Signals and Communication Technology

For other titles published in this series, go to
www.springer.com/series/4748

Ramjee Prasad • Sudhir Dixit
Richard van Nee • Tero Ojanpera
Editors

Globalization of Mobile and Wireless Communications: Today and in 2020

Springer

Editors
Ramjee Prasad
Center for TeleInFrastruktur (CTIF),
Aalborg University
Denmark
prasad@es.aau.dk

Richard van Nee
Qualcom
Waalre
The Netherlands
rvannee@Qualcomm.com

Sudhir Dixit
Hewlett-Packard Labs
Bangalore
India
sudhir.dixit@hp.com

Tero Ojanpera
Nokia
Espoo
Finland
tero.ojanpera@nokia.com

ISBN 978-94-007-3440-1 ISBN 978-94-007-0107-6 (eBook)
DOI 10.1007/978-94-007-0107-6
Springer Dordrecht Heidelberg London New York

Printed on acid-free paper

Springer is part of Springer Science+Business Media (www.springer.com)

To the organizers of Ramjee Prasad's 50 PhD Graduates' Celebration Event held on April 11, 2008, in Aalborg, Denmark

Preface

नियतं सङ्गरहितमरागद्वेषतः कृतम् ।
अफलप्रेप्सुना कर्म यत्तत्सात्त्विकमुच्यते ॥१८- २३॥

niyatam sanga-rahitam
araga-dvesatah kritam
aphala-prepsuna karma
yat tat sattvikam ucyate

That action which is regulated and which is performed without attachment, without love or hatred, and without desire for fruitive results is said to be in the mode of goodness

The Bhagavad-Gita (18.23)

This book is the outcome of the panel discussions held on the special event honoring first 50 PhD students of Prof. Ramjee Prasad. Several of his PhD students are today worldwide telecommunication leaders themselves. Over 60 post-docs, PhDs, colleagues and the scientific staff were present at the event to celebrate the research and development achievements in the field of mobile and wireless communication. The topics of the two panel discussions held were 'Globalization of Mobile and Wireless Communications' and 'Beyond 2020'. Therefore the title of the book is 'Globalization of Mobile and Wireless Communications: Today and in 2020.

The book reflects on how rapidly mobile communication has spread and touched almost every person on this planet, may that person be rich or poor, man or woman, or child or adult. The book also attempts to provide a window into the future and speculates how the wireless world might look like at the end of this decade, and presents some of the key enabling technologies that would drive the process to get there. Since naturally the business models drive the investment decisions and the adoption rate of any new technology, we have also included a chapter on that.

This book serves as a good starting point for casual readers, engineers and researchers who would like to have access to a book that provides in one single place what is there to know about the reasons behind the huge success of the wireless communication around the world.

As globalization becomes more and more important, we foresee more readers (including the policy makers and decision makers in the governments) being interested in the topic of this book. The reader would benefit from a wide range of topics that are discussed in this book, some of which are listed below.

1. The book introduces the notion of globalization of wireless communication
2. The book covers the key technologies (past, present, and future) and their enablers to make mobile wireless a global phenomenon
3. The book not just covers technology aspects, it also addresses applications and techno-economic aspects
4. The book covers standards and international fora and their critical role in globalization
5. The book sheds light on the future developments in wireless communication, and accompanying potential business opportunities
6. Potential impact of wireless communication on efficient management of energy resources and global warming

Because each topic can easily expand into a book of its own and it is difficult to have in-depth knowledge in all of these domains, we chose to invite the various thought leaders in their fields to contribute to this book. The book is written in a style to provide a broad overview of the various technologies with a special emphasis on the user as the center of all activities. It has been our objective to provide the material in one single place to enable quick learning of the fundamentals involved in an easy-to-read format.

Finally, we (including the chapter authors) have tried our best to ensure that each and every chapter is as accurate as possible; however, some errors in any manuscript are inevitable. Please let us know of any errors and ideas to improve the book – such comments will be highly appreciated.

June 2010

Ramjee Prasad, CTIF, Denmark
Sudhir Dixit, Hewlett-Packard Laks, India
Richard van Nee, Qualcom, The Netherlands
Tero Ojanpera, Nokia, USA

Acknowledgments

We are indebted to the contributors of this book for their hard work that made this book possible. All throughout this project, they were patient and forthcoming with any revisions we requested of them.

We also thank the organizers of the 50 PhD Graduates' Celebration Event that was held on April 11, 2008, in Aalborg to commemorate the attainment of Professor Ramjee Prasad to have graduated 50 PhD students. Most chapters published in this book were first presented as the invited talks and were later expanded into book chapters.

Credit for the major efforts in helping us to put the material together and shape it into a final version goes to Kirti Pasari from CTIF.

About the Editors

Professor Dr. Ramjee Prasad Fellow of IEEE (USA), The IET (UK) and IETE (India), has obtained B.Sc. Engineering in Electronics and Communication from the Bihar Institute of Technology, Sindri, India in 1968 followed by a M.Sc. Engineering from the Birla Institute of Technology (BIT), Ranchi, India in 1970 and a PhD from BIT, India in 1979. Ramjee Prasad is a world-wide established scientist, who has given fundamental contributions towards development of wireless communications. He achieved fundamental results towards the development of CDMA and OFDM, taking the leading role by being the first in the world to publish books in the subjects of CDMA (1996) and OFDM (1999). He is the recipient of many international academic, industrial and governmental awards and distinctions, huge number of books (more than 25), journals and conferences publications (together more than 750), a sizeable amount of graduated PhD students (over 60) and an even larger amount of graduated M.Sc. students (over 200). Several of his students are today worldwide telecommunication leaders themselves. Recently, under his initiative, international M.Sc. and PhD programmes have been started with the Sinhgad Technical Education Society in India, the Bandung Institute of Technology in Indonesia and with the Athens Information Technology (AIT) in Greece. Ramjee Prasad has a long path of achievements until to date and a rich experience in the academic, managerial, research, and business spheres of the mobile and wireless communication area. Namely, he played an important role in the success that the Future Radio Wideband Multiple Access Systems (FRAMES) achieved. He was the leader of successful EU projects like the MAGNET and MAGNET Beyond, among others, as well as the driver of fruitful cooperation with companies in projects, like Samsung, Huawei, Nokia, Telenor, among others. He started as a Senior Research Fellow (1970–1972) and continued as an Assistant Professor (1972–1980) at the Birla Institute of Technology (BIT), Mesra, Ranchi, India. He was appointed as an Associate Professor in 1980–1983 and head of the Microwave Laboratory there. From 1983–1988 Ramjee Prasad worked at the University of Dar es Salaam (UDSM), Tanzania, where he became Full Professor of Telecommunications in the Department of Electrical Engineering in1986. From February 1988 till May 1999 Ramjee Prasad worked at the Delft University of

Technology (DUT), The Netherlands at the Telecommunications and Traffic Control Systems Group. He was the founding head and program director of the Centre for Wireless and Personal Communications (CWPC) of the International Research Centre for Telecommunications-Transmission and Radar (IRCTR) at DUT, The Netherlands. Since June 1999, Ramjee Prasad has been holding the Professorial Chair of Wireless Information and Multimedia Communications at Aalborg University, Denmark (AAU). Here, he was also the Co-Director of the Center for Personal Communication until December 2002. He became the research director of the department of Communication Technology in 2003. In January 2004, he became the Founding Director of the Center for TeleInfrastruktur (CTIF), established as large multi-area research center at the premises of Aalborg University. CTIF at Aalborg University was inaugurated on January 29, 2004. Under Ramjee Prasad's successful leadership and due to his extraordinary vision, CTIF turned into CTIF-Global by opening four divisions, namely: CTIF-Italy (inaugurated in 2006 in Rome), CTIF-India (inaugurated in 2007 in Kolkata), CTIF-Copenhagen and CTIF-Japan (inaugurated in 2008).

Ramjee Prasad is the founding chairman of Global ICT Standardization Forum for India (GISFI).

Dr. Sudhir Dixit is the Director of HP Labs India. The principal focus of the laboratory is on creating new technologies to address the information technology needs of the next billion customers of HP. Prior to joining HP Labs, Dixit held a joint appointment as a CTO at the Centre for Internet Excellence and a Research Manager at the Centre for Wireless Communications, at the University of Oulu, Finland. From 1996 to 2008, he held various positions with Nokia and Nokia Siemens Networks: Senior Research Manager, Research Fellow, Head of Nokia Research Centre (Boston), and Head of Network Technology (USA). Before that he was with NYNEX Science and Technology and GTE Laboratories (both now Verizon Communications) from 1987 to 1996. He also held the position of Senior Director at Research In Motion for a brief period in 2008.

He has published over 200 papers in journals or conferences and edited 4 books, and holds 19 patents. He is on the editorial boards of IEEE Communications Magazine, Cambridge University Press Wireless Series and Springer's Wireless Personal Communications Journal. He has organized numerous conferences and has also served in other capacities. He is Chairman of the Vision Committee and Vice Chairman of the Americas region of the Wireless World Research Forum.

Dixit is also an Adjunct Professor of Computer Science at the University of California, Davis. Dixit received a PhD in electronic science and telecommunications from the University of Strathclyde, Glasgow, and an M.B.A. from the Florida Institute of Technology, Melbourne, Florida. He received his M.E. degree from BITS, Pilani and B.E. degree from MANIT, Bhopal. He is a Fellow of IEEE (USA), IET (UK) and IETE (India).

Dr. Tero Ojanperä who heads Nokia's Services business, is responsible for the company's portfolio of location, messaging, entertainment and context-based services.

He has been a member of the Group Executive Board since 2005, and was appointed to his current position in 2009.

Tero has played a defining role in driving Nokia's evolution since joining the company in 1990, holding several senior management positions at Nokia Networks.

In 2003–2004, he headed the Nokia Research Center, and was appointed chief strategy officer a year later. From 2006, Tero served as chief technology officer, responsible for corporate and technology strategy, strategic alliances and partnerships, research and intellectual property rights.

Tero has a master's degree from the University of Oulu, Finland, and a PhD from Delft University of Technology in the Netherlands.

He was born on Nov. 12, 1966, in Korsnäs, Finland. Tero is married and has three children. In his spare time, he enjoys boating and skiing.

Dr. Richard van Nee received the M.Sc. degree in Electrical Engineering from Twente University in Enschede, the Netherlands, in 1990, followed by a PhD degree from Delft University of Technology in 1995. From 1995 to 2000, he worked for Lucent Technologies Bell Labs on wireless LAN transmission techniques. He invented the CCK codes that are used in the IEEE 802.11b standard – which was the standard that led to the formation of WiFi. He co-developed the OFDM-based proposal that was adopted by the IEEE 802.11a wireless LAN standard in 1998. In 2001, he cofounded Airgo Networks – acquired by Qualcomm in 2006 – that developed the first MIMO-OFDM modem for wireless LAN and which techniques form the basis of the IEEE 802.11n standard. Together with Ramjee Prasad, he wrote a book on OFDM, entitled 'OFDM for Mobile Multimedia Communications.' In 2002 he received the Dutch Veder award for his contributions to standardization of wireless communications. He holds more than 40 patents related to various WiFi standards and served as an expert witness in several WiFi related lawsuits. He is currently a principal engineer at Qualcomm where he is responsible for WiFi algorithm design and for developing proposals for the new 802.11ac standard.

Contents

Abbreviations

ACL	Adjacent Channel Leakage
ADSL	Asymmetric Digital Subscriber Line
AES	Advanced Encryption Standard
AI	Artificial Intelligence
AKE	Authenticated Key Exchange
AMPS	Advanced Mobile Phone Systems
ARQ	Automatic Request Control
BIS	Bidding Strategy
BS	Base Station
BWA	Broadband Wireless Access
CAGR	Compound Annual Growth Rate
CALA	Context Access Language
CDMA	Code Division Multiple Access
CHESS	Communication Using Hybrid Energy Storage System
COMET	Cooperative Mobile Positioning
CR	Cognitive Radio
CSI	Channel State Information
DAMPS	Digital AMPS
DoD	Depth of Discharge
DPI	Deep Packet Inspection
DSA	Dynamic Spectrum Auction
DSA	Dynamic Spectrum Access
DS-CDMA	Direct Sequence CDMA
DSL	Digital Subscriber Line
DSS	Dynamic Spectrum Sharing
DVB-SH	Digital Video Broadcasting - Satellite services to Handhelds
EAS	Electronic Auction System
EDGE	Enhanced Data rates for GSM Evolution
EEHF	Environmental Energy Harvesting Framework
EHWSN	Energy-Harvesting Wireless Sensor Network
EIRP	Equivalent Isotropic Radiated Power
ETSI	European Telecommunications Standards Institute

E-WME	Energy-opportunistic Weighted Minimum
FCC	Federal Communications Commission
FDE	Frequency Domain Equalization
FFT	Fast Fourier Transform
FPM	Fixed Price Model
FSK	Frequency Shift Keying
FSU	Flexible Spectrum Usage
3GPP	3rd Generation Partnership Project
GDP	Gross Domestic Product
GMSK	Gaussian Minimum Phase Shift
GPRS	General Packet Radio Service
GPS	Global Positioning System
GSM	Global System for Mobile Communications
HESS	Hybrid Energy Storage System
HIPERLAN	High Performance Radio LAN
HSPA	High Speed Packet Access
IBE	Identity-Based Encryption
ICI	Inter-carrier interference
ICI	Inter-channel interference
ICT	Information and Communication Technology
IMT-A	International Mobile Telecommunications-Advanced
IP	Internet Protocol
IPR	Intellectual Property Rights
ISI	Inter-symbol interference
ISM	Industrial, Scientific and Medical
ITU	International Telecommunication Union
KB	Knowledge Base
LR-AKE	Leakage-Resilient AKE
LTE	Long Term Evolution
LTE-A	Long Term Evolution - Advanced
MAC	Medium Access Control
MAGNET	My Personal Adaptive Global Net
MIMO	Multiple Input Multiple Output
MISO	Multiple Input Single Output
NFFT	Nonequidistant Fast Fourier Transform
NGN	Next Generation Networks
NLOS	Non-Line- Of-Sight
NMT	Nordic Mobile Telephone
OFDM	Orthogonal Frequency Division Multiplexing
OFDMA	Orthogonal Frequency-Division Multiple Access
OPEX	Operational Cost
OSI	Open System Interconnection
OWL	Ontology Web Language
PAPR	Peak-to-Average-Power Ratio
PDF	Probability Density Function (PDF)

PHY	Physical Layer
PIR	Private Information Retrieval
PLL	Phase Locked Loop
PNC	Public Network Computing
PPM	Pulse Position Modulation
PSTN	Public Switched Telephone Network
PU	Primary Users
QKD	Quantum key distribution
RAT	Radio Access Technology
RRM	Radio Resource Management
RSS	Received Signal Strength
SDR	Software-Defined Radios
SIM	Subscriber Identity Module
SIP	Session Initiation Protocol
SLA	Service Level Agreement
SMC	Secure Multi-party Computations
S-OFDM	Shaped OFDM
SVCs	Switched Virtual Connections
SAA	Simultaneous Ascending Auction
TACS	Total Access Communication System
TCO	Total Cost of Ownership
TDD	Time Division Duplex
TDMA	Time Division Multiple Access
TDOA	Time Difference Of Arrival
TRAI	Telecommunication Regulatory Authority of India
UMTS	Universal Mobile Telecommunications System
UTRAN	UMTS Terrestrial Radio Access Network
UWB	Ultra Wideband
VSAT	Very Small Aperture Terminal
VSB	Vestigial Sideband
W-CDMA	Wideband CDMA
WiMAX	Worldwide Interoperability for Microwave Access
WLAN	Wireless Local Area Network
WLL	Wireless Local Loop
WSN	Wireless Sensor Network
WTO	World Trade Organisation
XML	Extensible Markup Language

Chapter 1
Introduction to Globalization of Mobile and Wireless Communications: Today and in 2020

Sudhir Dixit, Tero Ojanpera, Richard van Nee, and Ramjee Prasad

1.1 Introduction

Mobile communication has expanded the basic need for humans to communicate to "from anywhere at any time," which was unthinkable only 25 years ago. It has touched almost every person, rich or poor, child or adult, around the globe. Increasingly, new developments in wireless data and mobile internet are augmenting the voice communication experience with much richer multimedia content consisting of text, video, and images. In not too distant a future, wireless data connectivity between the machines and sensors is destined to reshape the environment we live in and impact our lives in ways that we can hardly imagine. The emerging economies are already skipping legacy technologies, going for the next generation, and will be the hotbed of some of the innovations that would find their way to the affluent economies.

This book reflects on the recent developments in wireless communications from the global perspective and attempts to understand the reasons and provides an insight behind a phenomenal rate of adaptation by the wealthiest to the poorest nations around the globe, and what can we expect in the year 2020 time-frame.

S. Dixit (✉)
Hewlett-Packard Labs, India
e-mail: sudhir.dixit@hp.com

T. Ojanpera
Nokia, USA
e-mail: tero.ojanpera@nokia.com

R. van Nee
Qualcomm, the Netherlands
e-mail: rvannee@qualcomm.com

R. Prasad
CTIF, Aalborg University, Denmark
e-mail: prasad@es.aau.dk

R. Prasad et al. (eds.), *Globalization of Mobile and Wireless Communications: Today and in 2020*, Signals and Communication Technology,
DOI 10.1007/978-94-007-0107-6_1,

1.2 Mobile Devices

In a short span of about 25 years, mobile devices have evolved from a luxury business communications tool to a necessity that is literally used by everybody in the world. What used to be the exception to the rule is now the rule. The world is rapidly approaching 4 billion mobile subscribers, a figure that continues to grow leaps and bounds each day, but a new sea of change is being fuelled by the proliferation of smartphones, advanced mobile devices that put the power of a desktop computer into the palm of your hand. Smartphones are currently the fastest growing category in the mobile communications industry. According to Gartner, smartphone volumes will represent 14% of total mobile devices sales in 2009, growing by 23.6% from 2008. Fig. 1.1 illustrates the total mobile device sales worldwide.

Different people around the world have different needs to suit their personal and cultural tastes. There are numerous success stories in the development of highly-personalized, complementary services that further enable people to stay connected with their families, friends and their passions. Similar to the proliferation of 'voice' in the past, today, being online whenever and wherever we choose, is becoming a reality for the masses.

1.3 The Current Landscape

Internet and mobile communications are on a collision course, which is enabling the development of completely new types of applications. Examples of these new applications impacting the vast populations in the emerging markets are agricultural and educational services designed especially for the people in small towns and rural areas who have limited or no access to the internet. These services, which may cost less than $1/month, allow hundreds of millions of people the opportunity to access essential information to improve their businesses, potential prospects and their way of life. In more developed markets, the growing adoption of mobile applications is also having a significant impact.

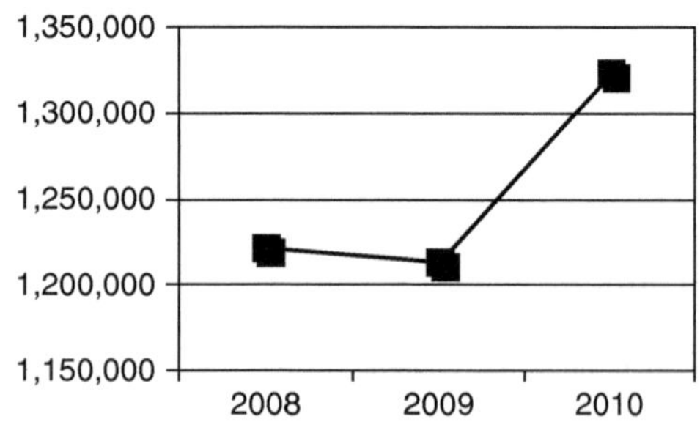

Fig. 1.1 Forecast: Worldwide Mobile Terminal Sales to End Users 2008–2010 (thousands of units) (Source: Gartner, December 2009)

1.4 'Sharing' is the New 'Sending'

Social networking has transformed the manner in which people share information. New possibilities are emerging as people link all of the different features of their mobile devices with their social connections and physical location. In this new age, 'social location' has a new meaning, now 'where you are' is just as relevant as 'what you are doing.' For example, to build on the transformation, Nokia is working on a feature called "Life casting," which provides zero-effort status updates to Facebook from phone and allows consumers to easily share where they are and what they are doing with a community of friends, as broad or as narrow as desired.

1.5 Sensors and Machines Become a Catalyst

Integrated sensors and global positioning systems (GPS) in mobile devices are serving as the catalyst for new technology solutions that can learn and anticipate a consumer's needs. The mobile industry is looking for ways to capitalize on the boom of GPS-enabled smartphones. According to Canalys, the number of GPS-enabled phones grew globally by more than 250% from ca. 10 million units in 2008 to ca. 27 million units in 2009 as depicted in Fig. 1.2. Several service providers and equipment manufacturers are already providing guided voice navigation, including 3D landmarks for drivers and pedestrians. Additionally, more and more mobile phones are beginning to offer RFID (radio frequency identification) sensors, providing easy and concrete access to services, content and repeat functions by a simple touch. Mobile workforces can launch services by simply touching RFID tags (transponders) with their mobile phones.

Increasingly machines are being connected with the web through the wireless interfaces, at homes, in offices, and outdoors. This is particularly true in locations where there is no wirleline infrastructure and in hard-to-reach locations. Some examples of these are security cameras, environment monitoring, dispensing machines, bus stops, etc. In the long-term we anticipate that every electronic device would come with its own IP address and would be web enabled and connected to the cyberworld.

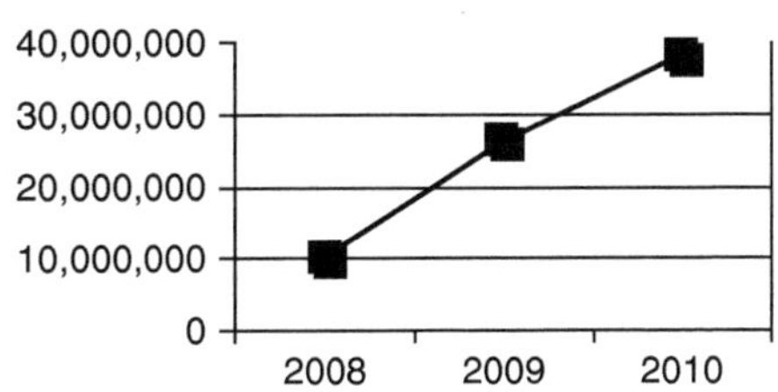

Fig. 1.2 Worldwide Mobile GPS Navigation Installed Base Forecasts (Source: Canalys, November 2009)

1.6 A Look to the Future

Until now, devices, applications and services, connectivity and radio have been treated as separate domains interconnected by layered architecture, such as the 7-layer OSI or TCP/IP stack. However, the future is about the convergence and layerless architectures. Additionally, over time, most radio and networking functions would be cognitive, and increasingly applications would be context-aware.

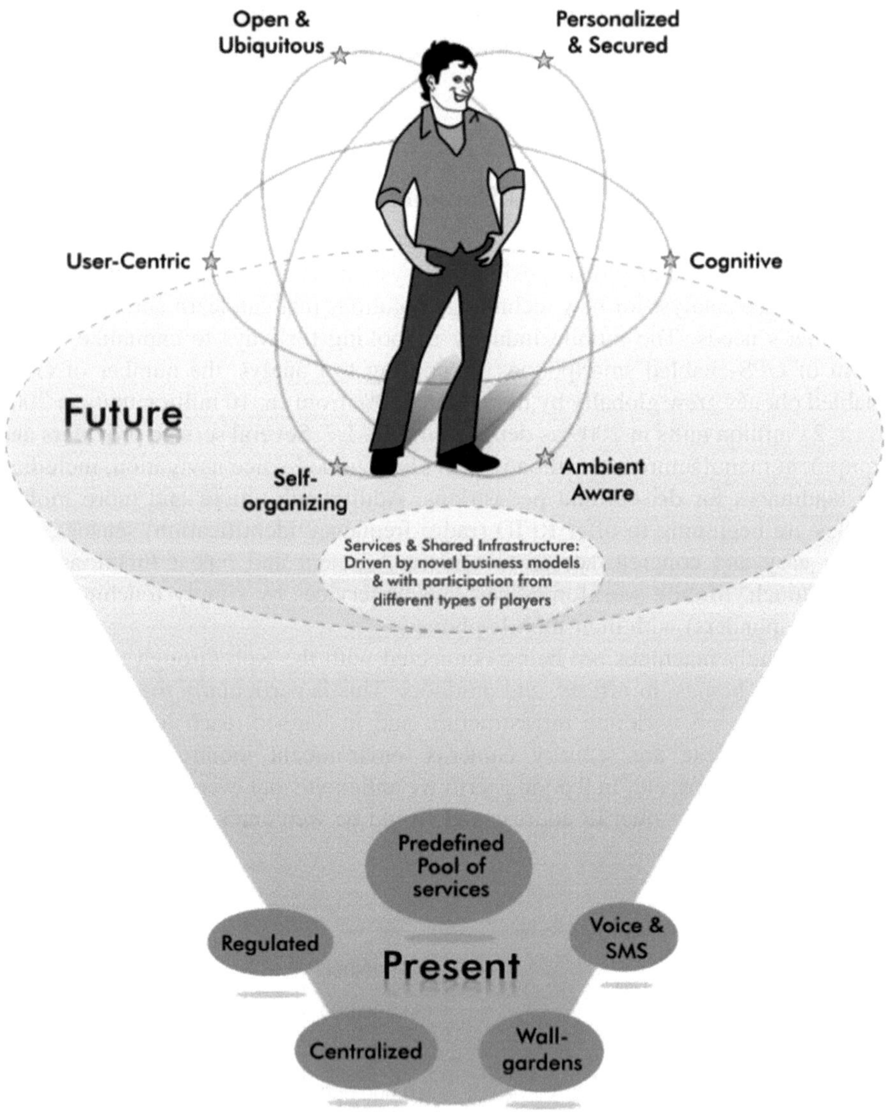

Fig. 1.3 Evolution of the world of wireless communication into the future

The service providers are already pushing for zero-configuration, self-management and autonomous operations in order to bring down the operating costs. The world of communication will be open, ubiquitous, secured, personalized, context-aware, and the majority of the wireless users would come from the emerging new markets consuming multimedia, typically video. Fig. 1.3 provides a window into the future. Connecting the virtual world – of location-aware and socially-responsive mobile solutions – to the real-world is the future of mobile multimedia. The next transformation is slowly coming with the mobile handsets becoming intelligent, automatically interacting with and sensing the world around it to provide truly personalized and useful services that help the user get the most out of life.

1.7 Brief Overview of the Book

This book has been organized to envision how the world of future wireless might look like and what enabling technologies will get us there with the principal goal of improving the users experience, quality of life and ease of use for everyone. This has meant covering a broad range of topics from technologies to business models. Additionally, to put the objectives in perspective, we have included chapters that present the current environment as illustrated in Fig. 1.4. A list of topics that have been covered in this book is provided below.

- Role of networks operators and manufacturers in building a globalized wireless world.
- Globalization of wireless networks and its impact the world over with emphasis on bridging the digital divide.
- Emerging business models.
- Access security and personal privacy and how it might impact globalization in the future.
- Wireless LAN going global.
- Radio resource management and quality of service.
- What would drive the wireless standards in the future: CDMA or OFDM?
- Future vision of wavelet radio and could it revolutionize radio interfaces.
- Designing future communication systems from clean-slate.
- Flexible and dynamic allocation of spectrum and its usage.
- Location determination in hybrid WiMAX-WiFi networks.
- Energy harvesting using wireless sensor networks.
- Role of coherent transmission in taming global warming and climate change.
- Role of intelligent software to drive novel and modern services.
- The challenges of synchronizing systems today and in the future.

The book has been organized into several distinct but interconnected parts: role of wireless in globalization, privacy and security, broadband wireless radio and networks – today and in the future, future services (e.g., location identification),

Fig. 1.4 Illustration of the coverage of the book. The number in the branches denotes the chapter of the book

energy harvesting, global warming, intelligent software for novel services, and finally synchronization of systems.

Chapters 2–4 We present some background material on how the operators and the manufactures have played a key role to connect everyone on this planet from the richest to the poorest, which is rapidly closing the digital divide. The world is getting flatter and by the year 2020 the vast majority of the people would have access to the same information with the result that prosperity would spread the world over.

Chapter 5 The growth in the connected world would bring a lot consolidation in the telecommunications and computer networking industries with the result that numerous interesting and novel business models would be born. No technology can survive and reach the mass market unless it is accompanied by a viable business model. Therefore, we next present the theory of business models and discuss how they are likely to evolve over this decade.

Chapters 6–7 Since wireless poses rather unique challenges to privacy of the users and to the security of the information that it transports, we discuss this topic next.

The revolution in broadband wireless cannot continue and succeed unless the users feel that their data is secured. Consequently, the next two chapters are devoted to user authentication and ensuring that privacy is not compromised. Any user data that traverses the network must be free from eaves-dropping and tampering.

Chapter 8 We all are witnessing the pervasiveness of the WiFi networks that are based on the IEEE 802.11x family of standards as a means to deliver untethered broadband to the last tens or hundreds of meters; therefore, next we provide an overview of the IEEE 802.11x standards.

Chapters 9–11 Making limited radio resources available to billions of broadband users in some of the most densely populated cities is going to be a daunting task both for the service providers and the equipment manufacturers. Therefore, we address this topic in three ways. First, the chapter provides an overview of the existing methods of radio resource management and quality of service, and then extrapolates them into the year 2020. Second, we investigate which radio interface (CDMA or OFDM) would be most suitable in the long-term. Third, we present a completely new approach to wireless communication that is based on the theory of wavelet that could potentially bring a paradigm shift.

Chapter 12 When the Internet was first born over 30 years ago, no one anticipated its huge success, and, therefore, the approaches of the day were rather simplistic and did not consider the issues of traffic types, performance and scalability. The World Wide Web in the early 1990's further accelerated the spread of the Internet. This raised an interesting question: were we to design a new network from scratch today with the knowledge that we have today, what and how we would go about designing an optimum network? We endeavor to answer this question in this chapter.

Chapters 13–15 The laws of physics limit the spectrum that is available thwarting efforts to offer higher and higher bitrates to everyone who wishes to access the internet using wireless. Over the past decade there has been a growing interest in studying and employing cognitive radio and dynamic spectrum allocation schemes to efficiently use spectrum. Chapters 13 and 14 provide an overview of these technologies, and Chapter 15 presents a case study in India where it is anticipated that by year 2013 there would be about 1.2B mobile users, the largest mobile market anywhere in the world.

Chapter 16 Since in the future we anticipate that location will play a significant role as mobility does, this chapter is devoted to methods for locating a user when he or she is indoor using WiFi in conjunction with the wide area radio access such as WiMAX. Locating a user when outdoor is a solved problem because of the success of the GPS, cellular and other radio access technologies; therefore, the focus in this chapter is mainly on indoor location determination.

Chapters 17–18 It is a widely accepted view today that wireless can play a significant role in energy conservation and reducing global warming. Therefore, Chapters 17 and 18 focus on these issues. Chapter 17 describes the role wireless sensors can play in reducing energy consumption. Chapter 18 makes a case that coherent transmission significantly reduces transmit power for the signal to reach the same distance as the traditional methods.

Chapter 19 Nothing will drive the usefulness and adoption of technology than the applications and services. Therefore, this chapter is dedicated to discussing the need to develop and use intelligent software to enable development of novel and user-centric applications and services that are personalized, context-sensitive, and help user to offload the mundane and routine tasks that can otherwise be done efficiently by a virtual assistant.

Chapter 20 An often overlooked topic is that of system synchronization. With the solutions becoming complex and global, there is a growing trend towards systems on chips and distributed architectures. This chapter provides an interesting insight into the synchronization issues in communication systems and proposes a number of solutions.

Since the various subject matter experts already cover the above topics in depth, it would be redundant to repeat them here. We, therefore, stop here and wish you happy reading!

Chapter 2
Role of Networks Operators and Manufacturers in Building a Globalized Wireless World

Sofoklis Kyriazakos

2.1 Introduction: Trends and Developments

One of the most popular situation that the telecommunications market is experiencing is the merging of operators vendors and manufactures. The list of famous mergers is not limited to the list below; it is extended into a large number that has both advantages and disadvantages. It is difficult to judge the result of a merger as this should cover all aspects and sides, including financial results for the stakeholders, new workplaces (or number of laid-off employees) and future potential. The motivation is simple: Better financial results in a very competitive market, aiming for sustainability and more promising future. As it is proven, brand name of the interested parties is a key role for mergers; and this is often very expensive.

This chapter is organized as follows. Section 2.1 is an overview of current trends and developments in the area of telecommunications. More specifically, famous mergers are discussed presenting both the benefits, as well as the socioeconomic impact. In Section 2.2 two giant international operators are compared in terms of networks, management and results. Section 2.3 presents different aspects of globalization in wireless systems. This includes short description of technologies they share. Section 2.4 concludes the chapter, where the future of wireless telecommunications is predicted. The projection considers the current trends, as well as the technological enablers and barriers.

2.1.1 Nokia Siemens Networks

Nokia Siemens Networks is currently a leading enabler of communication services. The organization comprises the former Networks Business Group of Nokia and the carrier-related businesses of Siemens. The net sale for the calendar year 2009 was

S. Kyriazakos (✉)
CTIF Aalborg University, Denmark
e-mail: sk@es.aau.dk

R. Prasad et al. (eds.), *Globalization of Mobile and Wireless Communications: Today and in 2020*, Signals and Communication Technology,
DOI 10.1007/978-94-007-0107-6_2,

approximately €12.6 billion. The company is one of the top vendors in the telecommunications infrastructure industry and continues the legacy of two industry champions, namely Nokia and Siemens.

The target group definition for NSN is based on the assumption that 5 billion people, around 70% of the world's population, will be connected by 2015; while billions will be connected by broadband. In the current market only 400 million out of the 2.6 billion people who are connected also have a broadband connection. The estimation is very promising and NSN has the background, the customers and the potential to be a major stakeholder in the coming years [1].

2.1.2 Deutsche Telecom – OTE

On March of 2008, Deutsche Telekom – Europe's biggest telephone company – acquired 20% of Hellenic Telecommunications Organisation (OTE) shares from Marfin Investment Group (MIG) for €2.5 billion. A few weeks later, Greek Government came to an agreement to sell another 3% of the Greek state's share to the German buyer in the context of its decision to secure the German provider as a strategic partner. Deutsche Telekom will have the management of OTE, while the Chairman of the Board of Director will be from the Greek Government and will have a say in critical decisions.

This deal is not only for the 3%, as the Greek State has supported the negotiations and is an interesting case study, as it involves the State and it bridges the mentality gap of two nations, namely Greece and Germany. On the other hand side, OTE is a key player in the Balkans with presence in several countries, owing both fixed and wireless networks, e.g. RomTelecom – Romania, Armentel – Armenia, Cosmote – Greece, Cosmote – Romania, Globul – Bulgaria, Cosmofon – FYROM, AMC – Albania and other infrastructures and assets. OTE has the potential to make its position even stronger the coming years; while the support of a strategic partner as Deutsche Telekom, will minimize the risk of facing issues with the competitors that still have privileges compared to the incumbent operator [2].

2.1.3 AT&T – Cingular

In the beginning of 2007 AT&T, the standard bearer of communications excellence for more than a century in the US, fold the 6 year-old Cingular wireless name into the iconic AT&T brand. AT&T Inc. launched a multi-media campaign to begin transitioning the Cingular brand to AT&T in advertising and customer communications, throughout Web sites and nationwide retail stores, and on company buildings and vehicles. "Around the world, our customers recognize the AT&T brand for meaningful innovation, a commitment to customer service, high quality and exceptional reliability," said Edward E. Whitacre Jr., chairman and CEO of AT&T. "AT&T, BellSouth and Cingular are now one company, and going to market with our services under one brand is the right thing to do" [3]. This is another example of cooperation between operators that aims to make both stronger base on the experience and the significance of the AT&T brand and the potential of Cingular.

2.1.4 Telia Sonera

TeliaSonera is the result of a 2002 merger between the Swedish operator Telia and Finnish operator Sonera. Before that, Telia has not succeeded the merger with Norwegian telecommunications company Telenor, that is now a major competitor in the Nordic countries. Telia is a telecommunications operator with famous as a state telephone monopoly. Sonera on the other hand used to have monopoly only on trunk network calls, while around 75% of local telecommunication was provided by telephone cooperatives. The brand names Telia and Sonera have continued to be used separately in the Swedish and Finnish markets respectively. Of the shares, 37% are owned by the Swedish Government, 13.2% by the Finnish Government, and the rest by institutions, companies, and private investors worldwide [4].

2.1.5 Alcatel Lucent

News of the proposed $13.4 billion merger with Alcatel sent some at the U.S.-based Lucent Technologies polishing up their French while others updated resumes. The new joint company worth $25 billion plans a European headquarters and an 8,000-person job cut. While CEOs of the two companies focused on the synergistic benefits of a "merger of equals" between wire-line Alcatel and wireless Lucent, analysts and others point out this new marriage could have a rocky start [5].

2.1.6 SONY – Ericsson

Sony Ericsson Mobile Communications, a joint venture between Sony Corp and Ericsson, began operations on 1 October 2001 with a capitalization of US $250 million each, an Ericsson spokesperson said. Sony and Ericsson agreed to merge their mobile phone units in April. In a statement today, Ericsson said that after "necessary approvals", both companies will start to consolidate their respective operations. The joint venture started with 3,500 employees in product, marketing and sales [6]. According to Ericsson Asia Pacific spokesperson, Arthur Huang, only the management team is based in London.

2.1.7 International Telecommunication Union – ITU

In the 1995–1999 planning period, "globalization" was more a slogan than a reality, since it referred mainly to alliances between major operators to provide end-to-end services to multinational enterprises. Public networks and residential customers were relatively unaffected by this kind of globalization, although various forms of "alternative calling procedures" provided consumers in countries, which allowed such practices a "poor-man's version" of the benefits enjoyed by big business users.

In the 1999–2003 planning period, globalization is likely became much more of a reality. The WTO agreement made it possible for foreign operators to have direct access through interconnection and interoperability to public networks in most of the world's major telecommunication markets, as well as to make direct investments in the development of those networks [7].

2.2 Case Study: Vodafone and Telefonica

An interesting study is the comparison of Vodafone and Telefonica, as both operators are widely spread; owing a large number of networks; while they have both similarities and differences.

Vodafone Group Plc is the world's leading mobile telecommunications company, with a significant presence in Europe, the Middle East, Africa, Asia Pacific and the United States through the Company's subsidiary undertakings, joint ventures, associated undertakings and investments. On 31 March 2008, based on the registered customers of mobile telecommunications ventures in which it had ownership interests at that date, the Group had 260 million customers, excluding paging customers, calculated on a proportionate basis in accordance with the Company's percentage interest in these ventures [8].

Vodafone holds interests in 33 licensed network operators located in 27 countries (Fig. 2.1).

Vodafone has centralized many of functions in Newbury in the United Kingdom. There is of course a distributed management mechanism; however networks that fully belong to Vodafone report to Newbury.

On the other hand, Telefónica is one of the world's leading operators with presence in Europe, Africa and Latin America. Telefónica has more than 60% of its business outside its home market and a reference point in the Spanish and Portuguese speaking market. In Spain the Group is providing services to more than 46.4 million customers while in Latin America, Telefónica gives service to more than 134.1 million customers as of the end of December 2007 becoming the leader operator in Brazil, Argentina, Chile and Peru and has substantial operations in Colombia, Ecuador, El Salvador, Guatemala, Mexico, Morocco, Nicaragua, Panama, Puerto Rico, Uruguay and Venezuela. The Group stands in 4th position in the sector worldwide in terms of market capitalisation the 1st as an European

Fig. 2.1 Vodafone's presence world-wide [8]

integrated operator and 2nd in the Eurostoxx 50 ranking, composed of the major companies in Europe (31 December 2007). The Group is listed on the main Spanish and foreign stock markets and has over 1.5 million direct shareholders according to separate records in favour of individuals and corporations [9]. As of December 2007, Telefónica's total number of customers amounted to 228.5 millions.

The contrasting approaches of Vodafone and Telefonica show the importance of getting the globalization model right. This requires the right balance between global integration and local management.

In Fig. 2.2 the stock of Vodafone and Telefonica are compared for a period of 4 years. One can see that even the pattern of their stock behavior has similarities, although the figure shows a higher increase for Telefonica (Fig. 2.3).

2.3 Aspects of Globalization

Aspects of globalization are discussed in the previous section, mainly with examples and success stories. In this section the technical aspects of achieving globalization of wireless network is presented. The technical aspects are classified into following categories:

- **Network sharing**: Describes the situation where two or more operators are sharing networking resources, whether these refer to antennas and base stations, or even core components.

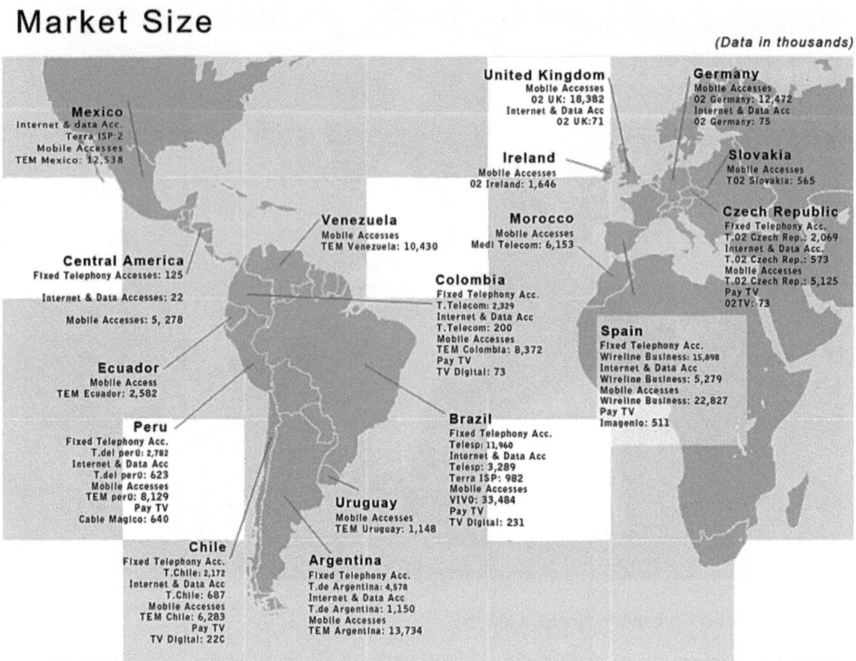

Fig. 2.2 Telefonica's presence world-wide [9]

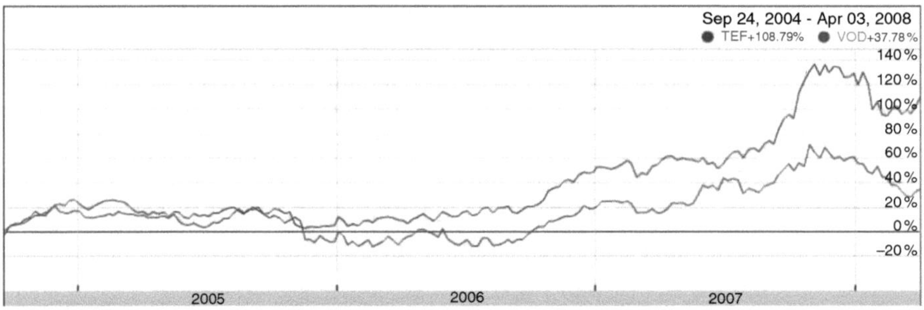

Fig. 2.3 Comparison of Telefonica and Vodafone stock for the last 4 years [10]

- **Site co-location**: Use of the same site for more than one operator to host its antennas and base stations. This is a typical situation, especially in urban areas, where sites are difficult to be found.
- **Base station hotels**: This is a new trend, where a third party offers space for base station subsystems and can offer resources based on SLAs to interested operators. This is mainly applicable in 3G systems.
- **Intersystem handover**: Handover between heterogeneous systems that belong to the same or different operator. The handover should be as seamless as possible; in any case the sessions/call should never be interrupted.

- **Roaming agreements**: Roaming agreements plays a significant role in the globalization world, as it is mainly linked with prices that the end user pays when using networking resources of different operators in the converging world of telecommunications.

In the following section, one of the technical aspects, namely network sharing, is discussed.

2.4 Network Sharing

As presented in Section 2.3, network sharing, in any way it is applied, is a situation that is required to bring operators closer, allowing them to use their resources in a more efficient way and achieve their goals. Often this is the beginning of a stronger relationship that leads into mergers as described in Section 2.2.

Network sharing has already been applied in several cases and according to [11]:

- "The impact is uncertain"
- "The benefit is a reduction in CAPEX"
- "Sharing does allow operators to potentially realize the value of their assets but by sharing they are degrading the value of a potentially valuable tactical asset"
- "Sharing could potentially be negative for an incumbent with a strong tactical asset in terms of the network which is costly and increasingly difficult to replicate"
- "In general the pros and cons of sharing is not clear although the immediate city reaction to a sharing agreement is likely to be positive

In general, the objectives of such collaboration are to reduce network cost or spread investment in time to improve balance sheet and enhance ability to raise finance and to reduce time to market and start to recoup from investment as early as possible. This helps to rollout "environmentally friendly" networks and enhance 3G footprint even if at present the industry emphasizes services instead of geographical coverage as for 2G.

Following figures present different approaches to network sharing (Figs. 2.4–2.7).

2.5 Future of Wireless Telecommunications

The future of wireless telecommunications is unforeseen. From a technological point of view it is very promising and definitely "globalized". This will involve all combinations presented above, in order to provide services by exploiting new technologies in the most efficient way.

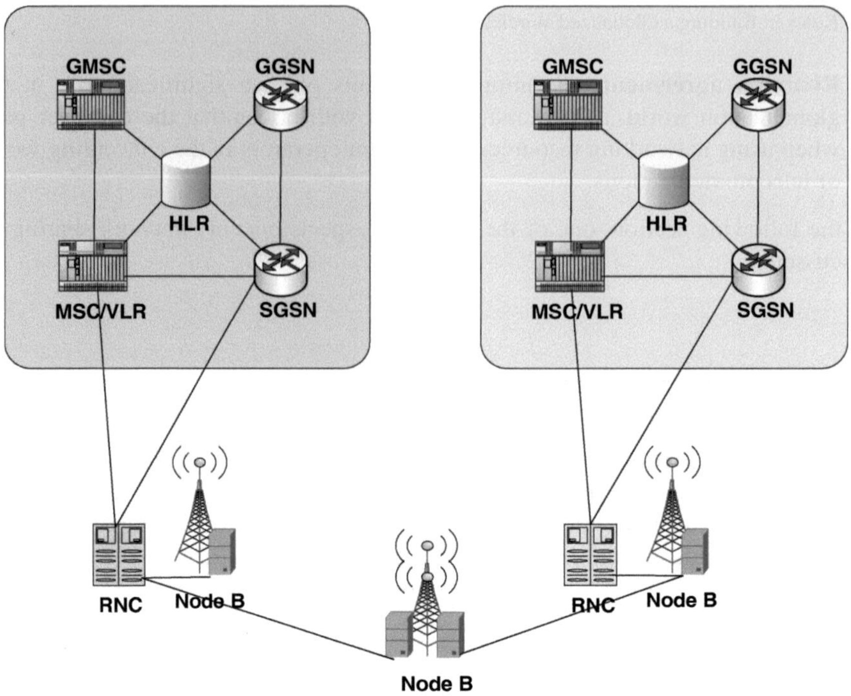

Fig. 2.4 Site sharing

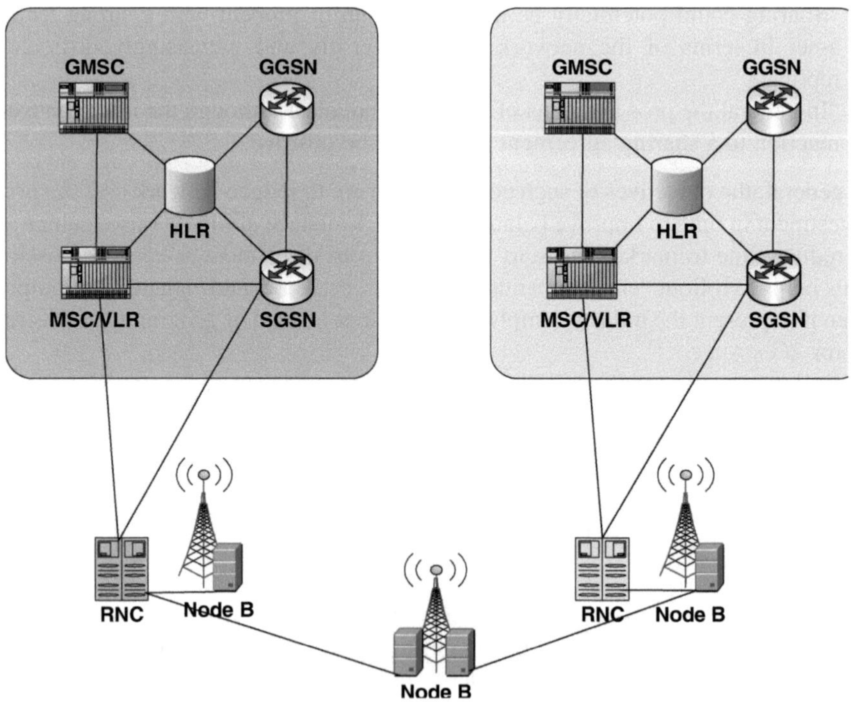

Fig. 2.5 Antenna sharing

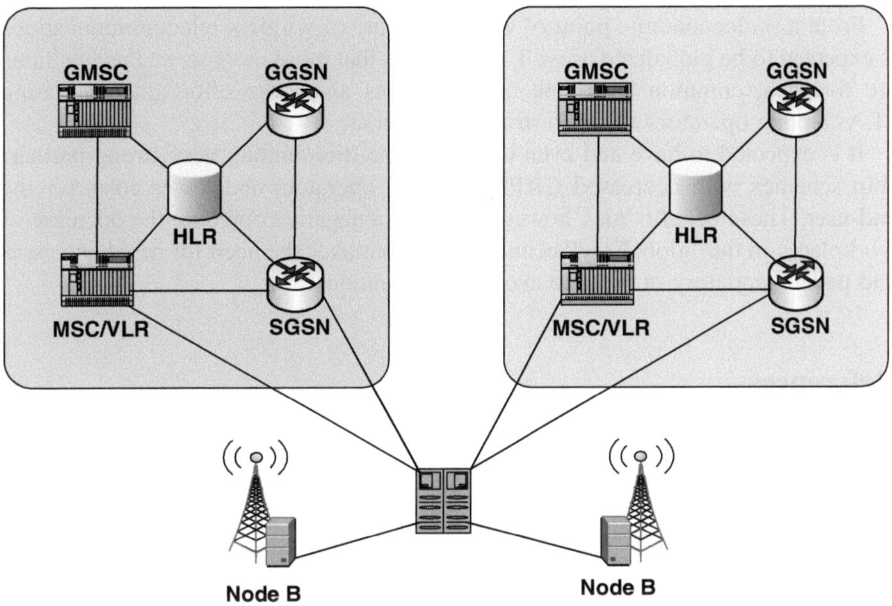

Fig. 2.6 RAN sharing (NB and RNC)

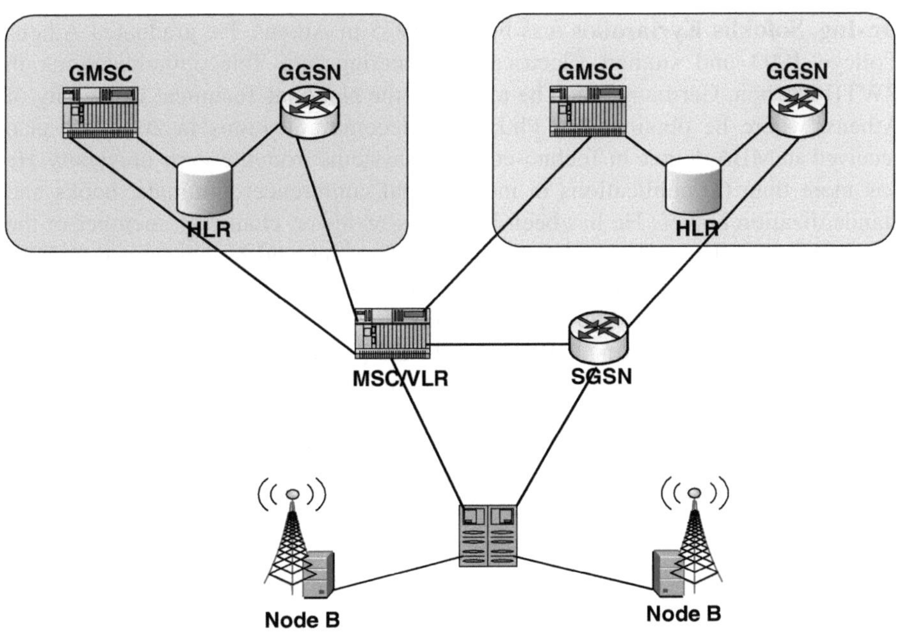

Fig. 2.7 Full MVNO scheme

From a socioeconomic point of view the future of wireless telecommunications is expected to be globalized as well. This means that more mergers and acquisitions are foreseen, common backbone infrastructures and more effort on negotiating SLAs among operators and infrastructure owners.

It is expected to have and even higher competition among very strong partnership schemes with decreased CAPEX for the operators and lower costs for the end-user. These benefits may however have as a negative outcome the decrease of workplaces in the "globalized" economy, which makes the need for new businesses and paths mandatory in order to avoid such situations.

References

1. www.nsn.com
2. www.ote.gr
3. http://www.mobiletracker.net/archives/2007/01/12/cingular-becomes-att
4. http://en.wikipedia.org/wiki/TeliaSonera
5. http://www.internetnews.com/bus-news/article.php/3596246
6. http://www.zdnetasia.com/news/communications/0,39044192,38008998,00.htm
7. www.itu.int
8. www.vodafone.com
9. www.telefonica.com
10. Google Financials
11. www.coleago.com

Dr.-Ing. Sofoklis Kyriazakos was born in 1975 in Athens. He graduated Athens College 1993 and studied Electrical Engineering and Telecommunications in RWTH Aachen, Germany. Then he moved to the National Technical University of Athens, where he obtained his Ph.D. in Telecommunications in 2003. He also received an MBA degree in Techno-economic systems from the same university. He has more than 60 publications in international conferences, journals, books and standardization bodies. He has been invited as reviewer, chairman, member of the committee, panelist and speaker in many conferences and he has served as TPC chair in two International conferences. Currently he holds the academic position of Assistant Professor in the University of Aalborg.

Sofoklis has managed, both as technical manager and coordinator, a large number of multi-million Telecom and IT projects in NTUA. During the last 7 years he also worked as external consultant in the area of Telecommunications. In 2006 he founded Converge S.A., an ICT company that is member of the PRC Group, that he is now the Managing Director.

Chapter 3
Globalization of Mobile and Wireless Communications: Bridging the Digital Divide

Marco Moretti

3.1 Introduction

According to Wikipedia the term *globalization* refers to "economic integration on a global scale, into a global economy, which blurs national boundaries". Globalization is a phenomenon involving the integration of economies, cultures, governmental policies, and political movements around the world. Depending on audience and context, the term globalization may have positive or negative connotations. For example, globalization of trade, markets and production has led to a steep reduction of prices of most goods worldwide. On the other end, it has also dealt a lethal blow to those companies that have not been up to the challenge, resulting in unemployment and loss one of economic stability for many communities and countries. The recent financial crisis, that has affected the world economy on a global scale, has shown that the globalization of the financial markets has paved the way to speculations out of control of any regulatory institution. What is happening to the mobile communication market is a good example of positive globalization: in few years mobile phones have spread all around the world becoming a low-cost tool capable of providing wireless connectivity almost everywhere. Nevertheless, until now most of the efforts have targeted developed countries and yet too little has been done for the rest of the world.

The term *digital divide* refers to the gap between those people with access to digital and information technology and those without access to it. The importance of solving the digital divide has been highlighted by the former Secretary-General of the United Nations Kofi Annan, who has said

> The new information and communications technologies are among the driving forces of globalization. They are bringing people together, and bringing decision makers unprecedented new tools for development. At the same time, however, the gap between information

M. Moretti (✉)
Department of Information Engineering, Università di Pisa, Italy
e-mail: marco.moretti@iet.unip.it

R. Prasad et al. (eds.), *Globalization of Mobile and Wireless Communications: Today and in 2020*, Signals and Communication Technology,
DOI 10.1007/978-94-007-0107-6_3,

> haves and have-nots is widening, and there is a real danger that the worlds poor will be excluded from the emerging knowledge-based global economy. (Annan, 2002)

In modern economy, information and communication technology (ICT) is a booster to the potential growth and development of any country and the problem of closing the digital divide has been identified by the United Nations as one of universal challenges for the future [1]. From 1995 to 2006, the EuropeAid, the European Commissions co-operation office, invested some 300 million in ICT-related activities, mainly targeted at African, Caribbean and Pacific countries [2].

Mobile communications have had an unprecedented growth and they are probably one of the best examples of the potentials of globalization. In less than 30 years they have changed the meaning of communications so that a large part of the world population now has wireless access to voice and/or data communications. Undoubtedly, such a result has been made possible by the forces of globalization. The amount of resources invested has been enormous. The investments in research necessary to develop completely new systems, the effort required for building the infrastructure, and the coordination necessary for standardization were possible only for multinational organizations aiming at the conquest of international markets supported by global institutions. The driving force of the mobile communication growth has been to try to reach the largest number of potential customers. Now that developed countries have reached full penetration, all the efforts and investments made in the past can be used with little cost to help the less developed countries to build their own telecommunications infrastructure. In fact, as the developed countries have carried the burden of accumulating global critical mass and resolving uncertainties related to standards and dominant designs, developing countries can benefit from being late entrants in wireless communications and thus reduce the digital divide.

3.2 Globalization of Mobile Communications: GSM and WLAN

The history of mobile communications ideally begins with the experiments on radio propagation performed by Marconi at the end of the nineteenth century. The first mobile radio telephones were employed in 1921 by the Detroit Police Department. Nevertheless, the costs of the wireless systems remained prohibitive for any commercial application for many years to come. A major step forward into the development of mobile communications took place in 1947 when the concept of cellular telephony was invented at the Bell Labs. In 1973 Dr. Martin Cooper filed for Motorola the first patent describing a mobile phone and the first commercial applications of mobile radio communications were deployed at the beginning of the eighties. While initially wireless communications were perceived as a niche application for rich consumers, they rapidly evolved into a global phenomenon beating even the most optimistic forecasts. In particular, we will focus on two of the most successful examples: GSM, which accounts for mobile telephony, and wireless LANs. The two cases are extremely different: GSM networks need a very expensive infrastructure and transmit on licensed bands. On the other end, WLANs offer wireless access to the Internet by transmitting on license-free bandwidths and require almost no infrastructure. In both cases we will try to outline how globalization has affected their development.

3.2.1 The GSM Case

According to the GSM Association (GSMA), in April 2008, just 17 years after the first GSM network was launched in 1991, the impressive landmark of 3 billion GSM connections has been surpassed. A 2009 survey reports 4.3 billions mobile connections of which approximately 3.5 billions are GSM users. To date, more than 700 mobile operators are spread across 222 countries and territories of the world. In this scenario, the worlds biggest GSM markets today are China and India, which together account for more than a quarter of the growth worldwide. Already in 2006 cellular services accounted for 1.8% of the global economy. Many reasons of this success can be found in the capacity of standardization bodies and manufacturing companies to take advantage of globalization. Some aspects of this process are:

- *Global standards: ETSI*. The standardization of GSM has been the result of the coordinated effort of the regulatory institutions of several three European countries under the umbrella of the European Telecommunications Standards Institute (ETSI). The fact that GSM equipment was the same all over Europe has been one of the most important reasons for its initial success. It has allowed the manufacturers to reduce development costs and the users to roam seamlessly across all Europe. This has given GSM a competitive advantage over all other mobile standards that were developed at about the same time in the USA, Japan, and Korea.
- *Global markets*. Since its first deployment, GSM has been a truly globalized system: its goal was to be the first pan-European communication system. International roaming is nowadays a normal aspect of mobile communications but it has been one of the most innovative services provided by GSM. On the financial markets, the success of mobile communications has led to the creation of big international groups, such as Vodafone, T-Mobile, and Orange, for example. The aggregation process of service providers of different countries has resulted in the creation of global networks with benefits evident in terms of international roaming and economies of scale.
- *Global manufacturers of mobile phones*. All the largest mobile manufacturers are multinational companies that assemble their products with components manufactured in several different parts of the world. For certain specific parts,

there are just one or two producers worldwide who supply all the mobile phones manufacturers. This determines economies of scale that contribute to reducing the equipment costs and thus helping the spread of GSM.

3.2.2 Meanwhile, WLAN Has Conquered the World…

With an evolution parallel to cellular networks wireless LAN modems have also become a commodity present in approximately 90% of the laptops sold worldwide. Also in this case, the commercial success of the system is due to the globalized approach of its development.

- *Global standardization bodies and regulatory agencies: IEEE and ITU.* In the early nineties the IEEE 802.11 Wireless Access Methods and Physical Layer Standardization Committee started the standardization phase for WLANs. The Institute of Electrical and Electronic Engineering (IEEE), although US-based, is universally recognized as a global organization that attracts the best academic and industrial researchers from around the world. The fact that the IEEE 802.11 WLAN transmits on the free industrial, scientific and medical (ISM) bands has probably contributed to a great extend towards its enormous commercial success. The ISM radio bands are a set of frequencies allocated by the International Telecommunication Union (ITU), another international institution, for unlicensed operations globally in the world.
- *Global manufacturers of computers.* The adoption of the IEEE 802.11 standard by many computer manufacturers all over the world has reduced development as well as production costs. Nowadays almost every laptop has a WLAN modem so that the potential market is almost the whole world.

3.3 Bridging the Digital Divide: What Has Been Done and What Will Come…

It is a fact that access to basic telecommunications infrastructures in general and to the Internet in particular is fundamental for the development of the economy of any country. The ICT infrastructure of a country has importance comparable to that of the main roads at the time of the Roman Empire. Thus, together with reducing the spread of AIDS and eradicating extreme poverty, the UN has set among its millennium goals to "make available the benefits of new technologies, especially information and communications" for everybody. In its Nobel Lecture, the 2006 Nobel Peace Prize Muhammad Yunus stated:

> Information and communication technology is quickly changing the world, creating distanceless, borderless world of instantaneous communications. Increasingly, it is becoming

> less and less costly. I saw an opportunity for the poor people to change their lives if this technology could be brought to them to meet their needs.

As Fig. 3.1 shows, the ICT penetration rates in the different regions of the world are very unbalanced. One of the main reasons for the digital divide is the lack in poor or under-developed countries of a telecommunication infrastructure. Mobile communications, thanks to the effect of a globalized process of development, are probably one of the most effective means to close the digital divide. Figure 3.2 shows the stunning growth of mobile phone subscriptions worldwide in comparison with other types of connections.

A recent study by the London School of Economics showed that an increase of ten mobile phones per 100 people boosts GDP growth by 0.6 percentage points [3]. According to a McKinsey report [4], in 2005 the total economic impact of wireless in China was $108 billion, or the equivalent of about 5% of GDP. Even if not all the scientists agree on such a definite impact of mobile communications over developing economies, in recent years there has been an enormous effort to boost rapid growth of mobile phone networks in developing countries. The result is that nowadays developing countries account for about two-thirds of the mobile phones in use, compared with less than half of subscriptions in 2002. In Africa, mobile networks reach 91% of the urban population and rural coverage is steadily growing; the number of users has grown from 10 million in 2000 to 302 million in 2008 [5]. The reasons for this success, as illustrated by the steep growth-rate shown in Fig. 3.3, are manifold:

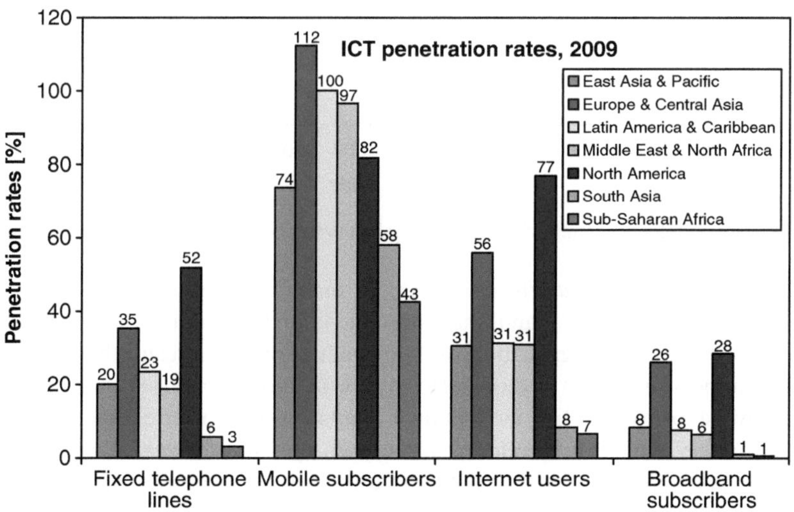

Fig. 3.1 ICT levels around the world, by region, 2009 (ITU statistics)

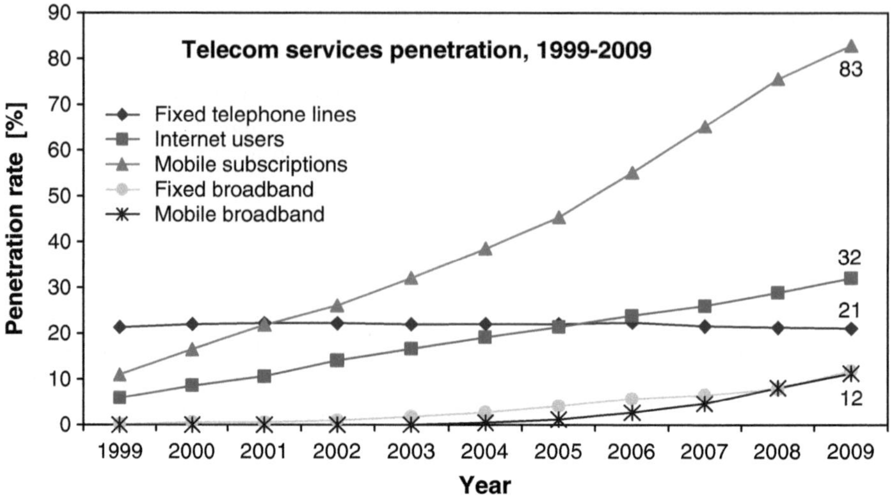

Fig. 3.2 Growth of telecom services globally, 1999–2009 (ITU statistics)

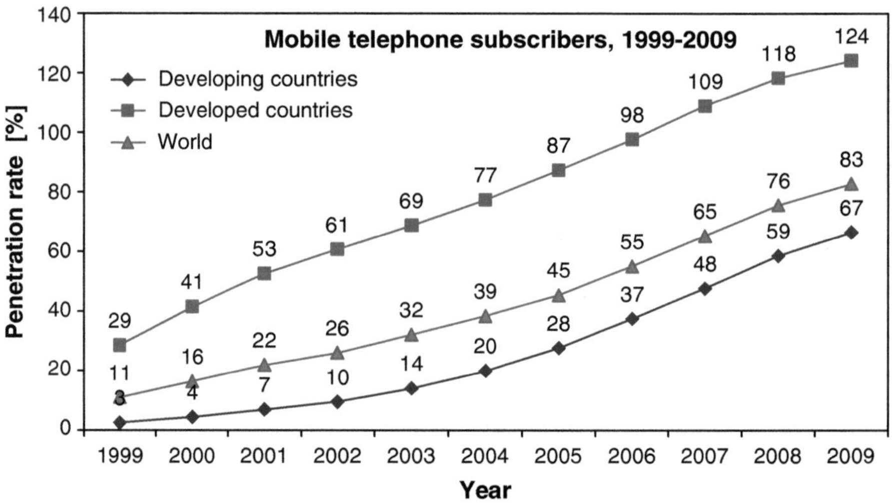

Fig. 3.3 Mobile telephone subscribers per 100 inhabitants, 1999–2009 (ITU statistics)

- *Competition among mobile operators*. In many countries, the privatization of state-owned companies and the consequent end of monopolies has greatly increased the pace of growth of mobile penetration. As a result there have been large improvements in sector performance in those countries that have taken decisive steps to establish independent regulators and foster competition [6].
- *Introduction of pre-paid services and innovative pricing schemes*. Prepaid services have encountered the favor of both end-users and operators so that, for example, in Sub-Saharan Africa 97% of the consumers are prepaid users. The benefits of prepaid strategies are twofold: prepayment reduces credit risks of the operators and encourages to plan and control the spending behavior of the users.

- *Offer of new innovative services*. In this case innovation occurs on a small scale, by either adapting or applying existing technologies and services and targeting them to developing countries. For example, the Bangladesh operator Grameen Telecom (GT) offers the village phone (VP) initiative in its poverty alleviation program. Village phones work as owner-operated pay phones. Anybody, but mostly poor people, otherwise unemployed, can take a loan to buy a handset and a mobile subscription and are then trained by GT on how to operate it. The VP operators then retail the mobile phone service to their fellow villagers, thus earning a living. Recently, mobile banking and financial services such as money remittance have gained a steadily growing interest among the poor.

The main requirement for the telecommunication infrastructure in a poor country is to keep low the cost of its deployment and its maintenance. Therefore, mobile systems that are easy scalable and require a small number of base stations to provide coverage for large areas are particularly suited for bridging the digital divide in underdeveloped countries. Nevertheless, we would like to underline some key aspects closely related to the phenomenon of globalization that greatly contribute to the success of mobile communications.

- *Cost of infrastructure*. Mobile communication infrastructure cost is kept low by the global competition among the largest producers and by the sheer size of the market that adds up to the whole world.
- *Global markets*. Just as in the rest of the world, in the under-developed areas too there are large multinationals with networks that span several countries. For example, in the first quarter 2008 more than 60% of the operators present in Africa were affiliated in some way with major international telecommunication groups.
- *Mobile terminal costs*. As for the infrastructure equipment, the price of mobile terminals has been steadily falling. The cost of a handset is often seen as the most important entry barrier. Consequentially, the manufacturers have been trying hard to reduce prices. Moreover, lowcost handsets have lower power consumption than their full-feature counterparts. This is particularly important in developing countries where electricity is not always available. The global competition between global players such as Nokia, Motorola and Samsung to win the market in under-developed countries resulted in the offer of a generation of ultra low cost terminals. Among these, the worlds best selling phone is the Nokia 1100, which has sold more than 200 million units. Mobile equipment costs are not the only variable to consider, anyway. Mobile handset price, the service fee and taxes form the total cost of ownership (TCO). In a GSMA report [7], penetration of mobile communication is significantly related to TCO; i.e., the lower the TCO, the higher the penetration. Thus, also local parameters, such as service fee and taxes are equally important for the successful spread of mobile subscription.

While in the developing countries the penetration of fixed-line connections is still low, the growth of mobile telephony partially compensates for it. The largest concentration of mobile phones is in the urban areas but affordability is still a problem in rural areas where the penetration rates are on average much lower. The benefit for the population is not only economical. On the contrary, the main perceived value of mobile use among the poor is improved communication with family and friends as illustrated by the results of a recent study [8]: in many cases family relationships and social purpose represent the most important use of mobile phones.

3.4 The World Is Not Online Yet…

Unlike the extraordinary success of mobile voice services, access to Internet is growing slowly in developing countries as shown in Fig. 3.4. For many practical reasons, a broader and more diffused access to Internet will contribute to reduce the digital divide more than just mobile telephony. In fact, broadband communications could greatly help developing countries to raise the standards of living by the use of ICT in many key sectors such as, for example, healthcare, distance learning and disaster intervention [9]. In this case, even the introduction of a very successful technology as WLAN has yielded limited success. In fact, it is the lack of computers and of broadband connections that limits the spread of Internet. To date, among the factors that slow down the expansion of Internet are:

- *High costs of computers*. The cost of personal computers is still very high when compared to that of a mobile phone. Several companies and international institutions are currently addressing the problem of reducing the cost of laptops so that they can be distributed to the poor. A good example is the "one laptop per child" project. This project aims at producing low-cost laptops with a WLAN interface with the main goal of promoting children's education in developing nations. In the first quarter of 2008 there were an increasing large number of countries participating to the project, including some very poor nations like Ethiopia,

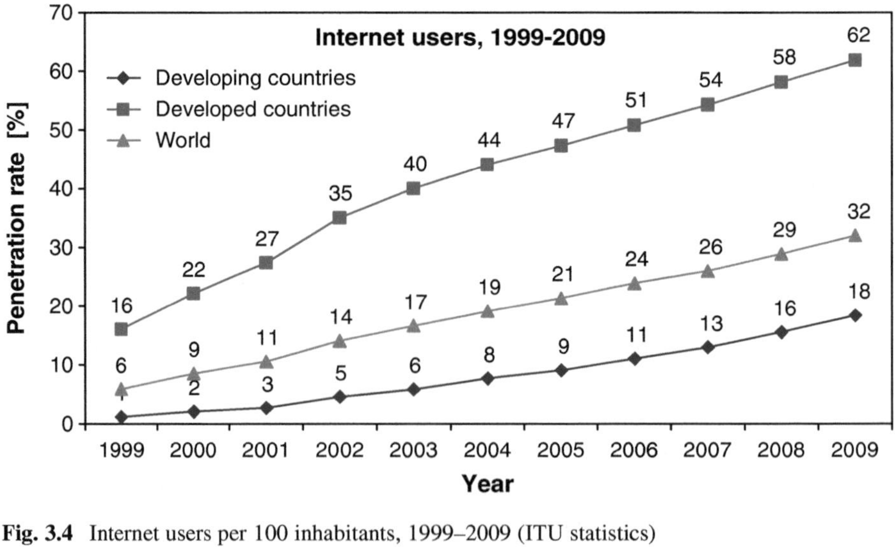

Fig. 3.4 Internet users per 100 inhabitants, 1999–2009 (ITU statistics)

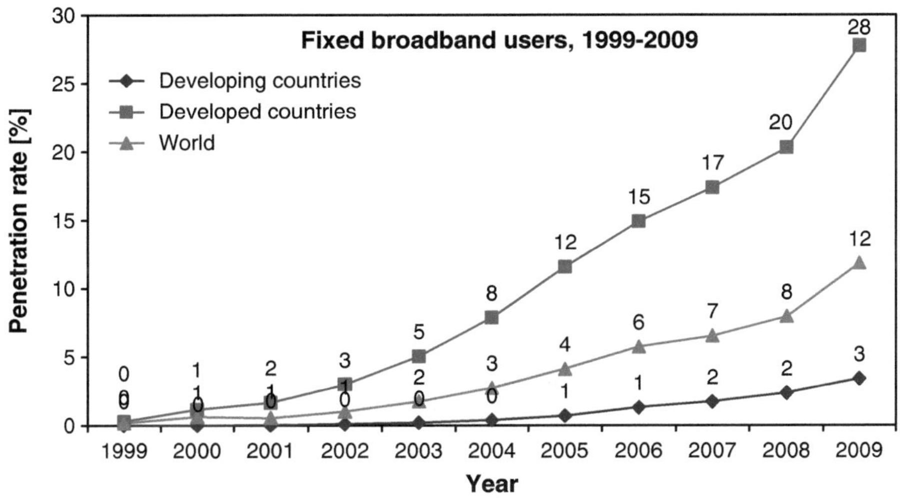

Fig. 3.5 Fixed broadband penetration per 100 inhabitants, 1999–2009 (ITU statistics)

Rwanda, Haiti, Afghanistan, Cambodia, Mongolia and Papua New Guinea. As a result of this and other projects, by various estimates, more than a billion PCs will be connected to the Internet by the end of 2010.

- *High costs and limited availability of bandwidth.* Figure 3.5 shows the penetration of fixed broadband services. The difference between developed and developing countries is increasing rather than decreasing. In general, developing countries have a much less developed fixed infrastructure, due to the costs of physically laying down new lines in areas where few people can afford to pay for a telephone subscription. Moreover, competition in the fixed-line sector is much less developed, and in many countries state-owned incumbents still control the market. Market reforms and liberalizations are slow and require complex legislations. Thus, even if many states seek to leverage mobile infrastructure boom into the broadband boom, few have so far succeeded. Nevertheless, WLAN, and its follow-up WIMAX, could play an important role in providing broadband access in many countries where fixed lines are scarce.
- *Low literacy levels.* Internet can be hardly accessed by those who can not read and high rates of illiteracy limit the spread of Internet. On the contrary, literacy is not a barrier for the usage of mobile phones. Moreover, low usage of Internet in a certain country is also strongly correlated with the lack of contents in local languages and vice-versa.

3.4.1 Africa Is Catching Up

What is now happening in Africa is a good example of how the world of communications is changing. For many years the penetration rate of fixed telephone lines in the continent was below 3% and the growth rate extremely slow. In just few years

Table 3.1 Main characteristics of most important sub-Saharan underwater cables

	SeaCom	TEAMs	EASSy	MainOne	GLO-1	WACs	ACE
Length (km)	13,700	4,500	10,000	14,000	9,800	14,000	17,000
Capacity (Tb/s)	1.28	1.28	3.84	1.92	2.5	5.12	5.12
Completion	2009	2009	2010	2010	2010	2011	2012

the penetration of mobile lines has boomed to an impressive 46% (ITU, 2009). In almost every African country the number of mobile subscribers has exceeded the number of fixed telephone subscribers, and only in 2009 76 million new mobile subscribers were added (amounting to a 20% growth rate). At the beginning of 2010, a total of 440 million Africans had subscriptions to mobile phones [5].

Moreover, also the broadband market is moving. The potential for growth of African telecommunication markets has been demonstrated by recent financial investments in underwater cables exceeding 3.5 billions USD. As a result of increasing demand for bandwidth, new undersea cables are being deployed by private and/or institutional consortia to provide low-cost bandwidth. As shown in Table 3.1, in the near future sub-Saharan underwater cables will be able to offer an overall capacity exceeding 21 Tb/s. As an example, African owned SeaCom has contributed to enable global broadcasting of FIFA Soccer World Cup 2010 by sharing its underwater fiber optic cable capacity, successfully meeting the heavy data requirements experienced during events of this magnitude.

Underwater cables will replace satellite connections as the main source of internet access in Africa, increasing speed, reliability and reducing cost. This should improve productivity and allow increased access at lower prices. Among many difficulties the continent is trying to close the digital divide gap!

3.5 Conclusion

We can conclude by quoting once again Muhammad Yunus 2006 Nobel Lecture.

> I support globalization and believe it can bring more benefits to the poor than its alternative. But it must be the right kind of 11 globalization. To me, globalization is like a hundred-lane highway criss-crossing the world....

Much has been done and even more needs to be accomplished but, in our view, the globalization of mobile communications is helping to bridge the digital divide by making *this hundred-lane highway* digital and available and open to everybody.

References

1. www.un.org/millenniumgoals
2. ec.europa.eu/europeaid/what/infrastructure-transport/index en.htm

3. L. Waverman, M. Meschi, M. Fuss, The impact of telecoms on economic growth in developing countries. *The Vodafone Policy Paper Series*, 10–23 (2005, 2 March), http://www.ictregulationtoolkit.org/en/Publication.3532.html
4. Wireless Unbound: The Surprising Economic Value and Untapped Potential of the Mobile Phone. *Mckinsey Report* (December 2006), http://www.mckinsey.com/clientservice/telecommunications/WirelsUnbnd.pdf
5. ITU World Telecommunication/ICT indicators (2010), http://www.itu.int/ITU-D/ict/
6. World bank, Information and Communications for Development 2009: Extending Reach and Increasing Impact. *Information and Communications for Development (IC4D)* (2009), http://www.infodev.org/en/Article.384.html
7. GSM Association, Tax and the Digital Divide: How New Approaches to Mobile Taxation Can Connect the Unconnected (2005), http://www.ictregulationtoolkit.org/en/Publication.3376.html
8. A. Zainudeen, in *What Do Users at the Bottom of the Pyramid Want?,* ed. by R. Samarajiva, A. Zainudeen. ICT Infrastructure in Emerging Asia: Policy and Regulatory Roadblocks (IDRC, Ottawa, 2008), pp. 39–59
9. A.T. Rashid, L. Elder, Mobile phones and development: an analysis of IDRC-supported projects. *Electron. J. Inform. Syst. Dev. Countries.* **36**(2) (2009)

Marco Moretti received the Laurea in Electrical Engineering from the University of Florence, Italy, in 1995. From 1995 to 2000 he was with the Department of Telecommunications and Traffic Control of the Delft University of Technology (Delft, The Netherlands), where he received the Ph.D. degree in electrical engineering with a dissertation on narrowband multi-user receivers. From 2000 to 2003 he worked at Marconi Mobile ACCESS in Genova, Italy as baseband receiver expert. During this period he managed as a team leader the research group involved on UMTS base stations. Moreover, he was in charge of coordinating scientific collaboration between Marconi Mobile and University of Pisa, University of Florence and other technical partners. From 2003 to 2005 he was employed as young researcher within the FIRB PRIMO project, an Italian government project aiming at the definition and implementation of a 4th generation communication system. Since January 2006 he has been with the Department of Information Engineering of the University of Pisa where he currently holds the *Ricercatore* position (full-time tenured position, roughly equivalent to Assistant Professor). His research interests are in wireless communication theory, with emphasis on synchronization, channel estimation and resource allocation for multi-carrier systems. He is author of several scientific journal and conference papers.

Chapter 4
Globalization of Wireless and Mobile Networks; the Impact to Africa

Stanley H. Mneney

4.1 Introduction

Globalization of wireless and mobile networks is making the network services ubiquitous; availability of any network service anywhere, at anytime and to anyone. The network should be able to transport interactive multimedia services and data streaming to facilitate both broadband and narrowband services such as video telephony and mobile TV, e-commerce, e-government, e-entertainment, e-education, e-health and other new services. The network must make these services accessible in both the developed and the developing world, in the densely populated urban centres and in the remotely located sparsely populated rural environments. The kind of network that is required should be a flexible with the ability to handle high traffic density and low traffic density depending on the requirements of the subscribers at a particular time. It should be adaptive and be able to maintain the same quality of service despite different radio environments. It should be robust and reliable. By globalization we also mean that the network should support both horizontal and vertical roaming and the information transported across such a huge network must be secured.

In this chapter we look at the evolution of mobile phone networks and related enabling wireless technologies followed by a discussion on the vision of globalization. In our discussion we focus on the African scenario and the impact of globalization on the African subcontinent.

4.2 Wireless Networks

In many parts of the world there are conditions that make wireless the only feasible way to provision of telecommunication networks. This may be due to factors such as large distances between communities, scattered and low population densities, hostile

S.H. Mneney (✉)
University of KwaZulu-Natal, South Africa
e-mail: Mneneys@ukzn.ac.za

R. Prasad et al. (eds.), *Globalization of Mobile and Wireless Communications: Today and in 2020*, Signals and Communication Technology,
DOI 10.1007/978-94-007-0107-6_4,

terrains and difficult vegetation. Yet many parts of the world that are already wired need the wireless links for additional flexibility, such as mobility, personalised services, and provision of new multimedia services. The traditional means of telecommunications, for over 100 years now, have been through the use of fixed-line telephony. Unfortunately, the continent of Africa and some other developing countries have seen a very modest growth of this service with the average penetration in Africa standing at three telephones per 100 persons in 2007. The introduction of the mobile phone in 1995 first found large acceptance in the urban areas of the developed countries, where the markets were attractive and gradually diffused to the developing world and even trickled to the rural areas. Mobile phones have provided the quickest and most economical solution not only to the problem of rural telecommunication but enhanced access to telecommunication facilities in the urban centres significantly.

Rapid development of the mobile phone occurred in the last 2 decades. Earlier work had the first mobile phone introduced in 1945, now classified as the 0G (zero generation), and this did not have the "handover" features. The handover feature was invented in 1984 at Bell labs. Motorola pioneered the introduction of the first mobile phone. In the next section we review the evolution of mobile phone networks and other wireless technologies and their impact to Africa.

4.2.1 The First Generation (1G) Mobile Phones

The 1G mobile phones were analogue. Though initially developed in the US the first commercial systems were launched by the Nordic Mobile Telephone (NMT). This was followed by the Advanced Mobile Phone Systems (AMPS) developed by Bell. AMPS was introduced in the USA but was used in many other countries. Another system known as Total Access Communication System (TACS) was introduced by Motorola in the UK and in many other countries. The various 1G systems were not standardised. All had a control channel to enable routing of the voice channels to available channels but channel spacing was different with 12.5 kHz for NMT, 30 kHz for AMPS, 25 kHz for TACS and 10 kHz for the narrow band version of AMPS referred to as NAMPS. The traffic demand on the 1G system soon became too high. These systems proved to be spectrally inefficient and were unable to meet the expected quality of service. The 1G system is still in use in some parts of Africa because of its simplicity and low cost [3].

4.2.2 The Second Generation (2G) Mobile Phones

A number of systems were introduced with the goal of solving the spectrum problem. These systems were called the 2G (second generation) mobile phones and they first appeared in 1991. The most successful system referred to as Global System for Mobile (GSM) was developed in Europe from the efforts of 26 Telecommunication utilities [4]. The system was fully digital and used TDMA to allow up to eight users in each of the channels that were spaced 200 kHz apart. The frequency allocated was in the 900 MHz band. Later other frequencies were added in the 1,800 MHz band. In the USA a digital system was developed to operate alongside AMPS and was referred to as Digital AMPS (DAMPS). DAMPS is defined under the standard IS-54. At the same time a company by the name of Qualcomm developed a more spectrally efficient mobile phone system using CDMA technology. This system had a more complicated receiver circuitry and was defined under a new standard IS-95 or cdmaOne. The GSM system is currently the most commonly deployed system in Africa. The problem with GSM is that it could handle a maximum data rate of 9.6 kbps which is too slow for data transfers in internet related services.

4.2.3 The 2.5 G Mobile Phones

Even though at this stage the 3G system was on the drawing board improvements in the 2G system were required to provide data services in the interim. A General Packet Radio System (GPRS) was developed for the GSM system [4]. The GPRS used packets in a packet switched system where packets selected the available routes. The practice of dedicating routes to a specific user was eliminated. This enabled channels to be used more efficiently and charges to be made in terms of data transferred. With GPRS data transfers of up to 171.2 kbps were theoretically possible. The GPRS was launched in South Africa and Egypt in 2003. The limitation in the deployment of GPRS was that in many parts of Africa used sets that were GPRS enabled were not available and in those countries where the GPRS service was available, it was only available to post-paid (contract) subscribers.

Further improvements in data rates were made by using 8PSK instead of BPSK modulation, where each symbol represents three data bits instead of one bit. This technique is referred to as Enhanced Data Rates for GSM Evolution (EDGE) [4].

At the same time enhancements were also applied to the CDMA system that started in the USA. CdmaOne evolved into CDMA2000 1x. The scheme retained the channel bandwidth of 1.25 MHz used in cdmaOne but by using additional channels it was possible to achieve data rates as high as 307 kbps. This is double the capacity of cdmaOne.

4.2.4 The Third Generation (3G) Mobile Phones [4]

The 3G system was designed to provide high speed internet access (burst speeds of up to 384 kbps), video telephony, real-time audio and video broadcast. There were three technologies that contributed to 3G mobile systems:

1. The Universal Mobile Telecommunications System (UMTS) was developed in Europe, using wideband CDMA, 5 MHz channel spacing and achieved data rates of up to 2 Mbps. The system used frequency division duplex (FDD) between the forward and backward channels.

2. CDMA2000 1xEV-DO (Evolution Data Only). This would provide a data service only and would therefore need a data connection to a PC or laptop for internet services over a mobile phone. For any service that requires both data and voice a standard 1x channel would be required as well. However, there was an evolution of the 1x system to CDMA2000 1xEV-DV which has full voice and data capability. The system was able to provide data rates of up to 3.1 Mbps in the forward direction. All these versions of CDMA systems used frequency division duplex (FDD) between the forward and backward channels.
3. Time Division Synchronous CDMA (TD-SCDMA) originated in China and used 1.6 MHz channel spacing.

The 3G mobile phone system is in its infancy in Africa. It has been introduced in 12 countries out of 55. The challenges of migration from 2G to 3G in Africa include

1. The mobile phone market is predominantly voice based.
2. There is very little relevant local content in the internet.
3. There is a lack of internet applications (airline ticketing, banking, e-learning etc).
4. The 3G capable handsets are expensive and are not easily available.
5. The mobile phone market is dominated by pre-paid customers.

4.2.5 The Fourth Generation (4G) Mobile System [5]

The goal of the 4G mobile phone system is to provide very high data transmission speed at lower cost. It is reported that the 4G system will provide wireless transmission speeds of up to 100 Mbps, substantially faster than the current 3G technology which only provides 2 Mbps at best, and also over four times faster than the fastest home broadband service. It will take a couple of years before this mobile system becomes available for deployment in the market. We believe that, in Africa, the 4G mobile will be ideal to provide broadband services for various applications like e-learning, e-government, e-health and entertainment for communities and individuals.

4.2.6 Other Wireless Technologies

Other wireless technologies include WiFi, WiMAX, WLLP and VSAT. In the next sections we will discuss how these technologies compete or complement the wireless mobile systems.

4.2.6.1 WiFi

Wi-Fi is a wireless technology based on the IEEE802.11 standard [6]. It is used to provide internet access, VoIP phone access, gaming, and network connectivity for consumer electronic products such as TVs, DVD players and digital cameras. Wi-Fi can access public Wi-Fi hotspots free of charge or after subscription making it possible to access internet services. It operates in the license-free 2.4 GHz spectral band offering connectivity at 11 or 22 Mbps over a range of up to 150 m. It has facilitated what is referred to as portable internet. There are several WiFi hotspots mushrooming in Africa with the majority of them in South Africa at the moment. WiFi is also used in the Wireless Local loop for the last mile telephone connectivity.

4.2.6.2 WiMAX

WiMAX, on the other hand, is a wireless technology aimed at providing data services over a range of up to 50 km using point-to-point communication or a fully fledged cellular type access [7]. It is a successor of Broadband Wireless Access (BWA) which has been deployed in many parts of Africa. WiMAX is based on the IEEE802.16 standard. It is a broadband service that can be used to interconnect WiFi hotspots and to connect them to the internet. It operates in bands that extend from 10–66 GHz and from 2–11 GHz without the requirement of line of sight. It can provide wireless alternative to cable and DSL access in the last mile.

4.2.6.3 Wireless Local Loop (WLL) [8]

WLL is used to link a subscriber to the local exchange. It uses wireless technology coupled with line interfaces and other circuitry to complete the last or the first mile of a telecommunication network. There are no defined standards for WLL and vendors can choose from fixed-access, mobile, or digital cordless technologies. The most common technologies include the following:

1. Analogue Cellular such as AMPS (69%), NMT (23%) and TACS (8%). These have limited capacity.
2. Digital Cellular such as GSM (72%), TDMA, and CDMA. GSM has an unnecessary overhead due to available roaming features.
3. Personal Communication Services/Personal Communication Networks (PCS/PCN). Possible standards include CDMA, TDMA, GSM, and PACS.
4. CT-2/DECT is a WLL when a public network operator provides wireless service directly to a user. DECT is capable of carrying higher levels of traffic, and transmit data at higher rates.
5. Proprietary implementations which include those networks that are customised for specific applications and are not available on public networks.

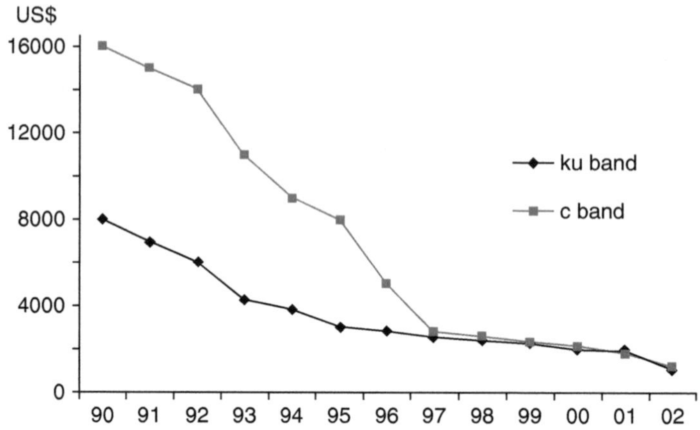

Fig. 4.1 VSAT terminals prices

WLL is a popular technology in Africa and is used in a number of countries. In section 3 we show the potential for Africa as an important wireless network market.

4.2.6.4 VSAT

In developing countries satellite technologies have been deployed for international telecommunication. However, VSATs have been deployed by private organisations, in development projects and for facilitating rural telecommunication. VSATs can be deployed fairly rapidly and almost anywhere. From 2002 to 2006 the market for VSATs has risen from 7,000 to 60,000 terminals in the African and Middle East Region [9]. VSATs can be used for voice, television reception and data transfers for internet services. The cost of the VSAT terminal equipment has been reducing drastically as shown in Fig. 4.1. At the moment the prices are $1,700 and 1,200 for a the C and Ku band terminal, respectively.

4.3 The Africa Scenario

4.3.1 The General Outlook

Africa has a total population of 923 million people or 14% of the world's population [1]. However, Africa accounts for only about 2% of the world's GDP. In comparison the Americas and Europe have similar shares of the world's population 14% and 12% respectively, and a share of the world's GDP of 35% each. The Americas and Europe are highly industrialised economies with well developed service sectors. In such countries the communication sector is pushed by advances in technologies and pulled and supported by market forces. On the contrary, Africa adapts technologies not optimised for its environment and the market support is limited due to the level and type of economic activities. In addition, the supporting services are severely underdeveloped and the majority of the people cannot afford to meet the cost of the state-of-the-art telecommunication services. In comparison, Asia has 60% of the world's population and share 25% of the world's GDP. Though the tele-densities may be relatively lower compared to the Americas and Europe the rapidly growing economies of Asia provide a huge market. The similarity with Africa is in the rural sector where the economies are mainly agricultural, and the basic services are underdeveloped and the economic activities cannot support an advanced communication system.

It is now a known fact that access to information and knowledge is essential for social and economic development and the regions that will benefit the most are the rural poor. An important requirement therefore is that the telecommunication service for such regions must be affordable. It should be possible to provide basic telecommunication services using low cost basic terminals. The African market is largely a voice market often complemented with simple text. A small percentage of the market would require broadband service for e-governance, tertiary institutions, industries, businesses and for the cyber cafés that have sprung up in the urban centres.

4.3.2 The Growth Stimulant

Africa's share is about 2% of the world's fixed-line telephone network. The fixed-line system requires a much more capital intensive infrastructure and is uneconomical to provide services over vast areas with a sparse population. In order to tap funds from investors the worldwide trend is to use private operators for both the fixed-line and the mobile networks. In 1995 there were only five private operators and 48 state owned operators of the fixed-line network in Africa. However, in 2007 the private operators (27) outnumbered the state-owned operators (26) for the first time [10]. Despite this positive trend Africa has still got the largest percentage of state-owned operators of the fixed-line networks compared to any other continent.

In order to make telecommunication services affordable many countries worldwide have allowed partial and full competition. In partial competition we have private instead of state-owned monopolies and in full competition we have more than one operator. In Africa the fixed-line service is the least competitive with 50% operated by monopolies, 25% partial competition and 25% full competition. The picture is very different with the mobile networks which are largely privately owned from the start; 7% are operated by monopolies, 44% partial and 49% full competition. This is partly the reason why the fixed-line networks have seen very little growth compared to the mobile networks which have been in existence over a much shorter period.

4.3.3 Regulatory Initiative

Where there is competition there is need to ensure there is order and fair play. The practice that is used worldwide is the deployment of independent regulatory authorities. By 2007, 83% of the African countries have introduced regulatory authorities to deal with matters related to spectrum management, competition policy, connecting tariffs, infrastructure sharing and license fees. The important requirement is to have skilled manpower to manage such regulatory authorities.

4.3.4 The Essential Infrastructure

Fifty years of the fixed-line in Africa has seen very little growth because the supporting network is capital intensive and is not economically feasible where the population density is sparse, the coverage area is vast or the physical features are hostile and vegetation is difficult. In addition there are problems related to lack of electrical power and a non-existence or a poor transportation infrastructure that should provide access for installation, management and maintenance. There have been efforts to provide wireless solutions using different wireless technologies and sometimes a hybrid of such technologies. A good example is a hybrid of the wireless local loops and satellite technologies. In fact, we believe the optimum solution for Africa, including its rural areas, is to use heterogeneous networks including all the technologies discussed in Section 2, optimised for the particular environment. Among the individual wireless solutions mobile networks have experienced great success in Africa compared to other networks. Base stations requiring very little maintenance can be placed in strategically accessible locations, powered by solar or wind. We have seen student projects adapting bicycle dynamos for charging mobile phones. It is quite possible to have centres in rural areas where solar or wind power is tapped to generate electrical power to charge mobile phones. In the sections that follow we review the existing wireless network for mobile phone and internet services.

4.3.5 Mobile Network in Africa

Compared to the rest of the world the number of broadband subscribers in Africa is only 0.4% of the Worlds total. Broadband services find applications in e-commerce, e-governance, e-learning, e-health, news and entertainment. It is interesting to note that in 2006 the number of internet users in Africa was 3.8% of the world's 1.1 billion. The mobile phone penetration in 2006 stood at 22 for every 100 inhabitants. This is about seven times higher than the fixed telephone penetration. This shows that the African telecommunication market is in the wireless networks.

Africa's mobile market has been the fastest growing compared to any region for a period of time as shown in Fig. 4.2. It also has a great potential to grow even more

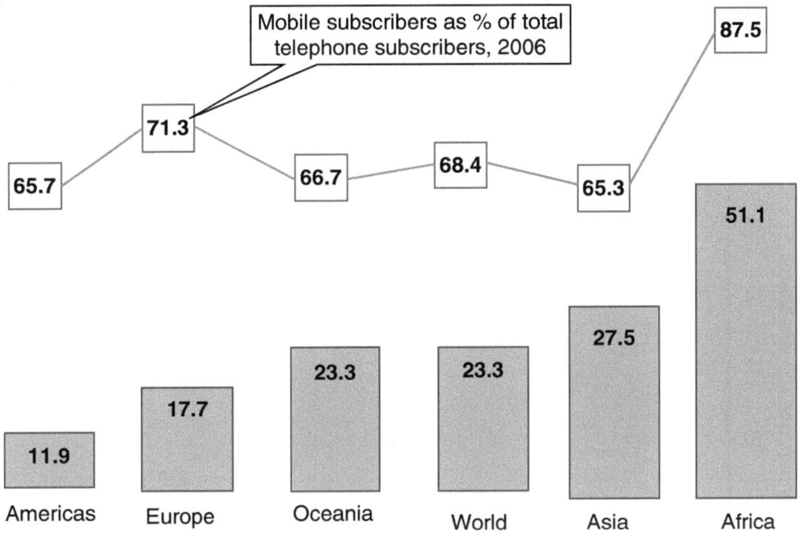

Fig. 4.2 Annual average growth rate in mobile subscribers, 2001–2006

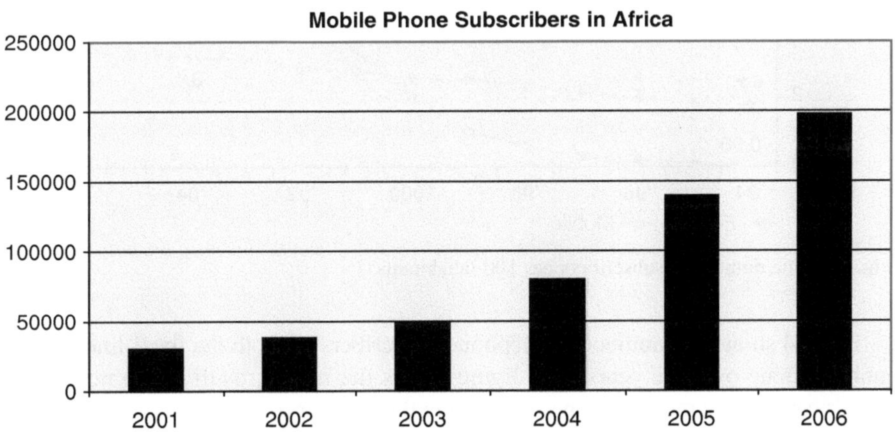

Fig. 4.3 Mobile Subscribers in Africa

as it is far from saturation. The total number of mobile phones at the end of 2006 stood at 198 million and it was expected to rise to 278 million by the end of 2007.

Fig. 4.3 shows the number of mobile cellular subscribers in Africa. In the year 2006 alone Africa added 55.3 million new mobile cellular subscribers to its subscriber base. The number of subscribers added was more in sub-Saharan Africa (excluding South Africa) than in South and North Africa. The Seychelles is the only country in Africa to have passed the 100% total penetration rate with everybody in the country having either a fixed-line or a mobile phone.

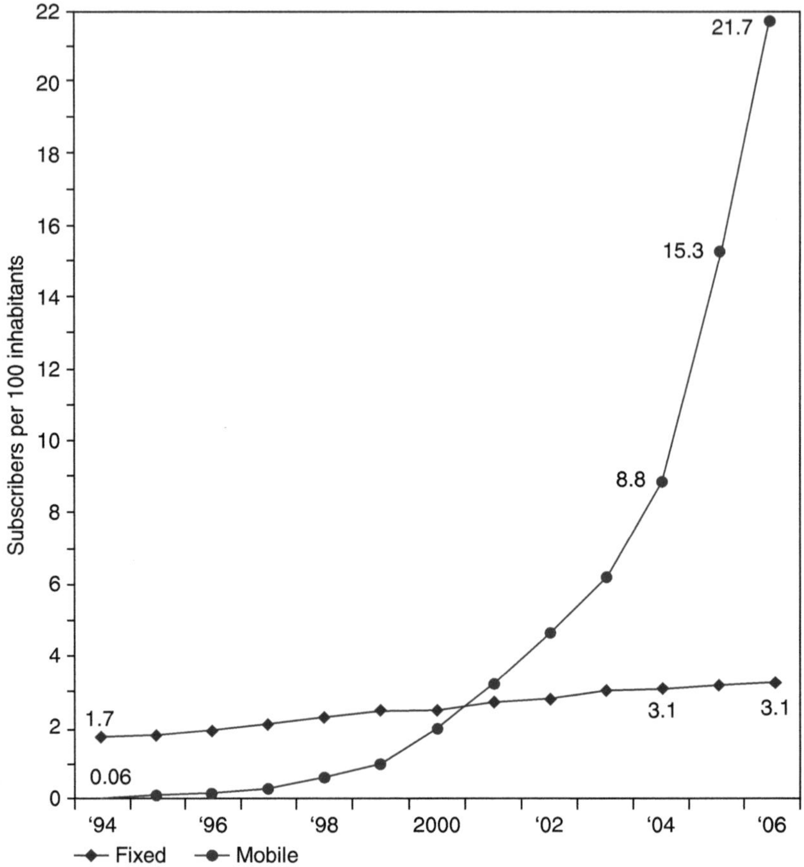

Fig. 4.4 The number of subscribers per 100 inhabitants

Fig. 4.4 shows the number of telephone subscribers for both the fixed-line and the mobile phone over the years. The figure shows the rapid growth in the number of subscribers for the mobile phones overtaking the number of subscribers for the fixed-line that has been around for more than 50 years, in just 5 years. It is also noted from the graph that there is almost exponential growth particularly in the last 2 years.

The rapid growth of the numbers of subscribers using the mobile networks is due to several reasons. Mobile phone networks, which are fully wireless, can be rapidly deployed as they do not require expensive cabling or trunking. Due to their wireless nature they can easily bypass most hurdles that the fixed-lines cannot and hence their coverage in many areas can be guaranteed. Connectivity of a mobile to a network is immediate as long as a SIM card is available and the subscriber has 'air-time'. There are no lengthy and frustrating waiting times typical of fixed-line telephones. Mobile system operators have been able to attract more subscribers because of their creative market strategies which have made it possible for people with low income to be able to afford a telephone service. The pre-paid model (in contrast to the contract model) also referred to as 'pay as you go' makes it possible for a customer to make a call at any time, anywhere and pay only the cost of that call. In addition the mobile phone service is more popular as it is a person to person service. Some operators, such as Celtel, have removed roaming fees across countries sharing a common boarder making it possible to make regional calls and pay local tariffs.

In offering licenses to operators some government have included a condition that the operators must provide some coverage to the non-lucrative markets like the rural regions and provide some connectivity to some of the villages in the rural areas. This is particularly important because since it is estimated that in sub-Saharan Africa about two thirds of the population lives in the rural areas. Whereas the fixed-line coverage is almost inexistent there is about 45 % coverage in the rural areas of sub-Saharan Africa and hence a good potential for mobile phone connectivity.

The largest mobile phone companies in terms of numbers of subscribers and therefore revenues collected include Vodacom (South Africa), MTN (South Africa), MTN (Nigeria), Glo Mobile (Nigeria), Maroc (Morocco), Djezzy (Algeria), Mobinil (Egypt), Vodafone (Egypt), Mobilis (Algeria) and Celtel (Nigeria) each with over six million subscribers. It was projected that the number of subscribers from the continent would be 278 million by 2007 with projected revenue of about US$36.4 billion.

4.3.6 Wireless Internet [5]

In the year 2007 Africa, with 14.2% of the world's population, had an internet penetration of 4.7%. The world average penetration stood at 20%. The reason for the low penetration is mainly due to lack of infrastructure to support the internet. The fact that there is lack of local content and local applications also contributes to its low penetration. Fig. 4.5 shows the top ten internet user-countries and Fig. 4.6 shows comparison of penetration with the rest of the world in 2006.

It is expected that the internet in Africa will be propelled by the following two technologies:

1. 3G mobile phones: The impediments here are that the 3G capable handsets are expensive, they are not easily available and the market is dominated by pre-paid customers.
2. VSAT technology: Though VSAT terminals have declined in price they are still too expensive for low income users.

4.3.7 Observation and Challenges of the African Market

From the discussion above the following observations are made:

1. When one compares the growth rate of wireless services and the fixed services one concludes that the African telecommunication market is a wireless market.

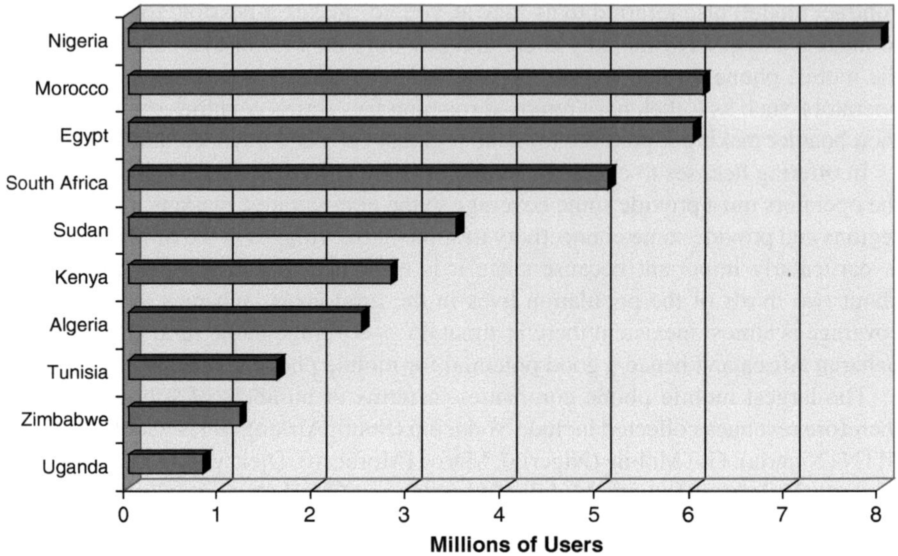

Fig. 4.5 Africa's top ten internet countries (December 2007)

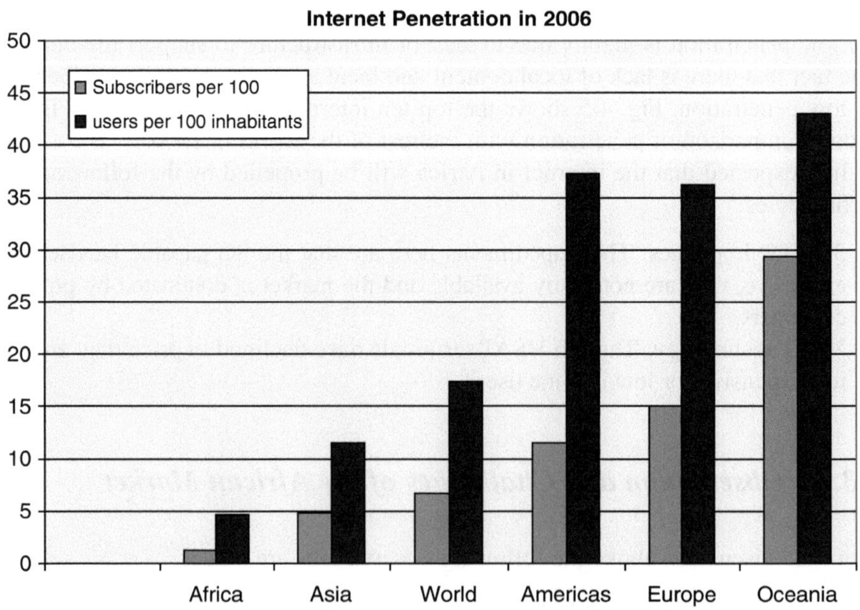

Fig. 4.6 Internet penetration: Africa versus the rest of the World

2. Despite having the fastest growth in wireless services in the world today Africa still has the lowest penetration of both the mobile phones and the internet compared to the rest of the world. This lack of penetration translates to a great business potential from the untapped market.
3. There are several wireless technologies being deployed in Africa and therefore any future network must be able to interface such technologies in a heterogeneous network system. It should be possible to seamlessly roam from one network to another while guaranteeing security and quality of the services.
4. The African market is mainly a voice based pre-paid market. This is created, to some extent, by the business models supported by the service providers. A prepaid market should not be a limitation to growth of the data services as it can be extended to provide such services in the form of data bundles. New applications that involve data services like e-government, e-education and e-health need to be created.
5. There is a lack of internet applications and also applications with local content. This is an untapped market. A demand can be created with the development of relevant applications.
6. The internet subscription per month is prohibitively expensive.
7. Alternative energy solutions, like wind and solar power must be deployed to provide electrical power for the base stations and to charge the mobile phones in the rural areas where electrical power is not available.

4.4 The Impact of Globalization to Africa

Globalization of the mobile and wireless networks will make accessible all types of services, anywhere and at anytime as long as one has the necessary devices or terminals that are connected. Current trends support an IP-based protocol for transport. We believe there will be integration between 3G, (or 4G), WLANs and other access networks to support the ubiquitous network. We believe that in order to maintain the same quality of service for networks under different channel conditions it may be necessary to deploy adaptive schemes; adaptive coding, adaptive modulation and adaptive equalisation. The impact to Africa will be as follows:

1. Services that are available elsewhere will be accessible from any point in Africa. At the moment there is limited internet access and almost a total absence of the broadband services.
2. It will be easy to extend services to even the remotest areas of the world by addition of ad hoc access networks to a network that can handle heterogeneous networks.
3. It will be possible to roam from one network using one type of platform to another type and may be from one service provider to another thus providing effectively more coverage.
4. The readily available access will stimulate growth of new applications with local African content which will create more demand for data services.

The most important end result will come from the use of the more effective global networks to stimulate growth and development of the African communities through applications that will help the people in health-care, education and in poverty alleviation.

4.5 Conclusions

In this paper we have discussed globalization of the mobile phone and the wireless networks. The objective is to make all types of services available anywhere and at anytime and to anyone. We have seen that in order to achieve such an objective we need to develop a system that can interface with the existing heterogeneous platforms, providing roaming possibilities across platforms meanwhile guaranteeing quality of service and security. We have looked at the evolution of the mobile phones and other wireless access networks, and have seen how they have been deployed in Africa to provide voice and data services. We have observed a tremendous growth in wireless networks in Africa over the past 10 years and yet there remains a great potential to grow even more. Globalization will benefit Africa with the added flexibility in the networks and the broader access to a wide variety of services.

References

1. Population Division of the Department of Economic and Social Affairs of the United Nations, *World Population Prospects: the 2006 Revision on World Urbanization Prospects* (February 2008), http://esa.un.org/unpp
2. ITU, *World Telecommunication/ICT Indicators*, http://www.itu.int/IFU-D/ict/statistics/at_glance/ictindicators_2006.htm
3. R.C.V. Macario, *Modern Personal Radio Systems*, IEE Telecommunication Series 33 (Short Run Press Ltd Exeter, 1996)
4. M. Schwartz, *Mobile Wireless Communication* (Cambridge University Press, Cambridge, 2005)
5. K. Daniel Wong, *Wireless Internet* (Artech House, 2005)
6. P. Brenner, *A Technical Tutorial on IEEE802.11 Protocol*, http://www.sss-maf.com/pdf/802–11tut.pdf
7. M.W. Thelander, *WiMax Opportunities and Challenges in a Wireless World* (Signals Research Group, LLC), http://www.cdg.org/resources/white-papers/files/WiMAX%20July%202005.pdf
8. Web-Forum Tutorial, International Engineering Consortium, *Wireless Local Loop*, http://www.iec.org/online/tutorials/acrobat/wll.pdf
9. Panos Media Toolkit on ICTs No.2, *Why Calls in Africa Cost More: the Need for VSATs*, http://wwwpanos.org.uk/files/wsistoolkit2.pdf
10. ITU, *Telecommunication/ICT Market Trends in Africa 2007* (Connect Africa Summit, Kigali, Rwanda, 29–30 October 2007)

Stanley H. Mneney obtained his B.Sc. (Honours) Engineer's degree from the University of Science and Technology, Kumasi, Ghana in 1976. In 1979 he completed his M.A.Sc. from the University of Toronto in Canada. In a Nuffic funded project by the Netherlands government he embarked on a sandwich Ph.D. programme between the Eindhoven University of Technology and the University of Dar es Salaam, the latter awarding the degree in 1988. He currently holds the position of Professor of Telecommunications and Signal Processing and is the Head of School of Electrical, Electronic and Computer Engineering at the University of KwaZulu-Natal in Durban South Africa. His research interests include telecommunication systems, services, networks and their performance. He has done extensive research in low cost rural telecommunications services and networks, channel modelling, mobile communications and Image and Signal processing.

Chapter 5
Globalization of Mobile and Wireless Communications: Today and in 2020: Emerging Global Business Models 2020

Peter Lindgren, Yariv Taran, and Kristin Falck Saughaug

5.1 Introduction

Mobile and wireless communication are a challenge to many innovators and established companies. Especially the old print media companies (Newspapers, book publishers e.g.) are indeed bleeding in these days due to that mobile and wireless technologies have entered the global market place. The strategic question is – should one leave the old technology and business model and join the new technology and market? Should one invest or just wait for technology to improve more?

Business leaders, when asked to explain their future company's strategy to business model in the mobile and wireless communication industry is very diverse. Some invest heavily – some just follows the development waiting for technology, standards and proved business models to come. Many are waiting for critical mass of customers and business. Others with quiet interesting running business models on behalf of the mobile and wireless communication technologies and markets (Apple, qq.com, Cisco) are often very closed or not particularly very communicative about how and where they do their business.

When they however do come up with an answer – Apple, HTC, Google, Nokia e.g. – they would most likely comment and present their new vision for new technology and business models together with their bid on future structure rather than comment on existing business. It is therefore very hard to see if they really are making a profit on mobile and communication technology.

Indeed much technology and many business models in mobile and wireless communication industry is still in the very introduction phase [52] – and many business opportunities are still question marks [52]. However nobody can deny that we are indeed moving into a mobile and wireless communication society – and fast – but do the mobile and wireless communication based markets represent one favourable business model? Many managers at the moment do not really know and few can

P. Lindgren (✉), Y. Taran, and K.F. Saughaug
Aalborg University, Denmark
e-mail: pel@production.aau.dk

R. Prasad et al. (eds.), *Globalization of Mobile and Wireless Communications: Today and in 2020*, Signals and Communication Technology,
DOI 10.1007/978-94-007-0107-6_5,

really imagine what the future business model with this technology will bring us and what the future business model will be about and will represents [17, 36, 46].

How will the moble and wireless communication technology and business models really function in 2020? Do they even have one explicit business model? And assuming they do, do we know how to commercialise and continually innovate these – to create successful businesses that will be able to grow and survive?

However lets take a look back on the history and development of the mobile and wireless industry and lets try to understand what this opportunity and industry is about Then lets have a look on what to become and where we expect to be in 2020.

5.2 The History and Evolution of the Mobile and Wireless Communication Business Models

What are the mobile and wireless communication technology and business model about. Basically today in 2010 the Mobile and Wireless communication technology are about any mobile and any wireless communication in our environment [32]. However in 2020 it will be about any human and business centric activity enable and often conducted by mobile and wireless telecommunication network.

The Mobile and wireless communication industry is not a variation on existing internet services – it is a natural extension of our mobility and living platform. Some would say we finally are given our opportunity of freedom back – because we now will be able to communicate anywhere, anytime, with anybody and with anything. Previously the technology bound us to home, offices and central points where technology allowed us to communicate. The future is opposite.

However it is not just about us getting our freedom back but it is also allowing us via our mobile and wireless devices finally to freely communicate and be mobile at the same time.

How did we reach this point. Well the journey to this stage game through several generations of wireless technology – from 1G to 4G and now towards 5G (Table 5.1).

Six billion people own a mobile phone today and we have been through various generations of cellular systems in the evolution of our mobile communication from first generation to fourth generation. Now almost all the service providers as well as the customers seek for availing 3G and 4G mobile and wireless services. Daily increase in the increase of mobile and wireless customers – increase the potential and reachable market. Mobile and wireless companies is launching high tech featured mobile into the market continuously.

In 2010 we are in the midst of 3G. In China, the 3G service came into existence only by February 2009. It may take time for exposing to other states, but we should also realize many other countries such as the Asian countries were using these services since last decade.

Thanks to heavy research and big investments in mobile and wireless technology we are now heading quickly towards 5G may which we expect will enter within few years. We finally reach the end of the beginning – 5G. We have started to create the

Table 5.1 From 1G to 4G and now towards 5G (First author, Lindgren 2010 inspired from [48])

Generations of wireless communication technology	Description	Time period
1G	The first generation of wireless technology. It was an analog-based technology used exclusively for voice	1979–1992
2G	The second generation based on digital radio technology and is able to accommodate voice and text messages (SMS)	
2.5G	An interim technology based on cell phone protocols such as GPRS (General Packet Radio Service and CDMA2000 (Code Division Multiple Access). This generation can communicate limited graphics, such as in picture text messages (EMS)	
3G	Digital wireless technology, which support rich media such asvideo. 3G utilizes packet switching in the high 15–20 MHz range	2001–
4G	Provide faster display of multimedia	2006–
5G	Real wireless world – completed WWWW: World Wide Wireless Web	–

first part and a new digital communication layer on our world. 5G – Real wireless world – completed WWWW: World Wide Wireless Web will be started from 4G technologies. The evolution will be based on 4G and completed into its idea – a real wireless world. Thus, 5G will make an important difference and add more services and benefit to the world. 5G will be a more intelligent technology and will create more intelligent business models that interconnect – but even more interconnect with the entire global market and without limits.

The technological platform leaves us with some tremendous amount of new applications and opportunity all based on what we did in the past on mobile and wireless communication technologies and we can possibly innovate in the time to come.

However we have not even touched the real potential of these technologies – seen in a business model perspective – because few of us can really imagine what we really can do with this new platform. Still we are playing with the technology and the business models on a rather simple level.

It leaves us with some important choice and decisions to take – which technology and business models should we go for. Just to give an overview of some of the classes of mobile and wireless application we are given were we stand in 2010 – this nice piece of work from [48] gives a brief overview of some of the classes of Mobile and the wireless applications (Table 5.2).

It leaves us with some new digital attributes, opportunities and benefits which we have just begun to investigate and understand. Mobility – is based on th fact that users and items carry a cell phone or other mobile devices everywhere they go. **Mobility** implies portability and thereby the users and items can initiate a "real-time" contact with e.g. commercial, human beings, devices and other systems wherever they happen to be on the globe. **Broad Reach** means that e.g. people, animals, buildings and devices can be reached any time and instantly. These two

Table 5.2 Classes of mobile and wireless applications

Class of application	Examples
Mobile financial applications (B2C, B2B)	Banking, brokerage, and payments for mobile users
Mobile advertising (B2C)	Sending user-specific and location-sensitive advertisements to users
Mobile inventory management (B2C, B2B)	Location tracking of goods, boxes, troops, and people
Proactive service management (B2C, B2B)	Transmission of information related to distributing components to vendors
Product locating and shopping (B2C, B2B)	Locating/ordering certain items from a mobile device
Wireless reengineering (B2C, B2B)	Improvements of business services
Mobile auction or reverse auction (B2C, B2B)	Services for customers to buy or sell certain items
Mobile entertainment services (B2C, B2B)	Video-on-demand and other services to a mobile user
Mobile office (B2C, B2B)	Working from traffic jams, airport, and conferences
Mobile distance education ((B2C, B2B)	Taking a class using streaming audio and video
Wireless data center (B2C, B2B)	Information can be downloaded by mobile users/ vendors
Mobile music/music-on-demand (B2C, B2B)	Downloading and playing music using a mobile device

Classes of mobile and wireless application inspired by [48]

characteristics break barriers of geography and time. They create value-added attributes which will drive the development of mobile and wireless technology and business models to somewhat we just can't imagine today (Table 5.3).

Different people and business will define the new opportunities and implication of the Mobile and Wireless communication field in dissimilar and differentiated ways. This is one of the major potential of mobile and wireless communication.

In order to simplify things however, it would probably be easier to separate the question we would like to touch upon in this chapter – namely focusing on two main components: firstly by looking into:

How the concept of business model has been defined until now? – **As is business models** and later on

What we can expect the 'business model' definition and the functions of the business model in 2020 will come to look like – **To Be – business models.**

5.2.1 Towards an Understanding of the Mobile and Wireless Communication 'Business' Model 2010 – from "As is Business Models" to "To Be Business Models"

Derek F. Abell in his well known 1980 book: "Defining the Business", argued that a business could be defined according to a three dimensional framework:

Table 5.3 Attributes, opportunities and benefits of mobile and wireless applications

Attributes and benefits	Opportunities	Examples of benefits
Ubiquity	Refers to the attribute of being available at any location at any given time	Real-time information and communication independent of user or device location. Easier information access in real-time environment
Convenience	Convenience for users and devices op operate in the wireless environment	Rapid progress, easier and faster access to information
Instant connectivity	Possibility for users and devices to connect easily and quickly to the internet, intranets, other mobile and wireless devices, and databases	Access to information and knowledge
Personalization	Enables information for individual users, devices purposes	Purpose and object related information
Location of products, services, human beings, animals, devices and business models	Enables knowledge about where anybody and anything at any particular moment	Targeting of anybody and anything

First author, Lindgren 2010 Attributes, opportunities and benefits of mobile and wireless communication inspired by [48]

1. Customer Functions: customer functions rendered by the company
2. Customer Groups – customer groups served by the company
3. Customer Technology – customer technology used to produce customer functions and serve customer groups

Abell argued that a company's core business, the Strategic Business Unit (SBU), was related to what is inside the box (Fig. 5.1). This is related to what the company could defined as its core business. Abell argued that it was particularly important to the company to know about and define the company's core business because of the strong relation of the core business to the company's mission, goals and strategy – the strategic perspective.

The SBU box represented therefore the core business of the company. The space external to the box represented at that time the potential for innovation and further business development. Abell argued that a company, in general, should stick to its core business (core business model). By doing so, the company could achieve optimisation of business processes and resource utilisations (operational effectiveness).

Derek F. Abell defined the SBU related to a physical business model world – because at this time Mobile and Wireless communication technology hardly existed.

Abell also argued that when innovating out of the core business area, Abell advised the focal company to carefully analyse its own competences and resources. Abell argued that moving into a new strategic new business field (the customer

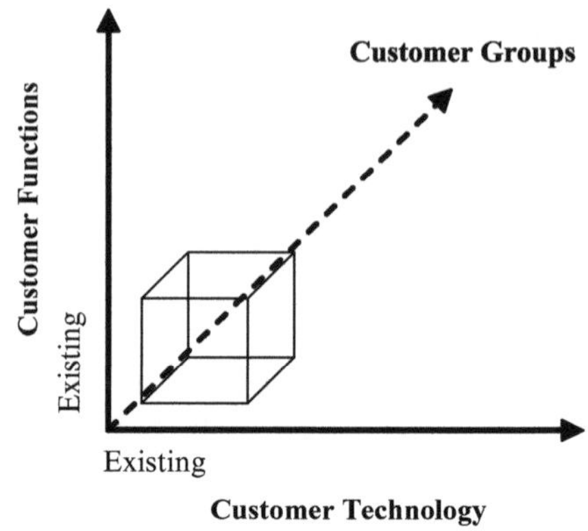

Fig. 5.1 Three Dimensional Framework for the 'Business' Definition [1]

function, customer group or customer technology) would cause a demand for other resources and new competences and, thereby, call for innovation into the field of product development, market development or different degrees of other diversifications.

However, Although Abell's model operated with the three dimensions – which indeed covers and prioritizes some of a company's BM components, it explained very little about how the business (model) structural design really looked like, how it was organized and how it functioned. Why because he was not able to "see" the business model. He was just able to imagine how it could look like and still in an physical context.

5.2.2 Definition of the 'Business Model' Concept

The term 'business model' (BM) became popular in the mid-1990s during the 'dot com era' and up in the 2000s several academics wrote about how the business model looked like and how it could be innovated in a incremental and radical perspective. Some also argued how the business model could be opened [10] (Table 5.4).

Due to today's 'hypercompetition' in a globalizing world, companies in all industries worldwide find themselves competing in ever changing environments. Those changes force companies in 2010 to rethink their operational business models more frequently and more fundamentally, as innovation based solely on single business models, just new products and aimed towards just the local markets is no longer sufficient to sustain their competitiveness and survival. Global competitors can relatively easily copy single business models, products and local market segments today and quickly capture business located elsewhere on the globe (Fig. 5.2).

As mobile and wireless communication technology and business ecosystems emerged many companies have started in 2010 to rethink their business model and business structure by shifting to include mobile and wireless communication into

Table 5.4 Examples of incremental and radical building block innovations (Adapted from [43])

Building block	Incremental innovation 'Do what we do but better'	Radical innovation 'Do something different'
1. Value proposition	Offering 'more of the same'	Offering something different
2. Target customer	Existing market	New market
3. Customer relationship	Continuous improvement of existing relationships	New relationship platforms (e.g. physical, virtual, personal)
4. Value chain architecture	Exploitation (e.g. internal, lean, continuous improvement)	Exploration (e.g. open, flexible, diversified)
5. Core competences	Familiar competences (e.g. improvement of existing technology)	Disruptively new, unfamiliar competences (e.g. new emerging technology)
6. Partner network	Familiar (fixed) network	New (dynamic) network (e.g. alliance, joint-venture)
7. Profit formula	Existing processes generating revenue followed by, or accompanied with, incremental processes of retrenchment and cost cutting	New processes to generate revenues followed by, or accompanied with, disruptive processes of retrenchment and cost cutting

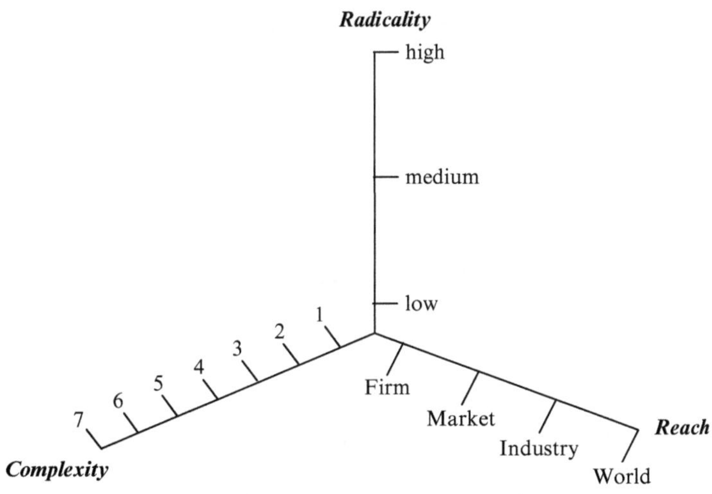

Fig. 5.2 A three-dimensional BM innovativeness scale (Adapted from [43])

their business model on an 'E-form' business basis [29] Companies saw during the 2000s the physical and E-form business model as separated – and many in 2010 still do. However more companies will turned out to see the two business models as integrated and it seems that most companies agree that the business model through 2010s will simply be a combination of the two: a physical and a digital 'business model'.

Accordingly, a company's 'business model' in 2010 serves as a building platform that represents the company's operational and physical manifestation but in

2020 it will serve as a building platform that represents the company's physical and digital operational manifest.

Thus, the challenge for the 2010s business model "designers" is firstly to identify the key elements and the key relationships between the physical, the digital and the virtual business model and secondly to describe and form the company's 'To – Be' business model as an integration of the three – before innovating it.

5.2.3 Components of 2020 Business Models

As part of the ongoing process of defining the physical, digital and virtual business model concept up to 2020 integrated with mobile and wireless communication technologies, it seems obvious that many companies will try to attempt and clarify the components and building blocks of such as well as the construction of a generic BM with an old perspective and view of business models. They will start thinking the business model with old business models framework – with defining it form a physical and single business model perspective. However because of the mobile and wireless communication technology development they will soon jump to the conclusion – that business models have to be thought about in other ways – with a multitude of dimensions – with a multi business model perspective.

Companies business in 2020 will consists of more business models and more layers of business models. They will differentiate to previous seen business models because they will have and consist of several layers of business models – the multi business model approach. A rather simple example of this will be a mobile company doing one business model on one customer and on the same data another business model to another customer – two revenue (R) streams but with just one cost (C) structure – R1 + R2 – C1 = Profit Formula.

The business models will also consist of both a physical, digital and a virtual layer [25], which means that e.g. the product or service will have a physical content (clothing and sales service), a digital content (the digital data on the clothing and extra service on e.g. the webside) and finally the product will just exist if there is a need for it – the virtual layer. All these three layers will be possible business model layers.

By definition all business models will in 2020 bee reachable and globally. Everybody, wherever and whenever will be able to reach the business model either physical, digital or virtually. The global market will be tremendously big for most business models – the long tail [4] theory will come into real life.

However another – and quiet interesting and new to many companies – business model dimension will be that business models in 2020 will always "be on the run" – they will be mobile.

The product lifecycle and customer lifecycle will as such be an integrated part of the business model. Products and services will be covered with sensors and sensor techniques. This means that products and services can be "tracked and traced" where ever, whenever and by who ever. A new business model potential which adds new possibilities to business models – what we called "the real time process dimension".

An overall view on 2020 business models will be that the BM will by definitions be interrelated business model frameworks with an integrated dimension of mobility. This will indeed set new standards to our understanding of business models. Companies will try to build their business models from what they know about in 2010 – what they called 'a unified perspective of business models' but they will soon come to the conclusion that this is not a possible road to drive and they have to change their mindset about business models and business model innovation when they take the real time process dimension in to consideration.

The indication for 2020 business models is that they will also have a cross-model perspective and there will be no single theory that can fully explain the value creation potential of these multi business models.

Consequently, it is very complex to identify a holistic business model and the building blocks of the business model framework for a mobile and wireless communication business model. However, we believe that they will be as an interactive framework between physical, digital and virtual layers of building blocks with multitude of business models placed simultaneously on many firms at a global level and with the mobile and wireless component built in.

Corresponding with the previous findings of business models, [1–54] nobody have yet come near to 2020 multi business model framework.

5.2.4 *Is the Mobile and Wireless Communication Business Model New?*

Having a conceptualization of what the Mobile and wireless communication business models can be about – another thing is how to innovate them. According to previous research Magretta (2002), new business models have until now been variations on a generic value chain underlying all businesses, which eventually can be divided into two categories:

1. All the activities associated with production; e.g. designing, purchasing and manufacturing
2. All the activities associated with selling something; e.g. finding and reaching customers, sales transactions and distributing the products/services

For that reason, according to her, a new business model can be seen as a new product for unmet needs (new customer segment), or it may focus on a process innovation and a better way of making/selling/distributing an already proven (existing) product or service (to existing and/or new customer segments). Or, formulated more generally, a business model until now have been define as new if one of the "building blocks" is new.

According to [3], business model innovation refers not only to products, production processes, distribution channels, and markets, but also to exchange mechanisms and transaction architectures. Therefore they proposed to complement the value chain perspective by concentrating also on processes that enable transactions. In view of that, they concluded that business model innovation does not merely follow the flow of a product from creation to sale, but also includes the steps that are performed in order to complete transactions. Therefore, the business model as a unit of analysis for innovation potentially has a wider scope than the firm boundaries, since it may encompass the capabilities of multiple firms in multiple industries. Also [11] and the IBM global CEO Study (2006) emphasized the importance of business model innovation to appear in the form of organization structure and network relationship changes, such as alliances, joint-ventures, outsourcing, licensing, and spin-offs.

Accordingly, we [25]concluded that any change can rightfully be called a business model innovation, but some changes are more radical and/or complex than others, and some (e.g. radical product innovation, incremental process improvement) reach farther and are less well understood than others (e.g. a holistic, new to the world departure from all business models known so far). Consequently, we get around the eternal discussion of when a BM innovation is indeed radical or incremental, simple or complex, far reaching or not, and, rather, portray the space in which any business model innovation can be positioned in terms of its innovativeness, defined as a composite of radicality, complexity and reach.

However the mobile and wireless communication business model concept was not in this context taken into consideration. Referring to Fig. 5.3: A four-dimensional BM innovativeness scale – the mobile dimension – and some would also call it – the process dimension – has now been taken into consideration.

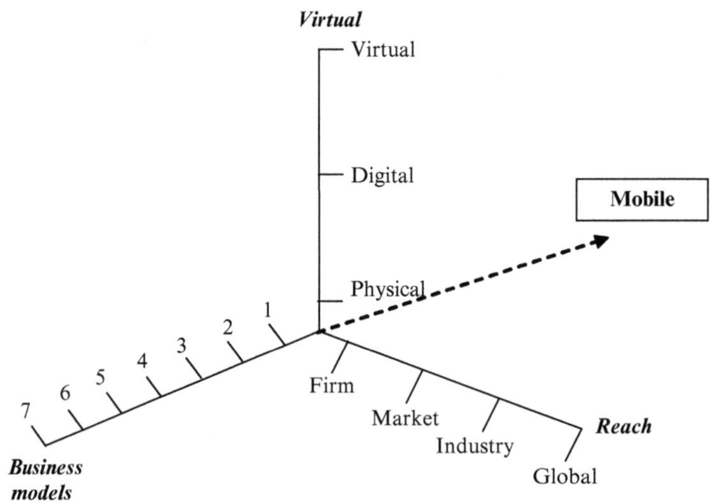

Fig. 5.3 The multitude dimensions of 2020 business models (Inspired by [43])

5.2.5 Levels of Mobile and Wireless Communication Business Model Change

The eternal question in the business model innovation literature is: when can we call something a business model innovation? Is it a new business model if it is new to the company but not within the industry as such? Is it new if the business model is used elsewhere but not in the local industry? Or should it be completely new to all, like e.g. the mobile and wireless communication shopping?

The debate on defining incremental or radical business model innovation [25] concerned the 'how new' question. Incremental business model innovation involves making small-steps on the number of building blocks change – the complexity, the radicallity dimension – defined as the dimension of change of the different building blocks (low, medium, high) and reach (firm, market, industry and world) parameter.

This framework model mainly used existing knowledge to improve, for example, existing products/services, customer practices and structures. It was very short term based and it focused on the focal company performing better than it already did. Radical business model innovation, also called quantum leap innovation, is however the ability to develop new knowledge and may even be based on competences of many companies – the network based business model [25]. It does not possess at the start of the process and result in products/services that cannibalize or remove the basis of existing products/services [45], but address new dimensions of all building blocks at both a physical, digital and virtual layer.

In order to understand how to innovate a business model, [40] argued that you first need to unpack it into individual components and understand how all the pieces fit together in holistic way. Furthermore, according to them, in order to build a breakthrough business model that rivals will find difficult to imitate, companies will need to integrate a whole series of complementary, value creating components so that the effect will be cumulative. Well this is exactly what the mobile and wireless communication business model innovation is about in 2020. This will take the business model innovation, suggested by [23] right to the level of radicality of business model innovation and presents what they called change models characterized as the extension models and journey models.

Extension models include radical changes by developing new markets – integrated physical, digital and virtual markets, new value chain and value chain functions with again physical, digital and virtual layers, and product/service lines which is both physical, digital and virtual. Most interesting considering the last building block in our seven building block business model [25] – the profit formular – will both include physical digital and virtual money. The *journey model* has taken its begin – involving a complete transformation of original business models – where numerous of companies moves deliberately and purposefully to new operating multi business models. The enabler to this development is indeed the mobile and wireless communication technology, which is believed in 2020 to enable and support this evolution.

5.2.6 Open Mobile and Wireless Communication Business Model

[11] introduced a whole new way of thinking about business models and innovation of business models in his book *Open Business models – How to Thrive in the New Innovation Landscape*. He argued towards open innovation (2005) and then to opening companies' business models (2008). Until then, all academic work [1, 23, 30, 27, 31] and practically all business thinking had been related to a 'closed' business model – sticking to your core business – where the business model was strongly related to a single and focal company. Chesbrough argued that 'a closed business model' was not efficient for future innovation and a global competitive environment. Much innovation was never even related to the core business model and companies who could have used the results of the innovation were prevented from this because of patents, unwillingness to network and even by lack of knowledge by those who had innovated.

Chesbrough argued that, in the future, companies would have to open up their business models and allow other companies to integrate them with their own business models and even take parts out of the model to use it in other business models. These innovations could be extremely valuable and more effectively used in other business models by other companies.

Because of 'a new global business environment' integrated with mobile and wireless communication technologies we claim that Chesbrough's theorem in 2020 will be taken even further. The business models of companies, their products and services, customers, customer relationships, key resources, competences, value chain and profit formulas [25] will have a much deeper structure than Chesbrough ever though about back in 2008. The business models to come will be managed in a much more open way than Chesbrough proposed and they will in 2020 have more and much more open layers than Chesbrough thought about. Developed and registered patents will be open to other companies or other business models of interest in order to support and develop other business models or the innovation of new multi business models. Previously developed innovations, IPR and competences will become much more value adding building blocks to both existing and new business models in 2020 because of the mobile and wireless communication technology.

Companies will play by different rules in 2020 than in 2010 using different and many business models simultaneously. This open approach – "internet in the clouds" – with "internet in all things" and later "human beings everywhere" will give new opportunities and challenges for both 'giver' and 'taker' of business models and business model innovation. We believe tt will also completely restructure the theory of mobile and wireless communication and related business models leading to a network oriented business model level. Networks – that at the same time, are physical, digital and virtual. These new multi business models will have more different partners' value equation included and will be completely open to all network partners of interest, both within and outside the value chain.

The open business innovation model has thereby in 2020 become not just a true and important concept for further work and development of business models – but has in major parts also be realized.

5.3 Mobile and Wireless Communication as a Key Enabler to New Business Models

Mobile and wireless communication technology we know already today will be of ever increasing importance to innovation of new business models and as the backbone of the new business models in 2020. The business models in 2020 will be based on mobile and wireless technology and operates within a mobile and wireless based market environment. This means that pretty much everything will be able to communicate with everybody due to advanced sensor technology and advanced wireless communication technology. Hereby the body of a human being, animals, devices, buildings, clothings e.g. will be able to communicate each other.

This opens for new business models. In 2010 we have only very limit possibilities to communicate from inside a body to outside a body of a human being. Pace makers and bigger instruments put attached to the body is today some of our possibilities but in future 2020 we will be able to communicate to and with practical every part of the body both on small and large scale perspective – organs, legs, arms, blood, brain can be objects for communication and measurements.

Digital devices of our "brains" "intellectual capital" – or devices that have learned our habits, interest and network partners will on an intelligent way communicate for us and thereby help us in a user centric world.

Virtual "avatars" of us and our business models including more advanced secondlife.com, google.com, facebook.com, linkedin.com and World of Warcraft liked environment will "play" or be allowed to "play" our virtual person and virtual life.

In a short term perspective, these business models will run at cost, speed, energy consumption vice, performance and change in the global market that are far above what we have achieved and would believe is possible today.

The development will reach a point of no return in 2020, where a even strong focus on mobility and "wireless life" of human beings and devices to cope with the increasing demand of slipping out of time pressure, being more agile, more flexible is the trend. Business models which are independent of time, place and people will be at the first generation. Integrated business models between physical, digital and virtual business models will be realized (Fig. 5.4).

The multi business models are created inside the cubic as presented in the model. However at the moment we do not know how they look like. Our research in the next years will focus on the multi business models and we expect to come up with advanced business models but we will probably not reach a standardized typology level before 2020.

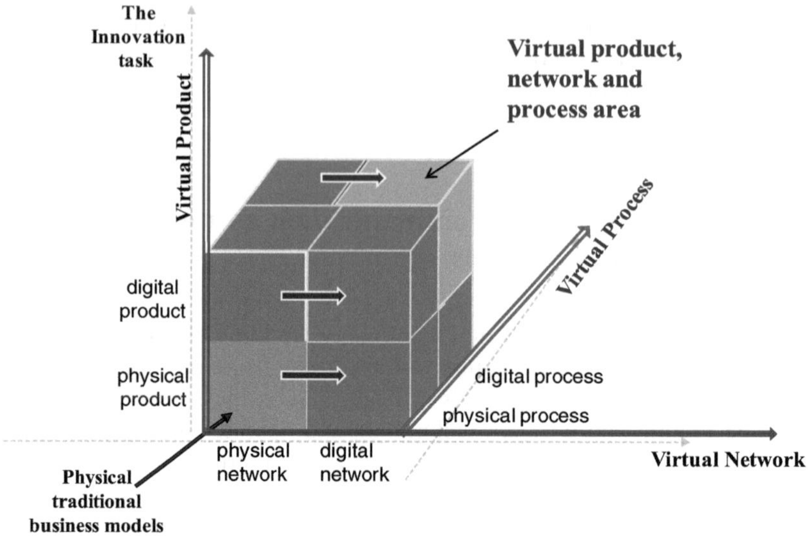

Fig. 5.4 Generation of physical, digital and virtual business model based on networks (First author, Lindgren 2010 inspired by [25])

5.4 Mobile and Wireless Communication a Move Toward a Process Enable New Business Models

An interesting synergy and spinoff of the above mentioned model is a first step towards a completely new and change in our understanding of the business model – taking us from a static understanding and measurement of the business model to a process based business model.

Business models in 2020 will not only consist of physical, digital and virtual layers but are also based on a process – integrated, connected, mobile, wireless in a continuously process of change – where ever and whenever the user or the customer demands it. All of "the 7 building blocks" in the business model will be in process. We predict that all building blocks can and will change simultaneously in the processes – both individual and across and together – creating continuously interesting new multi business models.

This is both a real challenge of the future multi business models but also the real potentials of multi business models above 2020.

In a long term perspective continuous improvement, learning and innovation [25], in combination with the continuous need to develop new multi business models, increases the importance of standards, security, legal rights and ethics of multi business models. The ability to include and create mega information fast, flexible and dynamically is not a question of whether it is possible. This will be the innovation platform to every company and network.

So – the answer to the question – What will be the global business models in 2020 – is that practical all business models of the future is global. We expected that all future business models will mainly be based on mobile and wireless communication technologies and will thereby be reachable from all parts of the globe. The majority of these future business models will include both physical, digital and virtual business components.

5.5 Discussion

By glancing through the overall research and practice that have been done so far in business model related to the mobile and wireless communication phenomena shows a rather fuzziness of the 'business model' concept and a very simple and single business model perspective. No matter why the conclusion is very simple – we are standing in front of a revolution of our understanding of the business model. Mobile and wireless communication technology have created several new layers and several new dimensions to our previous understanding of the business model.

Chesbrough commenced the trend on open business model but the dimensions of the mobile and wireless communication enabled business models to our opinion cannot be included in his framework. Our current research on business models and innovation has to be reinvented.

From a theory development perspective it is at the previous moment – 2010 not possible to agree on one argument what the phenomena will look like in 2020. Accordingly, in order to provide a solid ground for the multi business model (theory building) research to be built up, and based on the similarities/differences between the various explanations and perspectives of the multi business model, there is a need to narrow down and sharpen the large variation of potentials of mobile and wireless communication, as well as to develop new concepts, ideas and categories to the term 'multi business model' in general, and its innovation processes in particular.

Firstly, as to the components of, what we propose it to be in 2020, the *core* of the business model, despite the large variation in opinions we can still identify a strong integration between the different components. We proposed a multitude dimensional model including radicallity, reach, complexity and mobility as core dimensions for characterizing the business model to come in 2020.

The next issue concerns the question: when and what can we expect of the new models of business models in 2020? Three approaches have been proposed in this chapter. The first approach 'defines' the 2020 business model as a multi business model with integrated physical, digital and virtual parts. A multi business model in continuously change in any of the building blocks or the relationships between them as a form of dynamic multi business model innovation could be considered. The form and number of building blocks are changed simultaneously which we consider as the most radical form of business model innovation we expect will become in 2020.

If we combine the 'scale' of radicality (new to an individual, group, company, industry, world), and mobile a four-dimensional space emerged, which does not perhaps provide a precise definition of the multi business model we can expect in 2020, but helps indicate the radicality or, perhaps rather, the real complexity of business model and business model innovation in 2020.

This more accurate language for discussing mobile and wireless communication business model or better enabled business models will help us to better understanding of the the managerial challenges and issues involved using existing innovation and change theories. Not much of previously understood business model theory will be valuable I understanding business models in 2020. Maybe just the building blocks can be transferred – but even our previous customer definition and definition of money – and the profit formula – has to be revised.

5.5.1 Building a Process to Mobile and Wireless Communication Multi Business Model Innovation

Our findings so far has illustrated that the multi business model phenomenon is not well understood until now and innovation of these enabled by mobile and wireless communication technology is just partly understood, and the combination of the two leads to a high level of uncertainty and fuzziness, particularly to many company managers who will consequently be faced with tremendous challenges and obstacles when attempting to successfully innovate their multi business model (i.e. financial obstacles, strategic (choices) obstacles, operational obstacles, cultural obstacles, etc.).

Given this, and the understanding of radicallity in multi business model innovation seems to be indeed a 'risky business', is leading to the following tentative theory statement, namely: *Implementation of risk management processes within the overall innovation process might reduce the level of uncertainties, overall costs and (implementation) time, when developing business models of 2020.*

5.6 Futuristic Outlook on Emerging Globale Mobile and Wireless Communication Business Models in 2020

The business model concept will continue to evolve and embrace new perceptions, challenges and opportunities. As part of the authors' preliminary research, the following important trends and characteristics were found related to emerging globale mobile and wireless communication business models in 2020.

Table 5.5 above is clearly demonstrating that the changes in the economy will be the rules of the game and companies must in the future learn to adjust their mindset to a multi business model perspective.

Table 5.5 Future trends to Mobile and wireless BM's innovation

Context for Innovation	2010	2020
Market	Commencing global markets	Fully global market
	Unstable	Fragmented, dynamic, customised, mobile and process oriented
	Mainly physical but commencing digital and virtual market	New markets (Blue Ocean)
		Integrated physical, digital and virtual markets
Technology	Single technology	Mix of integrated mobile and communication technology or multi-technology based
	Expensive	Satellite based
	Data power low	Sensor based
	Stable	4G and preliminary 5G based
		Data power capacity challenged
		Stable technology
		Standardised technology
		Security technology based
		Stable new technology changes
Network	Closed networks, local networks, Fixed networks	Open physical, digital and virtual networks
		Dynamic networks, integrated physical, digital and virtual networks,
		Global physical, digital and virtual networks
Companies' competences	Unstable competences under development inside primary companies or in narrow networks	Dynamic – flexible and virtual competences
		Competences continuously under development
		Competences developed within a multitude of network and together with networkpartners – shared core competences and skills in the innovation process
Products	Mostly physical products to some extent immaterial products	A mixture of physical, immaterial, digital and virtual products and services – however with an overweight to digital and virtual products
	Unstable product – Short life cycle	Continuous development of physical, digital and virtual products and service – short life cycles of products and service
	Limited distribution and marketing channels	Multitude of physical, digital and virtual distribution and marketing channels
Business model innovation process	Un stable models	Many business model innovation models (flexible models, dynamic models, learning by doing, using, interacting)
	Slow, linear business model innovation process	Rapid prototyping models of business models
		Lean business model innovation process

(continued)

Table 5.5 (continued)

Context for Innovation	2010	2020
Success criteria	Individual success, business model innovation speed, business model time to market, business model cost and performance, business models to the local market. Emphasis on short term success criteria. More emphasis on continuous improvements and managing tangible assets efficiently	Network-based business model success, right-speed business model innovation, business model time to market, business model cost and performance optimisation, global business model markets Emphasis on sustainability in business models- short and long term success criteria are integrated More emphasis on radical business model innovation Most emphasis on managing intangible assets efficiently in business models Focus on Innovation leadership

The new multi business model based economy is fast coming. It is highly user and customer driven. Globalization are being strongly related to network and differentiation, learning and continuous innovation considered to be implemented via continuously new and open network formations.

5.7 Mobile and Wireless Communication Business Model Innovation Space in 2020

The dimensions of multi business model innovation space, presented earlier can be further analyzed into possible expected outcomes, when companies are considering how to innovate their multi business models, which eventually can result in (roughly) eight possible multi business model innovation outcomes (Table 5.6).

5.7.1 Future Thinking is also About Network-Based Mobile and Wireless Communication Business Model Platforms

Competition in 2020 will not be single company based and single business model will hardly exist. Consequently various ways of open network based multi business model created in network-based innovation processes will be the standard. A multitude of external competences and knowledge zones [4], among others, will be involved together with magnitude of customers and suppliers, knowledge consultants, venture capitalists and competing networks. Open network based innovation

Table 5.6 Mobile and wireless communication business model innovation space in 2020

Radicality	New to whom	Number of building blocks changed	Expected business model outcome
High	Low	Low	**'Me too' BM** Follow the others. Second mover strategy. Radical new value, but only to the organization
High	Low	High	**'Me too' BM** Despite the radical internal change processes, it is still a second mover strategy
High	High	Low	Potentially, **Highly competitive BM** Involved relatively low change management (and risks) processes to be implemented
High	High	High	**Radical new BM** Very risky, but also very competitive (blue-ocean)
Low	Low	Low	**Laggard BM** Not perusing any radical innovation, focus is on continuous improvements, risk of staying behind competition for the long run
Low	High	High	Potentially, **Highly competitive BM** Requires high levels of change management (and risks) processes to be implemented
Low	Low	High	**Laggard BM** Not perusing any radical innovation, stay behind competition for the long run, and also highly risky due to high levels of incremental innovations and change management processes that are needed to be implemented simultaneously
Low	High	Low	Potentially, **Highly competitive BM** Involved relatively low change management (and risks) processes to be implemented

assist to better leverage internal and external capabilities in the networks, as well as in opening continuously new markets and continuously experimenting with new cutting edge technologies and markets. Consequently, many networks are finding themselves increasingly tied to other networks either physical, digital and or virtual or all three integrated. The challenge each of them faces is to adjust their multi business model continuously meet their networkpartners' core competences and multi business models' value, so as to create a new platform for collaboration and innovation. This tendency, if continued, will have tremendous consequences on the way that we will see business, the company terminology – if it will keep on existing in 2020. Competition in 2020 will take new dimensions and we will learn new dimensions of competition.

Consequently, open multi business model innovation initiatives are expected to grow significantly from 2010 to 2020 in volume (i.e. joint ventures, spin-offs, licensing, outsourcing, and alliances). Open multi business model innovation, which can also be considered as network based business model innovation platforms, may therefore in 2020 either result from most possible equal partnership, or take place within loose, and more diversified physical, digital and virtual networks. In either case, some network partners will be forced to change their individual business model logic in the time to come – some more radically than others. In addition, the development of a network-based business model innovation may result in a radical change of our understanding of the customer, the customer and product lifecycle, since they involve and are integrates in the mobile and wireless technology, with new and a magnitude value propositions, new and magnitude of value chains, new and magnitude of network formations, and new markets.

Furthermore, the new network based market formation ties will not be characterised by industrial homogeneity, but rather by the large diversification of its network partners' identity for the purpose of pursuing radical multi business model innovation possibilities.

All in all, the present research and description has led the researchers to believe that the business model concept until 2020 will continuously evolve and embrace new perceptions, challenges and opportunities – beyond out 2010 imagination.

5.8 Conclusions

The first generation, 1G wireless mobile communication systems, was analog and the speeds up to 2.4 kbps introduced in the early 1980s and completed in the early 1990s. The second generation, 2G system, fielded in the late 1980s and finished in the late 1990s, was digital signal for voice and the speeds up to 64 kbps. the third generation, 3G wireless system provided the transmission speeds from 125 kbps to 2 mbps developed in the late 1990s and was well-done in the late 2000s. The fourth generation (4G) does not really exist yet – basically speaking – 4G is an evolution to move beyond the limitations and problems of 3G.

The speeds of 4G can theoretically be promised up to 1 gbps. The beyond will be 5G with incredible transmission speed with no limitation for access and zone size.

The differences between 3G and 4G are data rate (speeds), service, global roaming, interface with wire-line internet, qos and security. 4G will be supported by ipv6, ofdm, mc-cdma, las-cdma, uwb and network-lmds. c5G will be the completed version of wwww, world wide wireless web, to form a real wireless world with no more limitation with access and zone issue.

In 2010 Mobile and wireless technology is getting even more popular and important everyday. In this chapter we briefly introduced an overview of the evolution and the background of 1G–5G, compared the differences and illustrated how 5G may work for business models in the 2020. 4G just right started and there is still many standards and security technologies to come, which are still in developing process. Therefore, no one can really be sure what the future 5G will look like and what services it will offer to people. However, we can get the general idea about 5G from academic research; 5G is the evolution based on 3G's and 4G's limitation and it will fulfill the idea of wwww, world wide wireless web, offering more services and smooth global roaming with low cost.

A lot of buzzwords have come and gone in time, but it seems the business model concept will change tremendously until 2020. Despite its fuzzy definition and operationalization, it is capturing more and more attention of academics as well as company managers.

The objective of this chapter was to build a better understanding to global Mobile and wirless communication business models in 2020.

Through the chapter, we introduced and illustrated a framework of future and emerging new business models to come in 2020 and together the implementation process together with the challenges of these BM innovations.

It is the researchers' belief that the future BM concept, particularly with relation to its innovation will be one of a multi business model. A new understanding and theory of business models and what they are all about seems to be developing quickly at present. It is the researchers believe that the future BM concept should be considered and analyzed from a network perspective. A more network based business model moving from a closed to an open business models is developing, with much more focus on values, processes, cost – not only of partners inside the multi business models but also those outside – network partners who are either in the first, second or even third levels of the chain of the multi business model. Multi business models which are integrated physical, digital, and virtual models.

So when addressing a futuristic outlook on emerging multi business models and emerging business models based on Mobile and wireless communication technology – we need more research about the multi business models and how these will look like. This concerns both a theoretical and a practical perspective. Some short cases will point to some future examples and possibilities of multi business models.

The aim of this chapter was to open the discussion on multi business model and multi business model innovation as a important field of research, as well as in drawing future potential innovation possibilities, and expected outcomes, when considering how to innovate companies' business models.

References

1. D.F. Abell, *Defining the Business: The Starting Point of Strategic Planning* (Prentice-Hall, Englewood Cliffs, NJ, 1980)
2. A. Afuah, C. Tucci, *Internet Business Models and Strategies* (McGraw-Hill, Boston, 2003)
3. R. Amit, C. Zott, Value creation in e-business. Strat. Manage. J. **22**(6–7), 493–520 (2001)
4. Chris Anderson, *The Long Tail Hyperion*, 2008. ISBN 978 – 1 – 4013-0966-4
5. L.M. Applegate, E-Business Models: Making Sense of the Internet Business Landscape, in *Information Technology and the Future Enterprise: New Models for Managers*, ed. by G. Dickson, W. Gary, G. DeSanctis (Prentice-Hall, Upper Saddle River, NJ, 2001)
6. M. Baker, S. Hart, *Product Strategy and Management* (Prentice-Hall, Harlow, 2007), pp. 157–196
7. J.B. Barney, Firm resources and sustained competitive advantage. J. Manage. **17**, 99–120 (1991)
8. R.S. Burt, *Structural Holes: The Social Structure of Competition* (Harvard University Press, Cambridge, MA, 1992)

9. P.R. Carlile, C.M. Christensen, *The Cycles of Theory Building in Management Research Version 6.0*. Working paper (Innosight, 2005), www.innosight.com/documents/Theory%20Building.pdf
10. H. Chesbrough, R.S. Rosenbloom, The role of the business model in capturing value from innovation: Evidence from XEROX Corporation's technology spinoff companies. Indus. Corpor. Change **11**(3), 529–555 (2000)
11. H. Chesbrough, *Open Business Models. How to Thrive in the New Innovation Landscape* (Harvard Business School, Harvard, 2006)
12. C.M. Christensen, The ongoing process of building a theory of disruption. J. Prod. Innov. Manage. **23**, 39–55 (2006)
13. R. Cooper, *Winning at New Products: Accelerating the Process from Idea to Launch*, 2nd edn. (Addison-Wesley, Boston, 1993)
14. Y. Doz, G. Hamel, *Alliance Advantage* (Harvard Business Press, Boston, MA, 1998)
15. J. Dyer, H. Singh, The relational view: cooperative strategy and sources of interorganizational competitive advantage. Acad. Manage. Rev. **23**, 660–679 (1998)
16. J. Gordijn, *Value-Based Requirements Engineering - Exploring Innovative E-Commerce Ideas*. Ph.D. thesis, Vrije Universiteit Amsterdam, The Netherlands, 2002
17. Green Tech Media (2010)
18. G. Hamel, *Leading the Revolution* (Harvard Business School Press, Boston, 2000)
19. IBM, *Expanding the Innovation Horizon* (2006), http://www-35.ibm.com/services/uk/bcs/html/t_ceo.html
20. IBM, *The enterprise of the future* (2008), http://www-935.ibm.com/services/us/gbs/bus/html/ceostudy2008.html
21. T. Kuhn, *The Structure of Scientific Revolutions* (University of Chicago Press, Chicago, 1962)
22. R. Leifer, *Critical Factors Predicting Radial Innovation Success* (Technische Universität, Berlin, 2002)
23. J. Linder, S. Cantrell, *Changing Business Models: Surveying the Landscape* (Accenture Institute for Strategic Change, Cambridge, 2000)
24. P. Lindgren, Y. Taran, A.M. Schmidt, in *The analytical model for NEWGIBM*, ed. by P. Lindgren, E.S. Rasmussen (New Global ICT-Based Business Models, Aalborg Universitetsforlag)
25. P. Lindgren, Y. Taran, H. Boer, *From Single Firm to Network-Based Business Model Innovation*. Int.J. Entrepreneur. Innov. Manage. (IJEIM) (2009)
26. E. Loukis, E. Tavlaki, Business Model: A Perquisite for Success in the Network Economy, in *BLED Proceedings* (2005)
27. J. Magretta, Why business models matter? Harv. Bus. Rev. **80**(5), 86–92 (2002)
28. B. Mahadevan, Business models for internet-based e-commerce. Anat. Cal. Manage. Rev. **42**(4), 55–69 (2000)
29. J.F. Moore, The New Corporate Form, in *Blueprint to the Digital Economy – Creating Wealth in the Era of E-Business*, ed. by D. Tapscott, A. Lowy, D. Ticoll (McGraw-Hill, New York, 1998)
30. M. Morris, M. Schmindehutte, J. Allen, The entrepreneur's business model: toward a unified perspective. J. Busi. Res. **58**(6), 726–735 (2003)
31. A. Osterwalder, Y. Pigneur, L.C. Tucci, *Clarifying Business Models: Origins, Present, and Future of the Concept*. Commun. AIS No. 16, 1–25 (2004)
32. Paavilainnen (2002)
33. O. Petrovic, C. Kittl, R.D. Teksten, Developing Business Models for E-Business, in *Proceedings of the International Conference on Electronic Commerce,* Vienna, October–November
34. M.E. Porter, *Competitive Advantage: Creating and Sustaining Superior Performance* (Free Press, New York, 1985)
35. M.E. Porter, Strategy and the Internet. Harv. Bus. Rev. **79**(3), 63–78 (2001)
36. Printed Media (2010)
37. E.M. Rogers, *Diffusion of Innovations*, 3rd edn. (The Free Press, New York, 1983)
38. J.N. Rosenau, Turbulent Change, in *International Relations Theory: Realism, Pluralism, Globalism*, ed. by P.R. Viotti, M.V. Kauppi (Macmillan, New York, 1993), pp. 438–448
39. A.J. Schumpeter, *Capitalism, Socialism and Democracy* (Harper & Row, New York, 1942)

40. P. Skarzynski, R. Gibson, *Innovation to the Core* (Harvard Business School Publishing, Boston, 2008)
41. P. Stahler, *Geschäftsmodelle in der digitalen Ökonomie. Merkmale, Strategien und Auswirkungen*. Ph.D. thesis, University of St. Gallen, Switzerland, 2001
42. D.W. Stewart, Q. Zhao, Internet marketing business models, and public policy. J. Pub. Policy Market. **19**, 287–296 (2000)
43. Taran, B. Lindgren, *Managing Risks in Business Model Innovation Processes* (CInet Zürich, 2010)
44. D. Tapscott, D. Ticoll, A. Lowy, *Digital Capital. Harnessing the Power of Business Webs* (Harvard Business School Press, Boston, 2000)
45. J. Tidd, J. Bessant, K. Pavitt, *Managing Innovation. Integrating Technological, Market and Organizational Change* (Wiley, Chicester, 2005)
46. Times (2010)
47. P. Timmers, Business models for electronic markets. J. Electron. Market. **8**(2), 3–8 (1998)
48. T. Efraim, *Electronic Commerce a Managerial Perspective* (2004, 2006, 2008, 2010). ISBN-13 978-0-13-224331-5
49. K.T. Ulrich, S.D. Eppinger, *Product Design and Development* (McGraw-Hill, New York, 2000)
50. P. Weill, M.R. Vitale, *Place to Space* (Harvard Business School Press, Boston, 2001)
51. B. Wernerfelt, A resource-based view of the firm. Strat. Manage. J. **5**, 171–180 (1984)
52. O.E. Williamson, The economics of organization: the transaction cost approach. Am. J. Sociol. **87**(4), 77–548 (1981)
53. Y. Wind, A new procedure for concept evaluation. J. Market. **37**, October, 2–11 (1973)
54. A.B. Whinston, D.O. Stahl, S. Choi, *The Economics of Electronic Commerce* (Macmillan Technical Publishing, Indianapolis, IN, 1997)

Peter Lindgren, Associated Professor, Ph.D. is Associated Professor of Innovation and New Business Development at the Center for Industrial Production at Aalborg University. He is manager of International Center for innovation and he holds a B.Sc. in Business Administration and in M.Sc. in Foreign Trade and Ph.D. both in Network based High Speed Innovation. He has (co-)authored numerous articles and several books on subjects such as Product Development in Network, Electronic Product development, New Global Business Development, Innovation Management and Leadership and High Speed Innovation. His current research interest is in New Global Business models, i.e. the typology and generic types of business models and how to innovate these.

Yariv Taran, Ph.D. Student, M.Sc. Economics and Business Administration, is a research fellow at the Center for Industrial Production at Aalborg University. His current research interest lies in the area of business model innovation. Other areas of research interests include intellectual capital management, knowledge management, entrepreneurship and regional systems of innovation.

Kristin Falck Saughaug, Research fellow, M.A. Theology, is a research fellow at the Center for Industrial Production at Aalborg University. Her current research interest lies in the area of the creating human being. Focus is the potential of innovation defined via the connection of different knowledge and business domains. Other areas of research interests include innovation leadership, intellectual capital management, human resource and organizational theory.

Chapter 6
Access Security and Personal Privacy in Public Cellular Communication Systems: The Past, the Present and Beyond 2020

Geir M. Køien and Vladimir A. Oleshchuk

6.1 Introduction

In order to predict the future one needs to understand the past and then interpolate as best as possible. We expect this to work reasonably well for a "2020 Scenario", but we do not expect this approach to be valid for a "Beyond 2020" scenario.

- **24 Years ago: In the NMT450 years (1G)**
  Back in 1984 the 1G NMT system represents the state-of-the-art. It contained little or no security, but originally there were n006F intruder present. That changed soon enough, and the 1G systems started to suffer from subscription fraud and eavesdropping.
- **12 Years ago: In the midst of circuit-switched GSM (2G)**
  In 1996 the circuit-switched GSM system was the rising star. GSM had access security from day one, but one did not have a risk/threat analysis and so the security architecture is incomplete.
- **Present: Maturing 3G systems, approaching 4G systems**
  Today we have the 3G systems and the 4G systems are being defined. The 3G systems have relatively sophisticated security architectures, but there are important constraints inherited from the 2G systems. These both prevent effective personal privacy protection and it means that the outdated trust model found in the 2G systems is still in use.
- **In 2020: 5G system (whatever that means)**
  What will become of the cellular systems in 2020? We predict that the 4G systems will still be around and that the 5G systems will retain some compatibility

G.M. Køien (✉)
Telenor R&I, Norway
e-mail: geir.koien@uia.no

V.A. Oleshchuk
University of Agder, Norway

R. Prasad et al. (eds.), *Globalization of Mobile and Wireless Communications: Today and in 2020*, Signals and Communication Technology,
DOI 10.1007/978-94-007-0107-6_6, © Springer Science+Business Media B.V. 2011

with the 4G systems. In order to have credible personal privacy the subscriber addressing model will probably change and the limitations of the 2G/3G trust model will hopefully be removed.

- **Beyond 2020: Speculations about the future**
  Beyond 2020 we assume that compatibility with today's systems is no longer an issue, but there will nevertheless be practical constraints. We will likely see some very sophisticated cryptography being available, but it is not clear to which extent new security services can be supported in the overall system architecture. Furthermore, the intruder in a beyond 2020 scenario will be very powerful and entirely new attack types will be available to the intruders.

6.2 Assumptions

There are certain inescapable premises that a public cellular systems must take into account. Our investigation therefore postulates the following invariant assumptions:

Assumption 1: Public cellular systems. We only investigate public cellular systems. Other systems are not considered.

Assumption 2: Access part only. We only investigate security and personal privacy in an access related context. That is, we do not investigate application layer aspects in this chapter.

Assumption 3: Lawful interception. Like it or not, but public cellular systems are required to adhere to regulatory constraints, including support for *lawful interception* [1], *emergency calls* [6] and a set of "terrorist prevention" related measures (like the UE Data Retention Directive [10]). There is no reason to assume that this basic premise will be relaxed in the future, although the actual procedures may change. The standard lawful interception legislation seems to be relatively well justified, mature and stable. The present anti-terrorist legislation and the personal privacy legislation are much less mature and the justification is in many ways much more susceptible to change due to public perception of general validity and effectiveness of the measure to achieve the stated goals.

Assumption 4: A fundamental right to personal privacy. The issue is deeply philosophical, ethical and non-technical in nature [17]. However, even with an acceptance of personal privacy as a fundamental right one must always consider the limitations to personal privacy. This does not only cover the issues related to lawful interception vs. personal privacy or anti-terrorist measures vs. personal privacy, but also to the general laws concerning identity management and personal privacy [19]. With respect to basic cellular access the most prominent personal privacy requirements would encompass *data confidentiality* (both user plane and user associated control plane), *location confidentiality* and *identity confidentiality*. It is also common to include *untraceability* and *transaction confidentiality* as part of the standard requirements [20, 21].

6.3 Aspects and Issues

6.3.1 Security/Privacy Services

Several security services[1] will be important for a future cellular system. Amongst these are the following services.

- **Entity Authentication** of the Principal Entities
  A prerequisite for entity authentication is that one has identity structures for the principal entities. The principals may have multiple identities. Some of these identities may be public while others may be "anonymous". The identities may be long lived or they may be short-lived (down to a single transaction if necessary).
- **Data Confidentiality/Data Integrity/Message Origin/Destination Authentication**
  These are classical security services. *Data confidentiality* protects the data against eavesdropping, *data integrity* protects the data against illicit modification,[2] and *message origin/destination authentication* provides corroboration of the sender/receiver identities (or more often the associated routing addresses). Traditionally these services are provided by symmetric cryptographic primitives, but one may also use asymmetric (public-key) cryptographic methods and protocols.
- **Anonymity/Psedonymity**
  The subscriber may not want to disclose his/her identity or to remain anonymous. But the question then is: Anonymous with respect to whom? Traditionally (but it is not necessary should be the case in future systems), one would assume that the subscriber identity (the system identity, IMSI) is known both to the home operator and the serving network, but should be protected against eavesdropping on the radio interface by any third party. The name of the subscriber would normally only be known to the home operator and the telephone number (or SIP address etc) is normally seen as public information, but it should still not be visible over the link layer. It is also noted that currently the home operator generally is required to know the legal identity of the subscriber.
- **Location Confidentiality/Identity Confidentiality**
  These services are already defined for the existing systems, but no credible solution is yet provided for the current 2G/3G/4G systems. The issue is intrinsically tied to identity management and to solve the problem will require aligning the

[1] We make a clear distinction between the concept *security service* and methods to realize the service (the security methods and measures). That is, *entity authentication* is a security service. It may for instance be realized by a challenge-response security protocol (the method).

[2] By data integrity we here specifically mean cryptographic data integrity and not transmission oriented data integrity. That is, the data is cryptographically protected against willful modification. Only an entity that has the correct cryptographic key will be able to modify the data. Note however that an intruder may be able to delete data (this may or may not be detectable by the legitimate parties).

identity management (identity presentation in particular) with authentication at the access (link layer) level.

- **Untraceability and Transaction Confidentiality/Privacy**
  To provide the service "untraceability" implies that the intruder is unable to associate any point sampling of "a radio source at location (x, y)" with an identity. Identity in this context should be interpreted as any unique characteristic or trait that the radio source (mobile device) exhibit. That is, the concept "identity" can take on an emergent property. The intruder may not be able to derive the name or network address of the subscriber, but could successfully trace the subscriber based on radio transmission properties etc.[3] To conceal any emergent properties or characteristics is in principle very hard and may not even be possible against the most powerful intruder. However, the system should nevertheless be able to conceal identities, addresses and similar information elements. That is, a prerequisite is that the system provides credible Location Confidentiality/Identity Confidentiality.

  Transaction Privacy is a related concept and it relies on disassociation of the user identity and activities/transactions that the user performs. The *transactions* in question may be whether or not the subscriber accessed a particular website at a specific point in time, whether or not two specific subscribers were located at the same position at the same time or not, etc. Transaction Confidentiality relies on credible location and identity confidentiality. Untraceability is likely also required for an effective transaction confidentiality service.

6.3.2 Trust Relationships

Philosophically trust is a necessary evil. Therefore, a future system should ideally be designed such that the legitimate entities of a future system should not need to trust each other very much. Some trust is however inescapable. For instance, a home operator entity must have some information about the subscribers in order to provide services.

Trust is a difficult concept to capture and people are generally not good at evaluating security, trust and risk (see [18] for a recent and readable account). A good start is to identify the autonomous entities and physical nodes that are present in the system and to avoid trust decisions to be taken by the subscriber. The practical question often isn't so much which entities and node one trusts, but rather which entities and nodes one has to trust.[4] From a security perspective it is beneficial if the number of

---

[3] These may well be due to service usage at higher layers. A subscriber with particular browsing habits or other similar measurable parameters will exhibit statistically distinguishable patterns that can be detected and traced at the physical radio layer. This would be a particular case of traffic flow based tracing (i.e., lack of traffic flow confidentiality).

[4] Again, the decision should not normally be forwarded to the subscriber/user, but should be taken by the system (here the home operator) since subscriber/users are notoriously bad decision makers when it comes to assessing trust and risk [18].

trusted entities/nodes can be limited to as few as possible. That is, as few nodes as possible should be given cleartext access to user data, or be given access to cryptographic keys or be given authorization to make service provision decisions (which tend to imply charging decision authority).

One should also investigate the nature of the trust involved. Is the trust symmetric? That is, are the various entities on equal level or are there some entities with jurisdiction over other entities (implying a trust hierarchy). One should also differentiate between types of trust. The subscriber may well trust the home operator with respect to charging/billing, but may yet be reluctant to trust the home operator with respect to personal privacy unless when explicitly required.

Furthermore, with several trusted entities involved one should also look into whether one may allow the trust to be transitive. Today transitive trust is the norm for roaming (the serving network – subscriber trust is derived through transitive relations with the home operator[5]).

6.3.3 *Intruder Models*

The future cellular systems will face a number of attackers and adversaries. These intruders will attempt to stage large number of different attacks, ranging from naïve low-tech attacks to sophisticated attacks employing advanced equipment to attack the radio bearer and system protocols. Use of massively parallel computational attacks on "weak" cryptography should also be expected.

From a practical point of view it is important to consider scalability of attacks. Attacks that don't scale are much less worrisome from a system perspective, although the attacks may be damaging enough on per-subscriber level or similar.

From a theoretical modeling perspective one often wants to define the capabilities of the intruder, and the classical intruder model is the Dolev-Yao (DY) Intruder [5]. This intruder is very powerful and the DY intruder is able to delete, modify and insert messages to the communication channel at will. It will be able to do this for all available channels and will be able to store all previously transmitted messages. Thus any weakness exposed from previous communications can be used by the DY intruder. Still, the DY intruder is unable to break cryptography per see and it cannot physically compromise the principal entities. We note that a real intruder may be able to break "weak" cryptography[6] and that it may well be able to compromise the subscriber device (on smart phones the use of Trojans, Rootkits etc may facilitate this).

[5]Note that the relationship between the subscriber and home operator is asymmetric (the home operator have security jurisdiction over the subscriber in the traditional cellular architectures).

[6]We here distinguish between cryptoanalytic attacks on the cryptographic primitive and attacks on the composition of primitives in the protocols. Incorrect use of cryptographic primitives may result in a weak protocol even if the cryptographic primitive itself isn't broken. The opposite is also true, a provably secure protocol may be compromised by a weak cryptographic primitive (or from using inadequate key length etc).

In a wireless environment there will be practical constraints on the intruder capabilities. It may not be physically possible for the intruder to *delete* messages at will and likewise to insert and/or modify messages on the air may not always be physically possible. It must also be noted that in a wireless setting the intruder must be physically distributed to be able to operate as a true DY intruder.

Thus, an actual wireless intruder must consider spatial confinement in addition to processing and memory constraints. For some cases the constraints would involve real-time processing (i.e. to hijack a session) while for other cases the intruder can afford to do extensive prost-processing (eavesdropping etc where even access to aged data is of value).

A more detailed exposition of possible intruder models for wireless environments can be found in [27]. The conclusion in [27] is basically that the DY intruder is still a good theoretical model and that future access security protocols for cellular systems should be able to withstand the DY intruder. Any trusted entity/node should also have measures to protect against physical compromise. This is particularly important for the highly distributed subscriber devices, which for a number of reasons are the most vulnerable entity. Here one must be able to provide tools such that the subscriber/user is able to protect the integrity of the device. In practice this will as a minimum amount to disk/memory encryption, use of anti-virus packages, use of device firewalls etc.

6.3.4 Cryptographic Primitives and Key Sizes

One traditional way of classifying cryptography is to make a clear distinction between asymmetric and symmetric crypto-primitives. Computationally there are large differences between the two, and one would tend to use symmetric cryptography for cases when sheer performance is the decisive factor. That is why AES and similar symmetric primitives is almost always the choice for encrypting (providing data confidentiality) high throughput channels. However the public-key systems (asymmetric) are very useful, and are commonly used for entity authentication and for key distribution (of symmetric keys). Another service is data integrity (which in practice also includes message authentication). Again the implementation may be based on symmetric means or on asymmetric means, and again the most common today is to use symmetric methods for link layer protection while it is common to use asymmetric methods for single transactions etc (on the application layer) and for providing entity authentication and data signatures.

We do not foresee any large scale changes here, but we note that as processing power grows it will be feasible for more devices to support asymmetric cryptography. Still, for many devices the increase in transmission capacity outgrows the increase in processing power. For these devices the choice for channel protection will still predominantly be by symmetric cryptographic means. But we do expect entity authentication protocols to move towards use of asymmetric cryptography in the future.

A very important cryptographic service is to provide large random numbers. That is, normally the numbers don't have to be random (and sometimes we explicitly do not want them to be random), but they should appear to be random to any outsider (the intruder) in the sense that these pseudo-random numbers should be unpredictable (and uniformly distributed) and oftentimes non-repeatable. For the unpredictable property to be fulfilled it is important that the pseudo-random number is sufficient large to prevent guessing attacks. Typical sizes today would be between 64 and 128 bit, depending on actual requirements.

Sometimes it does not matter if the used number is unpredictable provided that it is unique. For instance, one may use a sequence number to uniquely distinguish between different protocol runs or one may use a timestamp to avoid replay attacks. In these cases the "random" number isn't random and one then uses the term Nonce (number used once). The pseudo-random numbers are also a type of Nonce number. The type of Nonce being used would depend on the overall design, but suffice to say that the number must be generated in a cryptographically secure way (for instance the clock providing the timestamp must be protected). For a cellular system it is likely that the bearer number, channel number and/or frame number be included as part of the Nonce for the link layer protection. This ensures a strong spatially associated real-time binding of the message payload and the actual message transfer frame. However, the system architecture must assure that there aren't any Nonce collisions (or that such events cannot otherwise be used to attack the system).

6.3.5 Near Future Options

What we know for sure is that future cryptography may be much more advanced than what we have today. We can foresee some of the future options, like Secure Multi-party Computations (SMC) and Identity-Based Encryption (IBE) [22], which may be important in the future. We know this since the basic ideas for SMC and IBE are already identified, but we also know that a lot more research is needed to make these crypto-systems practical in large scale systems. These issues will be elaborated with more details in the following sections.

6.4 Access Security and Personal Privacy 24 Years Ago

6.4.1 The NMT 450 Era

In 1984 the NMT450 had been operational for 3 years. The system consists of analogues PSTN like switches (MTX). The MTX'es comes with two databases: A home database for all subscribers in that area and a visitor database. The MTX directly controls the base stations (i.e. no BSC or RNC functionality present). NMT architecture is defined in [14].

6.4.2 *Security and Privacy in NMT450*

6.4.2.1 Security and Privacy

With respect to entity authentication the system originally had only an extremely weak password scheme. The mobile phone used a three digit "secret" password ($K_1K_2K_3$) which was transmitted in cleartext. With respect to subscriber identity we note that the NMT system did not distinguish between the mobile phone and the subscriber. That is, the identities used were in fact associated with the mobile station and not with subscriber (originally there were no subscriber identity module at all).

6.4.2.2 Cryptography

Originally there was no cryptography at all in the NMT450 system. In later phases one augmented the NMT450 system with the NMT-SIS functionality. NMT-SIS provided basic (uni-directional) subscriber authentication and signaling data encryption [16]. The idea of a functional entity for authentication in at the home operator was introduced with NMT-SIS [13].

6.4.3 *Trust in the NMT Ecosystem*

Inter-exchange (Inter-MTX) system signaling in NMT was by means of SS7 signaling through the HUP/MUP[7] user parts (one did not use SCCP in NMT and thus all system signaling had to be associated with a 64 kbps B-channel).

The SS7 system unfortunately provides no security services. This meant of course that the MTX'es had to trust any content it received over SS7 since there would be no way to verify/falsify the validity of the messages. Furthermore, the access signaling between the MTX and the Mobile Station was also transmitted without any security protection. We note that this was changed with the advent of NMT-SIS, and that the access signaling then got a measure of protection. But before NMT-SIS there were no way to corroborate claimed identities, and one was forced to naïvely trust all claims (and fraud level grew to become a significant problem!).

6.4.4 *The Intruder 24 Years Ago*

Originally there were no intruders present. But gradually it was realized that the system did not have any protection and inevitably this fact was exploited. Organized

[7]Handover User Part and Mobile User Part respectively.

crime in the Netherlands was probably amongst the first to exploit the weaknesses. It is interesting to note the primary goal for the drug traffickers probably weren't to steal airtime as such; the motivation was in all likelihood driven by a strong desire to attain anonymity from police monitoring.

Later the fraud became more focus on charging fraud and there were numerous accounts of "resellers" that sold call-time. In Norway this activity was primarily targeted at selling long distance calls and the problem became large enough for Televerket (the Norwegian NMT operator) to block all non NMT-SIS call to a set of countries.

In the fullness of time NMT scanner became more or less a commodity item and the scanners, while illegal to sell and use in many countries, were used so extensively that the credibility of the whole system was questioned. We note that although NMT-SIS was quite effective in combating charging fraud, it was also completely useless against eavesdropping on the user channel (the NMT-SIS encryption only covered system signaling; the analogue user channel was not protected).

6.4.5 Problems with NMT450

We can summarize the NMT450 security and privacy related problems in the following:

- No security proved to be a real problem
- Eroding the business model due to fraud
- Security provided by NMT-SIS was effective against fraud, but it was inadequate in that it could not provide data confidentiality for the users
- People lost faith in the technology due to missing speech privacy

6.5 Access Security and Personal Privacy 12 Years Ago

6.5.1 The GSM (Phase 2) Era

In 1996 the GSM system was well established. The basic service was circuit-switched speech and to a limited extent circuit-switched data. The need for security protection was recognized from the start and one also foresaw the need for subscriber Location- and Identity Confidentiality. The GSM did however have a fairly naïve assumption about inter-operator trust, and the authentication data (called the triplet or authentication set) is always transferred from the HLR (located at the home operator) to the VLR/MSC (which is located in the serving network (may be the home network but it may also be at a roaming network)).

6.5.2 Security and Privacy in GSM

The GSM system does provided some security and privacy services [7, 8]. In particular, the system provides entity authentication. The service is somewhat limited in that it is only providing uni-directional authentication. That is, the subscriber is authenticated to the network, by means of challenge-response mechanisms (GSM AKA protocol), but the subscriber (represented by the Subscriber Identity Module (SIM) smart card) has no mean of verify the validity of the network. As mentioned the entity being authenticated is the SIM and not the mobile station [9], and one has thus dissociated the radio equipment/mobile termination from the subscriber credentials.

6.5.2.1 Cryptography

GSM relies on a pseudo-random number function for its random-challenge protocol. This functionality, only presented in the HLR/AuC (in the home network), is not standardized. Key derivation and the response function are computed by means of Message Authentication Code (MAC) function. Collectively the function is called A38 (or A3 and A8 separately). GSM supports data Confidentiality and for the circuit-switched services one uses the A5 encryption interface. The actual A5 algorithm used originally was the A5/1 algorithm, with 54 significant bits (the nominal A5 size is 64 bit). Cryptographically speaking the A5 functions are stream-ciphers. The A5 functionality is located in the mobile equipment (ME) and so the SIM would transfer the key, Kc, to the ME. The GSM system provides no data integrity service.

6.5.3 Trust in the GSM Ecosystem

Core Network (CN) signaling in GSM is by means of SS7 (MAP, ISUP etc). The protocols used are more sophisticated than the NMT equivalents, but since the GSM protocols like their NMT cousins are based on SS7 they too inherit the SS7 weakness of not having any security services. In theory one might have added security services at the application part layer, but this never took place. The consequence is that all GSM CN nodes have to trust each other.

Another aspect, due to the uni-directional only authentication protocol, is that while the identity of the SIM can be corroborated by the MSC/VLR (with support from HLR/AuC) the SIM cannot authenticate the network. Thus, SIM has to trust network. Originally this was not a problem, but today masquerade attacks by means of "False-BTS" are both possible and feasible.

6.5.4 The Intruder 12 Years Ago

The intruder 12 years ago faced encrypted communication (but only over-the-air). Consequently, it would take less effort to eavesdrop on the communication on other interfaces. In particular, this meant eavesdropping attacks on the communication between the BTS and towards the BSC/MSC.

The GSM confidentiality protection isn't perfect and although system now permits the full 64 bit key length the original A5/1 algorithm shows its age. Cracking the A5/1 is now feasible and even the A5/3 algorithm cannot be said to be future proof (no real weakness has been found with A5/3, but system weaknesses in GSM and the limited key size means that even A5/3 can be cracked by a dedicated intruder).

However, in fairness, we note that the GSM system was for all practical purpose secure 12 years ago. Thus the intruder 12 years ago faced security that actually fulfilled its design criteria. The intruder would have found it very difficult to crack the security and indeed there were very low fraud levels in GSM (and from what we know the fraud attacks circumvented GSM security rather than attack it; most registered fraud was also very low tech (false identity, not paying the bill etc)).

The problem has rather been that the GSM system has outlived the initial security and that the initial design didn't adequately support a migration path to improved security (for instance it is very difficult to extend the key size of Kc and it is difficult to properly bind keys with good algorithms).

6.5.5 *Security and Privacy Related Problems with GSM*

Security in GSM has been hugely successful, but the success and longevity of GSM have outlived the basic initial security. The problems one see today can be summarized in the following:

- Too naïve trust model (delegating security credential)
- No risk/threat analysis (and no intruder model) – this resulted in an incomplete security architecture
- No data integrity protection
- Mutual Entity Authentication is missing – allowing False-BTS attacks
- Example AKA A3/A8 function (COMP128) was *very* weak
- Data confidentiality was too weak (64 bit not enough for system lifespan, >20 years)
- Location/identity protection not credible (TMSI method)

6.6 Access Security and Personal Privacy Today

6.6.1 *Maturing 3G (UMTS), Extensions to Provide 4G (LTE/SAE) Being Defined*

Today the basis for deployment is normally 3G with both CS and PS capabilities. The 3G systems have a reasonable set of 3G security services, but with some weaknesses and some omissions. For instance one still haven't got a credible subscriber location- and identity confidentiality service. This is largely due to the addressing- and subscriber identity model, which in principle is unchanged since the 1G systems. Furthermore, the delegation of authentication authority to the serving network has been retained from the 2G systems, and consequently the naïve assumptions about inter-operator trust are still in force. The forthcoming 4G system will provide some security improvements, but subscriber privacy is still problematic since the addressing scheme and identity management at the access level is basically inherited from the descendent 3G system.

6.6.2 *Security and Privacy in UMTS (And Quite Similar in CDMA2000)*

The security and subscriber privacy services provided is outlined and defined in [2, 11, 15]. In brief, one has a mutual entity authentication in the UMTS AKA protocol. The UMTS AKA protocol (a very similar protocol is used in CDMA2000) is a challenge-response protocol with authenticated challenge and use of sequence numbers to prevent replay attacks.

Subscriber identity and security credentials are stored in subscriber module (UICC/USIM) together with the authentication and key agreement algorithms. Data confidentiality is provided by the KASUMI and SNOW-3G encryption algorithms (mode-of-operation is to provide a stream-cipher service). The range of the service is between the mobile equipment and the Radio Network Controller (RNC) and essentially all data is encrypted once the service is initiated. A data integrity service is also provided. The coverage range is the same as for the data confidentiality, but it is only for signaling data. The basic crypto-primitives are the same as for data confidentiality, but this time the mode-of-operation obviously is to provide data integrity (though a message authentication code). The key length is 128 bits for both the data confidentiality and data integrity services.

The UMTS AKA algorithms are in principle operator specific, but an example set of algorithms, called MILENAGE, based on Rijndael (also known as AES) is provided. The MILENAGE set is complete except for the lack of a pseudo-random number function.

6.6.3 *Trust in the UMTS Ecosystem*

The inter-operator trust requirements are essentially the same as for the 2G systems since the model with distributed authentication authority is inherited from the 2G systems. Thus, the naïve 2G trust is retained through inheritance. Currently in the 3G systems the key material may be transported in cleartext in some protocols (examples are MAP and RANAP). This is of course naïve and it requires too much trust (gullible trust). We also note that the UMTS AKA protocol is not an online protocol. Thus, the USIM can authenticate the network, but only to the extent that the USIM knows that HLR/AuC was present at some time in the past (when the challenge was originally issued to the serving network). So, it is clear that the system architecture and the security architecture contain several weaknesses and minor flaws that put together require a fairly high level of trust between the legitimate entities. Only if that trust is present (and it should indeed be justified) can the system operate safely and securely.

6.6.4 The Intruder Today

The intruder today is in practice only able to attack the GSM cryptographic protocols. Furthermore, the current attacks don't scale very well and they consequently don't represent a serious threat towards the overall system. However, the reputation the system has as s secure system is damaged and a change in public perception can be very damaging to the longtime viability of the system.

The current intruder must be assumed to have considerable processing capability available. Even a single person intruder can buy commercial off the shelf computer systems with sophisticated and powerful co-processors, like the Cell processor or a dedicated FPGA, and with main memory in the range of tens of gigabytes. The ECRYPT yearly report gives a good overview of the capabilities of several intruder types [23]. A larger organization may use dedicated networks and the combined processing power can be quite formidable. Still, brute force cracking of cryptography is very much limited to attacks on a single conversation or inherently weak algorithms. Brute force attacks are still essentially facing exponential complexity measured against the bit length of the cryptographic key, and (as long as that remains what we have is that Moore's "law" etc) one only gains one bit of the key length for every doubling of processing power. So, by upgrading from the by now dangerously short 64 bit key length to 128 bit key length one has effectively killed off any brute force attacks (provided that the algorithm does not have serious flaws).

Another practical aspect for all wireless intruders is that they need to be physically distributed in order to eavesdrop or otherwise intercept the transmission. The opportunistic intruder does not itself have to be very mobile, but then the capabilities of the intruder will be limited to the air traffic in that area. However, we note that sensor technology is getting more mature and for a future intruder this may mean that the geographic coverage could be attained by deployment distributed sensor networks. We also note that the emergency of so-called zombie networks of hijacked computers that we see in today's personal computer arena could well be a phenomenon that is repeated for the next generation of smart phones. A network of zombie smart phones could potentially be utilized by an intruder to provide geographically distributed intruder capabilities. These possibilities are emerging today and in the near future one must assume that actual intruders will use them.

6.6.5 Security and Privacy Related Problems with UMTS and LTE/SAE

The 3G systems are evolving and the next generation is being defined. For UMTS this next generation is called LTE/SAE (also called EPS). There are strong requirements for coexistence and although there is no requirement for 100% compatibility, it is still clear that the LTE/SAE is becoming an evolved architecture. This, unfortunately, also means that many of the 3G problems will be migrated to the next generation (which, by the way is not officially called 4G). We summarize the main problem below and note that LTE/SAE will mitigate some of the problems.

- Too naïve trust model
- Intruder model is missing – Increasingly sophisticated intruders not accounted for
- UMTS: Data integrity protection is too limited (both in strength and coverage)
- UMTS AKA is not optimal (replay protection/seq.no.mngt)
- Location/identity protection not credible (TMSI method)
- UMTS: Backwards compatibility with GSM is problematic

6.7 Access Security and Personal Privacy in 2020 and Beyond

6.7.1 Unknown Territory

We assume that in 2020 compatibility with the 4G systems is likely still an issue. However we make no compatibility assumptions beyond 2020.

6.7.2 Trust in a Beyond 2020 Perspective

We do not expect large changes here, but we do foresee a development in which one requires less trust in the various network elements. That is, there will hopefully be fewer nodes that need to be trusted.

There exists different wireless system models and one may see a development towards less centralized systems, possibly without a central authority. Still, that requires large scale changes to the control model and it is not clear if there are economic models and/or regulatory space which will permit this to happen. That "internet" model with free access and the development towards fixed pricing structures may change the operator landscape significantly, but the regulatory requirements may still apply and then the trust scenarios will not change too much. However, subscriber trust in the serving network may be weaker in the future and this will be more pronounced if there is a development towards free access (with presumably weaker service guarantees).

6.7.3 The Intruder in a Beyond 2020 Scenario

In 2020 we expect the power of the intruder to have grown significantly. Moore's law alone would give an increase in processing power by about 256 times (eight bits), but the real increase in power will not come from an increase of power in any single processor.

Rather, one should expect the processing power to increase significantly faster than Moore's law, but with the proviso that it is achieved for multiple cores and data center technology (aka Google server parks). This would require computational parallelization of attack algorithm. Brute-force like attacks against symmetric algorithms is of course readily adapted to parallel processing, and so this is a viable approach. Still, to attack crypto-algorithm by brute force is only viable for algorithms with small key space or with known weaknesses that reduces the search space. Direct brute force against a 128 bit key space (or larger) will not be practical even in the beyond 2020 scenario for conventional computing methods [23].

Looking further ahead one may find that Quantum Computing (QC) becomes a reality. If the promise of scalable parallel computing made by QC proponents can be fully realized then obviously brute force attacks on will become feasible even for very large key spaces and cryptography as we know it today will have to be totally redefined.[8] However, it is not yet clear if QC can ever tackle real-life problems in this way and it may even be that some computations are inherently sequential in nature. QC may still provide orders of magnitude improvement in processing power and it may be highly effective against certain classes of problems. Another and less speculative promise of quantum cryptography is found in the area of key distribution. Quantum key distribution (QKD) has already been demonstrated over fiber optical channels and with optical switches it holds great promise. The basic promise of quantum key distribution is that QKD provides proof that the transferred key has not been eavesdropped on. That is, the "read" operation (observation) destroys/disturbs the state of the key. Thus, the key is successfully transmitted if and only if the key has not been observed while in transit. We note that there is nothing that prohibits the intruder from reading the key (if he/she has access to the channel), but doing so changes the data and the receiver would know that the key has been read. That is, unauthenticated QKD protocols do not prevent the possibility of the intruder inserting itself as a Man-in-the-Middle. For a readable account of the current state of quantum cryptography the reader is referred to [26].[9]

---

[8]Then the famous problem P = NP will be resolved and forced into the P domain by quantum computations.

[9]The SECOQC white paper provides a reasonable introduction to the field, without going into the details of quantum mechanics.

Another development is to produce massively parallel organic computer. These computers would operate on principles used in DNA replication etc. The field is hardly mature, but provides exiting glimpses of what the future may hold. Examples on solving the Hamiltonian path problem and the Boolean satisfiability problem (both NP-complete) are outlined in [24, 25].

Yet another key development will be an increase in the intruder's physical coverage. Sensors are becoming cheaper and cheaper while the capabilities increase, and mesh network technologies etc will develop to utilize this computational "cloud". This is an important development since many attacks on wireless systems will require a physically distributed listening-in capability. As previously mentioned, compromised wireless devices may well be part of this wireless computational cloud.

6.7.4 Cryptographic Tools and Security Services in 2020 and Beyond

In 2020 or may be slightly beyond security, privacy protection and trustworthiness would be embedded into network functionality due to ambient and sensing nature of telecommunication systems. Telecommunication systems will create tremendous opportunities to improve peoples' lives by their ability to perform joint cooperative tasks based on computations with inputs supplied by separate users.

Since computational power and digital communications will be embedded in almost all objects surrounding human beings, the new kind of ubiquitous applications will pervade people everyday's life and would be used to improve security and quality of their users' lives.

However such systems, by collecting information about people's daily activities, can easily become distributed surveillance systems [3]. Since information reflecting users' daily activities (for example travel routes, buying habits, and so on) is considered by many users as private it would be no surprising that one of the requirements to ubiquitous applications would be privacy preservation [12], and applications not conforming with the requirement may be prohibited by law (at least in some countries) as violating citizens' privacy. Thus if developers of new applications and services only focus on functionality without considering users' privacy they will limit usability and acceptance of the applications and services.

Since already nowadays private data are used to develop such important applications as traffic jam monitoring, monitoring of elderly people, anti-terror related monitoring of suspects, etc., we would expect growing number of such applications and services in the future and correspondingly growing the amount of private data collected by the systems.

Thus in 2020 the privacy concerns become a critical issue and questions like how exactly the private data will be used; how can the data possibly be misused to invade people's privacy; "who is watching the watchers", etc. is addressed and properly resolved. These questions arise many applications and services collecting information reflecting users' daily activities are cooperative computations performed among mutually untrusting parties on inputs containing private data. As result new security primitives based on secure multi-party computations (SMC) will become a relevant and practical part of deployed systems and services.

Generally, secure multiparty computations are cryptographic techniques that enable computations on data received from different parties in such a way that each party knows only its own input and the result of the computations [4]. Goldwasser [4] predicts that "the field of multi-party computations is today where public-key cryptography was 10 years ago, namely an extremely powerful tool and rich theory whose real-life usage is at this time only beginning but will become in the future an integral part of our computing reality".

Currently, the most common current drawback of SMC protocols, which substantially impacts their today's applicability, is their inefficiency with respect of both considerable computational and communicational resources will be overcome. We expect that in 2020 or slightly beyond this would be resolved both by tremendously increased computational and communicational capabilities of the systems and by developing new, more practical solutions based new security models that provide balancing security and efficiency. However even with increased computational and communicational capabilities of systems as a whole, it will still be required that SMC based crypto primitives are able to operate over narrow-bandwidth connections and must be able to operate on power constrained mobile devices. Furthermore, the algorithms will need to operate with real-time constraints and without the presence of any central (trusted) authority.

6.8 Conclusions

With respect to the past and the present of access security and personal privacy in public cellular communication systems we have learned that the following issues should be taken into consideration in order to make the future systems both secure and privacy preserving.

- Future systems should be designed to minimize trust
- Future systems must take into account the intruder
- The future intruder will be very powerful
- Access Security will become better and better
- The power of the intruder may outpace the improvements
- Actual access security could become weaker
- Past experience tells us that there is a lower limit, after which the business model breaks down. The past also tells us that there is a lower limit to personal privacy, after which the consumer trust in the product breaks down, But, Credible Personal Privacy requires that consumers care
- Society defines which privacy services are required and acceptable
- Personal privacy is difficult to achieve since technology makes privacy invasion very cheap

- "Privacy Intruder Models" must be developed
- Cryptographic methods must evolve to support credible personal privacy (SMC etc)
- The system architecture must provide pervasive support for personal privacy since personal privacy is a weakest link game

References

1. 3GPP TS 33.106, 3rd Generation Partnership Project; Technical Specification Group Services and System Aspects; 3G Security; Lawful Interception requirements (Release 6), 3GPP (2004)
2. 3GPP TS 33.102, 3rd Generation Partnership Project; Technical Specification Group Services and System Aspects; 3G Security; Security architecture (Release 6) (Sophia Antipolis, Valbonne, France, December 2004)
3. D. Brin, *The Transparent Society* (Perseus books, Reading, MA, 1998)
4. S. Goldwasser, Multi Party Computations: Past and Present, in *Proceedings of the Sixteenth Annual ACM Symposium on Principles of Distributed Computing* (Santa Barbara, California, United States, ACM, 1997)
5. D. Dolev, A. Yao, On the security of public-key protocols. IEEE Trans. Inf. Theory **29**(2), 198–208 (1983)
6. ETSI SR 002 180, Requirements for communications of citizens with authorities/organizations in case of distress (emergency call handling), ETSI (Sophia Antipolis, Valbonne, France, December 2003)
7. ETSI SMG, GSM TS 02.09 v6.0.1; Security aspects, ETSI (Sophia Antipolis, Valbonne, France, May 1999)
8. ETSI SMG, GSM TS 03.20 v6.0.1; Security related network functions, ETSI (Sophia Antipolis, Valbonne, France, July 1999)
9. ETSI SMG, GSM TS 02.17 v5.0.1; Subscriber Identity Modules (SIM); Functional characteristics, ETSI (Sophia Antipolis, Valbonne, France, April 1997)
10. EU Council/EU Parliament, *Directive of the European Parliament and of the Council on the Retention of data Generated or Processed in Connection with the Provision of Publicly Available Electronic Communications Services or of Public Communications Networks and Amending* (Directive 2002/58/EC, EU Directive, February 2006)
11. G.M. Køien, An introduction to access security in UMTS. IEEE Wireless Commun. Mag. **11**(1), 8–18 (November 2004)
12. M. Langheinrich, A privacy awareness system for ubiquitous computing environments. 4th international conference on ubiquitous computing (Ubicomp 2002), Springer, 2002
13. NMT-SIS, AUTHENTICATION REGISTER, Functional specification of an Authentication Register for NMT-900. The Post and Telegraph Administrations of Denmark, Finland, Norway and Sweden
14. Nordic PTTs, NMT Doc 450-1: System Description, Nordic PTTs, 1999, NMT Doc 450-1.
15. G. Rose, G.M. Køien, Access security in CDMA200, including a comparison with UMTS access security. IEEE Wireless Commun. Mag. **11**(1), 19–25 (2004)
16. SPECIFICATION FOR NMT-SIS KEY MANAGEMENT IN NMT-900 (1988-07-01). The Post and Telegraph Administrations of Denmark, Finland, Norway and Sweden
17. S.D. Warren, L.D. Brandeis, *The Right to Privacy.* Harv. Law Rev. **IV**(5) (Wadsworth, Belmont, CA, December 1890) (from "Ethical issues in the use of computers", 172–183, 1985). ISBN: 0-534-04257-0.
18. R. West, The psychology of security. Commun. ACM **51**(4), 34–40 (April 2008)
19. G.M. Køien, V.A. Oleshchuk, Personal privacy in a digital world. Telektronikk **103**(2), 4–19 (2007)

20. RFID and Privacy, G.M. Køien, Telektronikk **103**(2), 4–9/77–83 (2007)
21. S. Garfinkel, A. Juels, R. Pappu, RFID Privacy; An overview over problems and proposed solutions. IEEE Secur. Priv. Mag. **3**(3), 34–43 (2005)
22. D. Boneh, M. Franklin, Identity based encryption from the Weil pairing. SIAM J. Comput. **32**(3), 586–615 (2003)
23. IST ECRYPT (IST-2002-507932), ECRYPT Yearly Report on Algorithms and Keysizes (2006), Deliverable D.SPA.21, 29.February 2007
24. L.M. Adleman, *Molecular* computation of solutions to combinatorial problems. Science **266**(11), 1021–1024 (1994)
25. D. Boneh, C. Dunworth, R.J. Lipton, J. Sgall, *On the Computational Power of DNA*, DAMATH: Discrete Applied Mathematics and Combinatorial Operations Research and Computer Science 71, 1996
26. R. Alleaume et al., SECOQC White Paper on Quantum Key Distribution and Cryptography (22January 2007), www.secoqc.net
27. G.M. Køien, *Entity Authentication and Personal Privacy in Future Cellular Systems.* Ph.D. thesis, Aalborg University, Denmark, April 2008

Geir M. Køien is a postdoctor at the University of Agder, Norway, where he works with cellular security, personal privacy and similar topics. He holds a B.Sc. Honours in Computing Science from the University of Newcastle upon Tyne, England, a M.Sc. in IT from the Norwegian University of Science and Technology (NTNU, then NTH), and a Ph.D. from Aalborg University (AAU). Køien has previously worked for Telenor R&I, and participated in security standardization in 3 GPP during 1999–2009 as the Telenor delegate. He has worked extensively with access security in GSM/GPRS, UMTS and LTE.

Vladimir A. Oleshchuk is Professor of Computer Science at University of Agder, Norway. He received his M.Sc. in Applied Mathematics (1981) and Ph.D. in Computer Science (1988) from the Taras Shevchenko Kiev State University, Kiev, Ukraine, and his M.Sc. in Innovations and Entrepreneurship (2007) from the Norwegian University of Science and Technology (NTNU). From 1987 to 1991 he was Assistant Professor and then Associate Professor at the Taras Shevchenko Kiev State University. He has been working at University of Agder since 1992. He is a member of IEEE and a senior member of ACM. His current research interests include formal methods and information security, privacy and trust with special focus on telecommunication systems.

Chapter 7
Security and Privacy Challenges in Globalized Wireless Communications

Hanane Fathi

7.1 Introduction

Wireless communications added a disruptive value to communications by allowing location and movement independence into communications. This independence has brought wireless and mobile communications to be one of the most popular means of communications today. Many times, we have seen mobile phone users deciding not to get a fixed phone line for their homes, relying solely on the mobile phones. Most hotels do not wire all their rooms with Internet anymore but install wireless local area network (WLAN) access points for their customers to connect to Internet anywhere in the hotel premises. This is the trend worldwide, resulting in the convenience of wireless communications experienced by millions of users across the globe, as illustrated in Fig. 7.1. The impacts of wireless communication are not only limited to providing functional convenience in our daily lives and bringing the personal sphere in the public one. Wireless and mobile communications offer much wider scopes: as a globalization tool of information and communication technology (ICT), wireless and mobile communications play a vital role in empowering people worldwide. This vital role is however highly dependent and conditional on security of the wireless and mobile communications.

In this chapter, we discuss the opportunities and threats offered by globalization of ICT ranging from mobile, wireless communications to Internet. Globalization of wireless communications and Internet offers a variety of tools for people all over the world to combat or to help combat the essential challenges in developing regions of the world: ending hunger, poverty, disease, discrimination, establishing education, and reinforcing democracy.

We argue that security is a pillar that helps sustaining the benefits of globalization of mobile communications and Internet: protecting user-citizens against cyber-threats, guaranteeing user privacy against big brother (governments) and little

H. Fathi (✉)
National Institute of Advanced Industrial Science and Technology, Japan
e-mail: Hanane.fathi@aist.go.jp

R. Prasad et al. (eds.), *Globalization of Mobile and Wireless Communications: Today and in 2020*, Signals and Communication Technology,
DOI 10.1007/978-94-007-0107-6_7,

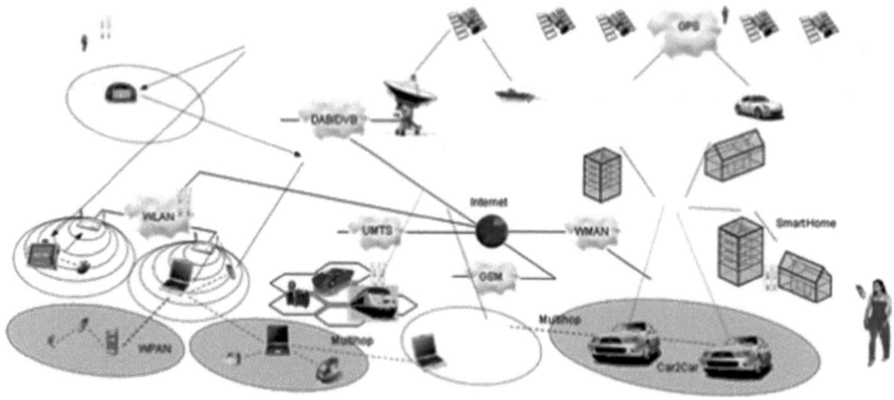

Fig. 7.1 The heterogeneity of wireless communication

brother (corporations), establishing trust between users, providers and organizations (governmental or corporate), allowing freedom of speech in hostile environments.

The paper is organized as follows. Section 7.2 describes how globalization of wireless and mobile communications can help closing further a telecom gap while Sect. 7.3 shows the globalization of the Internet. Then Sect. 7.4 shows the opportunities given in such a context. The threats that are mainly security oriented are exposed in Sect. 7.5. The security perspective introducing technical measures for a sustainable wireless communication system is given in Sect. 7.6. Concluding remarks are made in Sect. 7.7.

7.2 Globalization of Wireless and Mobile Communications

With four billion mobile phones already in use, and network coverage extending over 80 % of the world population [1], globalization of mobile communications has helped diminish the telecommunications divide between the world's richest and poorest nations. The telecom divide is defined here as the lack of access to telecom services. In developing countries, the penetration rate of mobile phones is much higher than that of the fixed landlines.

Fig. 7.2 shows that in Africa, from 1995 to 2004, the mobile telephone subscribers outnumbered the fixed telephone subscribers from year 2000. This is again confirmed in Fig. 7.3, which compares the average growth rate in mobile subscribers from 1999 to 2004 in different regions of the world.

In 2007–2009, the growth rates of monthly mobile subscriber in many developing countries are significant: every month India and China together add nearly 30 million new subscribers, Pakistan adds over three million new users per month while Egypt adds over one million new users per month [1].

In an increasing information society and connected era, wireless and mobile communications are powerful assets to use in bridging the telecom and digital

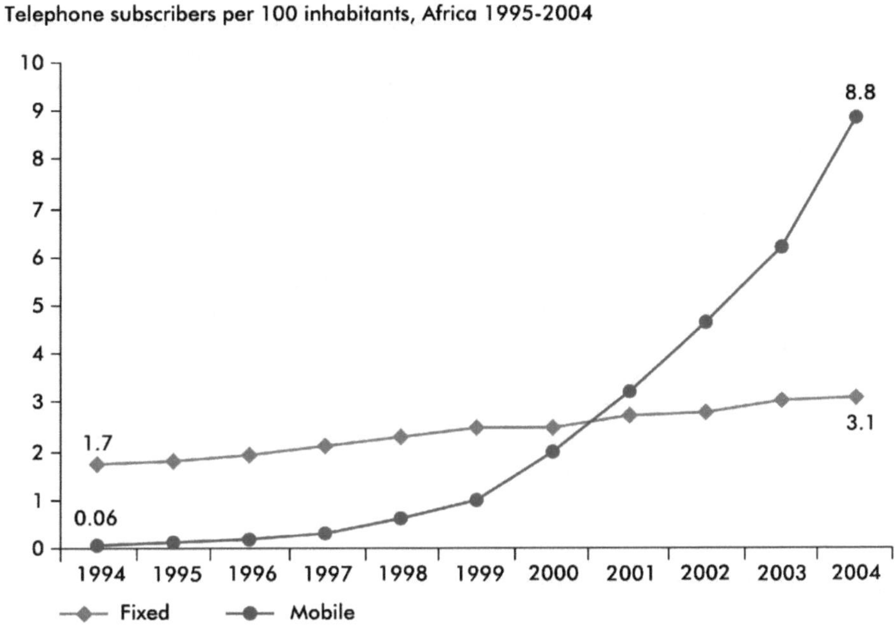

Fig. 7.2 Telephone subscribers per 100 inhabitants, fixed versus mobile, Africa 1995–2004

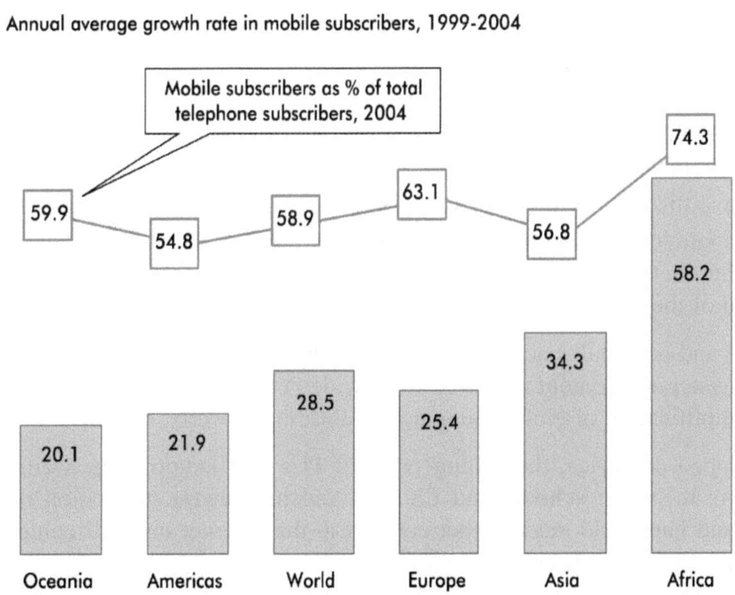

Fig. 7.3 Annual average growth rate in mobile subscribers 1999–2004

divide. The technology can help cutting the telecom infrastructure cost of reaching rural areas in developing countries and can thus facilitate the participation of all segments of the society in the economical growth, diminishing poverty and illiteracy. Global and affordable access to ICT has a direct correlation with the battle against hunger, poverty, disease, discrimination and for education and democracy. In certain regions of the world such as India especially in remote areas, a regular communication between family members and relatives represents a real value in keeping family relieved from the anxiety of not knowing what happens back in the hometown, and allows people to reach for help much faster.

The deployment of WLAN or WiMAX can close the gap between the connectivity of the inhabitants of urban areas and those of rural areas. It can also support the health organizations in providing online health advice, in collecting information on a disease status in a specific region, and in facilitating the help of humanitarian efforts. The deployment of WLAN and WiMAX is meant to be faster and cheaper than that of wired networks.

7.3 Globalization of Internet

Long ago, Internet was a tool mainly used by researchers. It has now become a cultural, political, and economic tool for many, and a workhorse of work, entertainment and social interactions.

Similar to wireless and mobile communications, the winning point of Internet is in the convenience it offers to users across the world. The Internet particularity is erasing territorial boundaries, or it creates its own territory: an e-world with

- E-mail
- E-commerce
- E-government
- E-ties to people across the globe

With 1.4 million worldwide users in 2008, Internet users represent 21% of the world populations, while they actually are only 5% in Africa [2].

The Internet is the tool for empowering people, independently of their social class and of the state of development of their country through, for example,

- Blogs and personal spaces: MySpace, Facebook, Twitter, LinkedIN
- User generated content available worldwide: YouTube, DailyMotion…
- An amplification of global campaign against e.g. poverty, hunger

In developed countries, the falling price of PCs and laptops, the availability of computers in public schools and libraries and the newest generation of mobile phones and hand-held devices that connect to the Internet have all contributed to closing the divide between social classes.

In developing countries, cyber cafes and used PCs allow a lot of mostly young people to stay connected to the highways of information: e-learning, online articles, online encyclopedias, and dictionaries. This way, both developed and developing

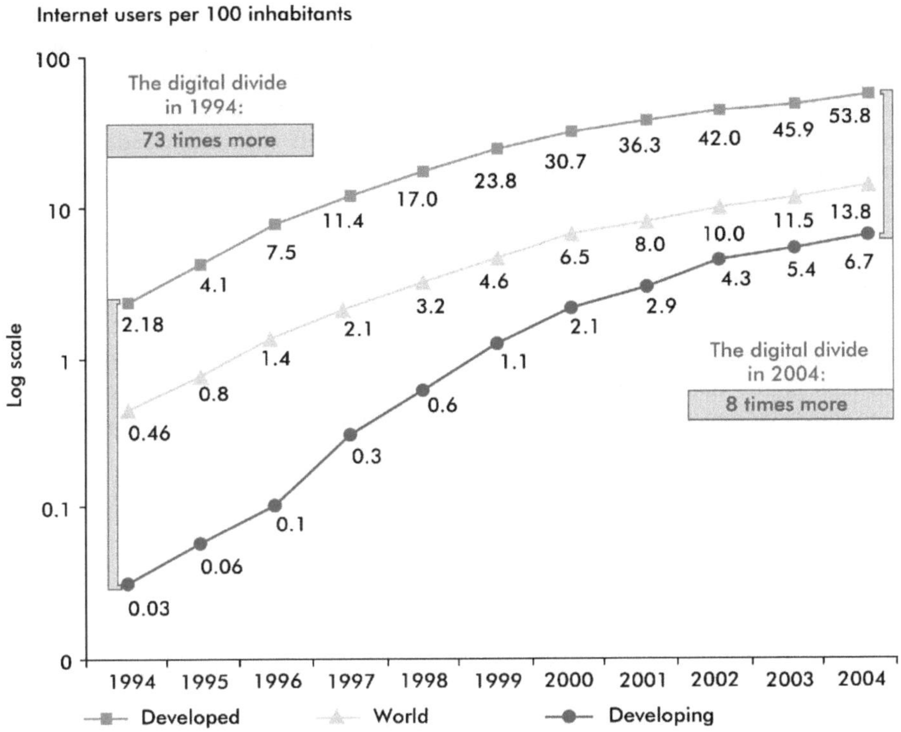

Fig. 7.4 Internet users per 100 inhabitants 1994–2004

countries exploit the very nature of the Internet as a communications, publishing, and transactions platform (Fig. 7.4).

In the context of globalization, the Internet has also opened the door to online Internet-based businesses (besides Google, Yahoo, E-bay, etc.) for millions of small enterprises that are for example shops selling hand-made traditional goods to worldwide consumers, artists advertising their artwork, etc.

The convergence of wireless and Internet allows for a faster deployment of infrastructures supporting Internet in rural areas, in countries with poor telecom wired infrastructures.

7.4 Opportunities

7.4.1 Driving Forces

The driving forces of wireless communication and Internet are mainly the cost and the availability of a global standard platform. Moreover, one should not forget that these technologies answer basic human needs in terms of communication, needs to feel linked to a community, a family, a friend and needs for entertainment.

These driving forces have lead to the globalization of GSM: the number of GSM subscribers has grown from one million in 1993 to 2.6 billion today across 193 countries [3]. This amazing worldwide growth is reinforced by the fast growing penetration rate in the rapidly developing countries. This tendency is also supported by globalization of platforms: iPhone/iPod, Microsoft software, vendor-specific hardware sold and operational worldwide.

7.4.2 Standardization

The globalization of GSM, Internet, and Wireless LAN has happened progressively through global standardization efforts, policies and regulations. But often when a few standards are established and available worldwide, only one is widely accepted regionally.

7.5 Threats

Wireless communication, Internet, and information systems have transformed our society into an information society. In such a context, the **threats are multiple** with a wide range of risks.

7.5.1 Denial of Service Attacks

Cyber-terrorism takes place in the form of denial of service attacks against high-profile websites. Such attacks happened to Yahoo, e-Bay, CNN in year 2000. Distributed denial of service attacks bring down the availability of the targeted website.

7.5.2 Masquerade Attacks

Cyber-crime takes advantage of the Internet and wireless communication to perpetrate, e.g., fraudulent financial profits. These can take the following forms: phishing, scams, or empowering organized crime. Phishing uses online masquerading of a trustworthy entity in order to criminally and fraudulently obtain sensitive information, such as usernames, passwords and credit card details. Scams aims at abusing the confidence of naïve people via, for example, emails, proposing quick money making deals. Also, mobile phones and Internet can be an efficient tool in the organization of crimes. This has also been seen multiple times in terrorist attacks.

Users have an entire personality and identity in the e-world that is actually susceptible to theft in order to gain credit or money, or for defamation. Furthermore, mobile communications users are susceptible to suffer from theft of service.

The Internet can be misused to entice, to invite, or to persuade an under-aged person to meet for indecent goals or to help arrange such a meeting.

7.5.3 *Leakage of Confidential Information*

Companies, organizations, governments, universities use and create information systems such as databases, intranets, websites, Internet and email servers. Some of the information stored in their information systems may be confidential. Leakage of such information represents a high risk as Internet and wireless communication are powerful tools for distributing information.

7.5.4 *Privacy Threats*

All these threats raise privacy concerns. Privacy protects the rights of the people to control what happens to their personal information. The major threats are Big Brother and Little Brother [4].

Big Brother as defined by George Orwell in his novel (1984) that describes a society where everybody is under surveillance of the main authority. In our information society, governments can use or control communication for foreign intelligence, law enforcement, tracking criminals, or tracking dissidents. Often such eavesdropping are undetected as governments do not want people to know that they are wiretapped. In democracies, big brother is not as major of a threat as in dictatorships that aim at monitoring internal traffic to find dissidents and deny access to information or certain websites.

Little Brother is the corporate controller. It concerns companies that aim at making money by tracking or profiling users, abusing and selling personal data. This can be done using cookies, third party cookies, cooperation with third party advertisement sites and web registration. Internet service providers monitor user traffic for traffic engineering but some providers monitor the traffic for disfavoring certain types of applications or for inserting advertisements in their traffic. Other corporations try to profile their users in order to sell users' personal data for targeted advertisements online or offline.

Privacy should also be taken into account with a perspective on where the major communication (Internet) pipes are.

7.6 A Security Perspective

Most security threats presented in the previous sections can be tackled with a technical security solution. The fundamental characteristic and strength of a security solution is to exploit what nature provides, i.e. human beings, their characteristics and their interactions.

7.6.1 Authentication Against Impersonation

The authentication that allows assessing the legitimacy of a user or of a website or of a correspondent is a strong property to hold against phishing, scam, theft of identity and service. Mutual authentication involves authentication of both communicating ends, as in Authenticated Key Exchange (AKE) protocols. The security goal of an AKE protocol is to establish secure channels between two parties, authenticating each other, and sharing a common session key (e.g., the key is used for confidentiality and/or data integrity) at the end of the protocol [5].

The advancement in user authentication leans towards biometrics exploiting human characteristics that are hard to forge such as vein patterns and iris.

7.6.2 Leakage Resilient Schemes

The leakage of secrets stored on a device may happen if the device is lost or stolen. There is thus a need for leakage resilient schemes. The strength of leakage resilient schemes is that they exploit human-memorable passwords to perform authenticated key establishment with mutual authentication, confidentiality and integrity protection and robustness against leakage of secrets stored on a device. In addition to mutual authentication and session keys generation as in any AKE, one could use the Leakage-Resilient AKE (LR-AKE) protocols proposed in [5] that provide additional security features for the protection of a party's short secret (e.g., password) against leakage of stored secrets from the both parties.

7.6.3 Identity-Based Encryption for Privacy

Encryption is the procedure of rendering a message into a concealed form so that it can be decrypted exclusively by one particular recipient or a group of recipients.

Identities and attributes publicly known can also be used for encryption purposes. ID-based encryption (IBE) exploits already available identities to generate verifiable signatures, or encrypting messages while attribute-based encryption (ABE) to allow a certain type of user to access.

IBE provides a public key encryption mechanism where an arbitrary string, such as a recipient's identity, is used as a public key. Such a property is useful, e.g., for simplifying public key infrastructure.

ABE enables an access control mechanism over encrypted data by specifying access policies among private keys and ciphertexts. One of the several applications of ABE is content broadcasting.

7.6.4 Anonymity for Privacy

For protecting humans/users' privacy, anonymous schemes can be efficient if not overloading the communication.

Anonymous authentication schemes and anonymous signatures proposed in [6] prevent the abuse of anonymity for performing illegal activities by allowing a server to be authenticated by a user and authenticate the fact that user is part of a legitimate group. The proposed protocol in [6] uses an anonymous Password-based AKE protocol extended for credential retrieval and stateless anonymous IPv6 address auto-configurations. This protocol has the advantage of being time efficient compared to the classical rerouting techniques for anonymous credential systems and channels. It can also prevent the abuse of the anonymous communication channels to perform malicious activities. The security analysis of our protocol shows that the anonymity can be guaranteed against the collusion of the credential server with the anonymous application service provider.

7.6.5 Private Information Retrieval for Privacy

A leakage is even more critical when it concerns for example competitive secret corporate information. This is the case when search information related to patents or patent applications of a company is leaked out, such a leakage can be very damaging to a company.

Private Information Retrieval is also a step towards full privacy protection. It permits a user to retrieve a piece of information from a database without the database knowing what item has been retrieved. It can be an efficient tool against little brother if it allows to retrieve user generated content stored in the database.

A private information retrieval (PIR) protocol allows a chooser to retrieve an item from a server (i.e. sender) containing a database without revealing the identity or the content of that item. The trivial solution is to let a sender send the entire string to a chooser, requiring n bits communication costs, where n is the number of bits in the database. In other words, the database, being the sender, sends the entire database to a chooser at every query so as to result in the database not knowing what is the item retrieved by the chooser.

7.6.6 Trust

The last point in this security perspective is how to establish trust in globalized communication environment. In mobile communication, a user should be able to trust a visiting network, or a foreign operator while only initially trusting and signing for the service of its home operator.

7.7 Conclusions

Security and privacy are the pillars, the guarantors and the guards of the sustainability and of the global success of future wireless and wired communication over the next decades. Privacy protection is a rising concern in the globally all-the-time connected era that we have entered. Along with security, it empowers solidly and consistently citizens of the world, in their usage of communications systems. As communications technology researchers in a globalized world, it is our mission to design, without compromising security and the environment, technologies that solve human issues in developing and developed countries. The first step towards this goal is to educate people about cyber-security threats through, e.g., comics; the second step could be to put in place zero-knowledge security techniques that would allow the users to reveal only needed personal information to obtain a desired service. The less personal information one provides to a third party, the stronger his/her privacy is protected. Beyond the "right to be forgotten", there should be the right "to give only necessary and sufficient personal information".

References

1. Ericsson, Can mobile communications close the Digital Divide? White paper found on http://www.ericsson.com/technology/whitepapers/Can_mobile_communications_close_Digital%20Divide.pdf
2. http://www.internetworldstats.com/stats.htm
3. http://www.gsmworld.com/news/statistics/pdf/gsma_stats_q4_07.pdf
4. S. Bellovin, in *Internet Privacy: Big Brother and Little Brother*, NICT Workshop, 2008
5. H. Fathi, S.S. Chakraborty, R. Prasad, *A Framework for VoIP in IP-based Wireless Heterogeneous Networks* (Springer, London, Nov 2008)
6. H. Fathi, S. Shin, K. Kobara, H. Imai, Protocols for authenticated anonymous communications in IP networks. Elsevier J. Comput. Commun. (Sept 2008)

Hanane Fathi (M'04) received her M.S. in Electrical Engineering and her Ph.D. in Wireless Communications Engineering from Aalborg University, Denmark, in 2002 and 2006 respectively. She also holds a M.S. in Information Technology received from Ecole Centrale d'Electronique, Paris, France in 2002. She visited Imai laboratory at the University of Tokyo, Japan, from October 2004 to March 2005. From April 2006 to May 2008, she was a research scientist at the Research Center for Information Security (RCIS) at the National Institute of Advanced Industrial Science and Technology (AIST), Japan. Currently she is a patent attorney at Plougmann & Vingtoft, Denmark. Her research interests include network security, secure network coding, wireless communications security, anonymity, and privacy.

Chapter 8
802.11n: The Global Wireless LAN Standard

Richard van Nee

8.1 Introduction

Two of the key factors playing a role in the globalization of wireless technologies are international standardization and global availability of spectrum. Both of these factors lined up almost ideally for WiFi wireless LAN technology. Worldwide license-free spectrum became available in the late eighties in the ISM bands, specifically the 2.4 and 5 GHz bands. International standardization took place in the IEEE 802.11 body, releasing the first global wireless LAN standard in 1997. The first IEEE 802.11 standard defined data rates up to 2 Mbps, which was as much as proprietary wireless LAN solutions had been capable of since 1990. In order to keep up with wired Ethernet speeds, new 802.11 groups started standardizing higher data rates. In 1999, two new standards emerged. The 802.11a standard defined rates up to 54 Mbps in the 5 GHz band using OFDM [1]. The 802.11b standard extended the rates in the 2.4 GHz band up to 11 Mbps while maintaining the same chip rate by using a set of complementary codes that was originally designed for use in OFDM systems [2]. Since then, data rates have kept growing exponentially as depicted in Fig. 8.1. Till 2004, the growth in data rate was achieved by going from single carrier direct-sequence spread-spectrum to OFDM using higher order constellation sizes up to 64-QAM. Unfortunately, this increase in rate came at the expense of a loss in range. The use of highly spectral efficient higher order modulations requires a significant larger SNR than the simple BPSK modulation used for the lowest 1 Mbps rate, resulting in a loss of range. In addition, the link becomes more vulnerable to co-channel interference, which reduces the total system capacity.

The solution to obtain significant higher data rates and increase range performance at the same time is MIMO-OFDM (Multiple Input Multiple Output

R. van Nee (✉)
Qualcomm, The Netherlands
e-mail: rvannee@qualcomm.com

R. Prasad et al. (eds.), *Globalization of Mobile and Wireless Communications: Today and in 2020*, Signals and Communication Technology,
DOI 10.1007/978-94-007-0107-6_8,

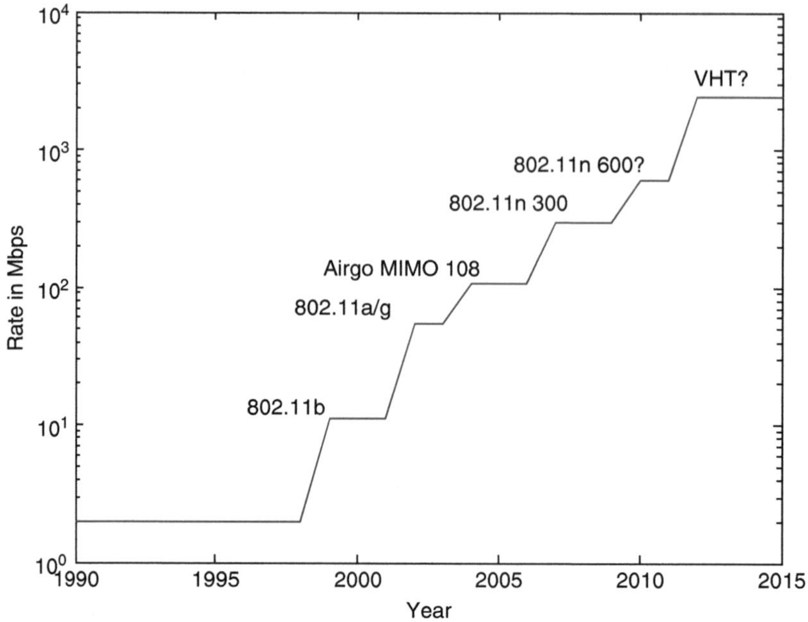

Fig. 8.1 Wireless LAN data rate growth

Orthogonal Frequency Division multiplexing) [3, 4]. MIMO-OFDM increases the link capacity by simultaneously transmitting multiple data streams using multiple transmit and receive antennas. It makes it possible to reach data rates that are several times larger than the current highest 802.11a/g rate of 54 Mbps without having to employ a larger bandwidth or a less robust QAM constellation [5]. With the introduction of MIMO-OFDM wireless LAN products in 2004 by Airgo Networks and the advent of the MIMO-OFDM based draft 802.11n standard in 2007, the performance of wireless LAN in terms of throughput and range is brought to a significantly higher level, enabling new applications outside the traditional wireless LAN area. The one time vision to replace wires in home entertainment applications, like TV cable replacement, has become a reality.

8.2 IEEE 802.11n

The 802.11n standard defines a range of mandatory and optional data rates in both 20 and 40 MHz channels. Table 8.1 lists the Modulation and Coding Schemes (MCS) and their corresponding data rates for the cases of 1 and 2 spatial streams [6]. For every MCS, four data rates are shown, as every MCS can be used in either a 20 MHz channel or a 40 MHz channel, using either a normal 800 ns guard interval

Table 8.1 Modulation and Coding Schemes (MCS) for 1 and 2 spatial streams

MCS	Code Rate	Modulation	Number of spatial streams	Data rate in 20 MHz, 800 ns GI	Data rate in 20 MHz, 400 ns GI	Data rate in 40 MHz, 800 ns GI	Data rate in 40 MHz, 400 ns GI
0	½	BPSK	1	6.5	7.2	13.5	15
1	½	QPSK	1	13	14.4	27	30
2	¾	QPSK	1	19.5	21.7	40.5	45
3	½	16-QAM	1	26	28.9	54	60
4	¾	16-QAM	1	39	43.3	81	90
5	⅔	64-QAM	1	52	57.8	108	120
6	¾	64-QAM	1	58.5	65	121.5	135
7	⅚	64-QAM	1	65	72.2	135	150
8	½	BPSK	2	13	14.4	27	30
9	½	QPSK	2	26	28.9	54	60
10	¾	QPSK	2	39	43.3	81	90
11	½	16-QAM	2	52	57.8	108	120
12	¾	16-QAM	2	78	86.7	162	180
13	⅔	64-QAM	2	104	115.6	216	240
14	¾	64-QAM	2	117	130	243	270
15	⅚	64-QAM	2	130	144.4	270	300
32	½	BPSK	1	N/A	N/A	6	6.7

Table 8.2 Optional MCS for 3 and 4 spatial streams

MCS	Code Rate	Modulation	Number of spatial streams	Data rate in 20 MHz, 800 ns GI	Data rate in 20 MHz, 400 ns GI	Data rate in 40 MHz, 800 ns GI	Data rate in 40 MHz, 400 ns GI
16	½	BPSK	3	19.5	21.7	40.5	45
17	½	QPSK	3	39	43.3	81	90
18	¾	QPSK	3	58.5	65	121.5	135
19	½	16-QAM	3	78	86.7	162	180
20	¾	16-QAM	3	117	130	243	270
21	⅔	64-QAM	3	156	173.3	324	360
22	¾	64-QAM	3	175.5	195	364.5	405
23	⅚	64-QAM	3	195	216.7	405	450
24	½	BPSK	4	26	28.9	54	60
25	½	QPSK	4	52	57.8	108	120
26	¾	QPSK	4	78	86.7	162	180
27	½	16-QAM	4	104	115.6	216	240
28	¾	16-QAM	4	156	173.3	324	360
29	⅔	64-QAM	4	208	231.1	432	480
30	¾	64-QAM	4	234	260	486	540
31	⅚	64-QAM	4	260	288.9	540	600

or an optional 400 ns short Guard Interval. The use of two spatial streams with a short guard interval in a 40 MHz channel gives a highest possible data rate of 300 Mbps. Even higher data rates are possible by using the optional MCS listed in Table 8.2 that use 3 and 4 spatial streams. For four spatial streams, the highest possible data rate becomes 600 Mbps. In addition to the MCS sets listed below, an 802.11n device also needs to support all mandatory 802.11g rates if it operates in the 2.4 GHz band, or all 802.11a rates if it operates in the 5 GHz band. This ensures full interoperability with legacy WiFi equipment.

In addition to the MCS values listed in Tables 8.1 and 8.2, there are other MCS up to number 76 that use a different QAM types across the spatial streams. These MCS do not provide higher data rates, but they can be used in combination with beamforming to provide a link budget gain over a non-beamforming case at the same data rate.

8.3 Preambles

Fig. 8.2 shows the packet structure of IEEE802.11a. One of the most important criteria for the choice of the new preambles for IEEE802.11n is compatibility with IEEE802.11a&g. To achieve this, a mixed-mode preamble is constructed as depicted in Fig. 8.3. The mixed-mode preamble starts with an 802.11a preamble with the only difference that multiple transmitters transmit cyclically delayed copies of the preamble. A legacy 802.11a receiver is able to receive this preamble up to the legacy signal field, which guarantees a proper defer behavior of legacy devices for 802.11n packets.

The short training field (STF) of 802.11n is the same as for 802.11a&g, except that different transmitters use different cyclic delays. The latter is done to avoid undesired beamforming effects and to get accurate power estimates that can be used to set the receive gain. For a proper AGC setting, it is important that the received power during the short training field is the same as the power during the rest of the packet. To achieve this, the short training fields from different transmitters must

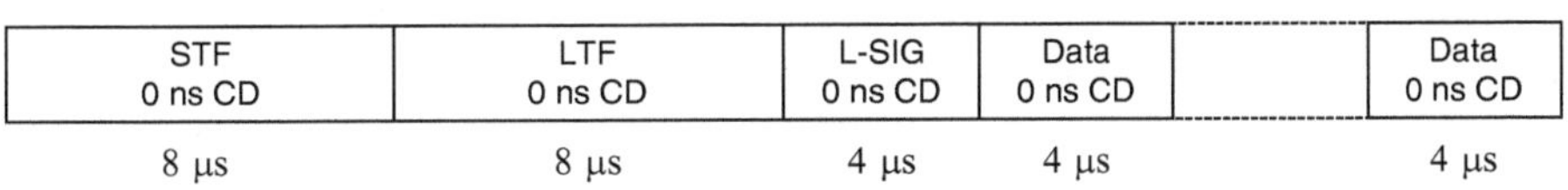

Fig. 8.2 IEEE802.11a/g packet structure

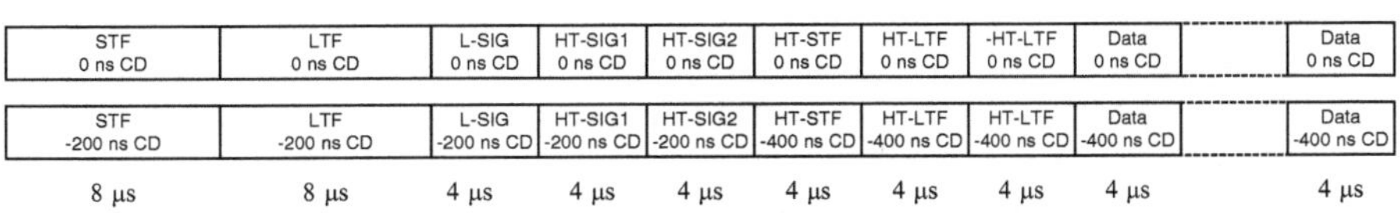

Fig. 8.3 IEEE802.11n mixed-mode packet with two spatial streams

have a low cross-correlation, also after being convolved with the wireless channel impulse response which has a typical rms delay spread in the order of 50–100 ns. This low cross-correlation is achieved by applying a different cyclic delay per transmitter. A greenfield preamble uses a cyclic delay of –200 ns, while a mixed-mode preamble uses a smaller delay of –200 ns for the case of two transmitters, because of a fear that legacy 802.11a/g receivers might not be able to deal with larger cyclic delay values. Legacy receivers are also the reason that only negative cyclic delays are used. To a receiver the presence of a second transmit signal with a positive cyclic delay appears like a multipath signal that arrives later than the signal of the first transmitter. If the receiver uses a correlation approach to set its symbol timing [8], then it will set the symbol time too late, which can result in inter-symbol interference. When the second transmitter uses a negative cyclic delay, the receiver will set its symbol timing too early, which will eat into the OFDM guard time without causing inter-symbol interference.

After the legacy signal field (L-SIG), a new high throughput signal field (HT-SIG) is transmitted that contains 48 bits with information including a 16-bit length field, a seven-bit field for the modulation and coding scheme (MCS), bits to indicate various options like LDPC coding, and an eight-bits CRC. To enable detection of the presence of a high throughput signal field, it uses a BPSK constellation that is rotated by 90°.

After the high throughput signal field, a second short training field is transmitted. This high throughput short training field (HT-STF) can be used to retrain the AGC, which may be needed for two reasons; first, the transmitter may employ beamforming for the high throughput part of the packet only, such that there may be a large power difference between the received signal before and after the start of the high throughput short training field. Second, there may also be a power difference because of non-zero cross-correlations between the cyclically shifted short training fields of the legacy part of the mixed mode preamble. This effect is small when using a cyclic delay of –200 or –400 ns like explained earlier. For the case of four transmitters, however, cyclic delays as small as 50 ns are used, which can result in a few dB difference between the received power before and after the high throughput short training field.

The high throughput short training field is followed by one or more high throughput long training fields (HT-LTF) that are used for channel estimation. The number of HT-LTF symbols is equal to the number of spatial streams. For the case of two spatial streams, the second HT-LTF of the first spatial stream is inverted to create an orthogonal space-time pattern. The receiver can obtain channel estimates for both spatial streams by adding and subtracting the first and second HT-LTF, respectively. The channel estimates can then be used to process the MIMO-OFDM data symbols that follow the HT-LTF. The only remaining training task after channel estimation is pilot processing. Every data symbol has a few pilot subcarriers – four in 20 MHz modes, six for 40 MHz modes – that can be used to track any residual carrier frequency offset.

In addition to the mixed-mode preamble, the 802.11n standard also defines a greenfield preamble. This preamble that is shown in Fig. 8.4 is 8 μs shorter, resulting

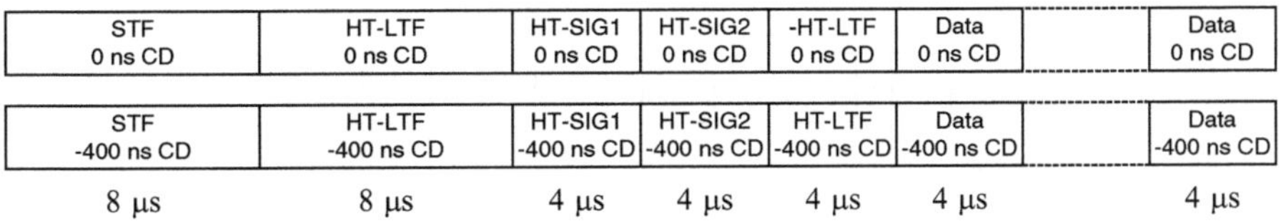

Fig. 8.4 IEEE802.11n greenfield packet with two spatial streams

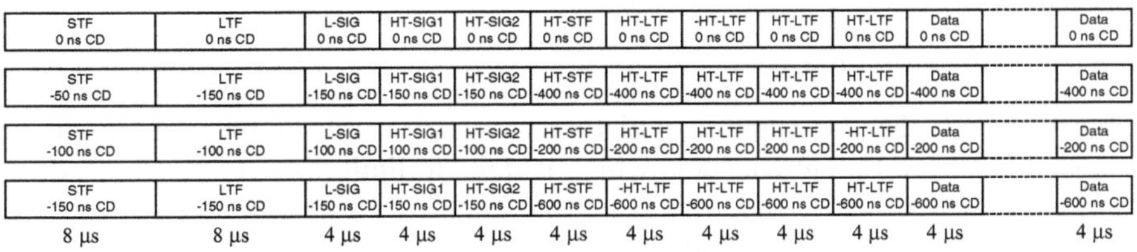

Fig. 8.5 IEEE802.11n mixed-mode packet with four spatial streams

in a larger net throughput. It is not compatible with legacy 802.11a or 802.11g devices as such devices will not be able to decode the signal field of a greenfield preamble. Because of this, the greenfield preamble is useful in two situations; first, in networks without any legacy devices, and second, in pieces of reserved time, also referred to as 'green time'. Green time can be reserved for instance by an RTS/CTS (Request-to-Send/Clear-to-Send). In the reserved time, the 11n standard allows a burst of packets to be send using a RIFS (Reduced Interframe Spacing) of 2 μs only. Using a greenfield preamble instead of the longer mixed-mode preamble minimizes the training overhead for such packet bursts, while a mixed-mode preamble can be used for the RTS/CTS to make sure that both 802.11n and legacy devices will properly defer.

It is mandatory for an 802.11n device to transmit or receive two spatial streams. In addition to this, optional modes are defined for 3 and 4 spatial streams. Fig. 8.5 shows the structure of the optional mixed-mode preamble for the case of four spatial streams. This preamble has four high throughput long training symbols that are encoded with an orthogonal pattern such that the receiver is able to obtain channel estimates for all four spatial streams. Together with the optional short guard interval option and the use of a 40 MHz channel, the four spatial stream mode gives a highest possible raw data rate of 600 Mbps.

8.4 802.11n Transmitter

Fig. 8.6 shows the block diagram of an IEEE802.11n transmitter. Input data is first scrambled using the same length-127 pseudo-noise scrambler that is used in IEEE802.11a. The convolutional encoder is also the same as IEEE802.11a, with the

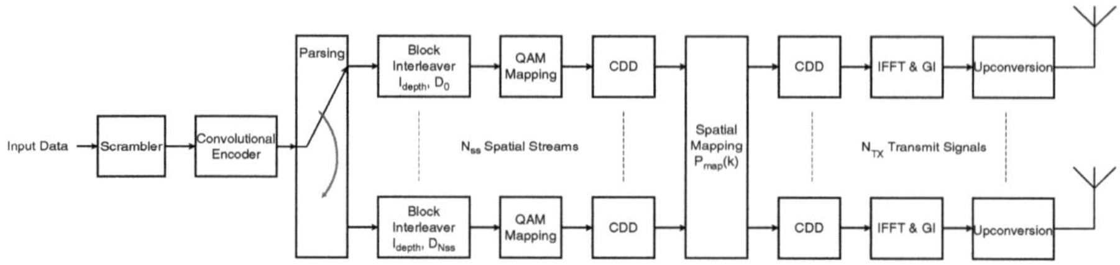

Fig. 8.6 Block diagram of an IEEE802.11n transmitter

only difference that for 3 and 4 spatial streams, odd and even bits are separately encoded by two different encoders which is done to limit the maximum decoding rate at the receive side.

After encoding, a parser sends consecutive blocks of $s=\max(N_{bpsc}/2,1)$ bits to different spatial streams, with N_{bpsc} being the number of bits per subcarrier. The bits are then interleaved by a block interleaver with a block size equal to the number of bits in a single OFDM symbol of the *nth* spatial stream, $N_{CBPS,n}$. By interleaving the bits across both spatial streams and subcarriers, the link performance benefits from both spatial diversity and frequency diversity. The interleaver for spatial stream n within its block of $\mathrm{N_{CBPS,n}}$ bits is defined by the following relations, where k_n is the input bit index for spatial stream n and j_n is the output bit index.

$$k_n = 0,1\ldots \mathrm{N_{CBPS,n}} - 1$$

$$\mathrm{s_n} = \max\left(\mathrm{N_{BPSC,n}} / 2,1\right)$$

$$i = \left(\mathrm{N_{CBPS,n}} / \mathrm{I_{DEPTH}}\right)\left(k_n \bmod \mathrm{I_{DEPTH}}\right) + \mathrm{floor}\left(k_n / \mathrm{I_{DEPTH}}\right)$$

$$j = \mathrm{s}_n \times \mathrm{floor}\left(i / \mathrm{s_n}\right) + \left(i + \mathrm{N_{CBPS,n}} - \mathrm{floor}\left(\mathrm{I_{DEPTH}} \times i / \mathrm{N_{CBPS,n}}\right)\right) \bmod \mathrm{s_n}$$

$$j_n = \left(j + \mathrm{N_{CBPS,n}} - \mathrm{N_{BPSC,n}} \mathrm{D_n}\right) \bmod \left(\mathrm{N_{CBPS,n}}\right)$$

The interleaving depth $\mathrm{I_{DEPTH}}$ and the subcarrier rotation $\mathrm{D_n}$ are defined in Table 8.3.

After interleaving, bits are mapped onto QAM symbols. Then, a spatial stream dependent cyclic delay (CD) is applied in the frequency domain. More details about this cyclic delay can be found in Section 8.3. At this point, a spatial mapping matrix is applied to each subcarrier to convert N_{ss} spatial stream inputs into N_{tx} transmitter outputs. If the number of transmitters is identical to the number of spatial streams, the spatial mapping matrix can simply be the identity matrix. To transmit legacy 802.11a/g rates that have only one spatial stream, the spatial mapping matrix reduces to a column of ones. After the spatial mapping matrix, an additional cyclic delay can be applied per transmitter to provide transmit cyclic delay diversity (CDD) and prevent undesired beamforming effects. Each transmitter subsequently applies an IFFT, inserts a guard interval, upconverts and transmits the signal.

Table 8.3 Interleaving parameters

N_{SS}	N_{SD}	I_{DEPTH}	D_0	D_1	D_2	D_3
1, 2, 3, 4	52	13	0	22	11	33
1, 2, 3, 4	108	18	0	58	29	87

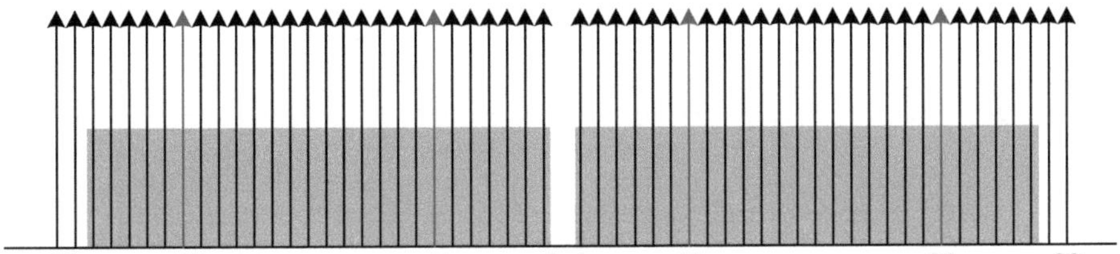

Fig. 8.7 Subcarrier allocations for a 20 MHz channel. Data tones are black. Pilots are dark grey. *Light grey blocks* are the subcarriers used for 802.11a, for the legacy part of the mixed-mode preamble, and for the HT-SIG field

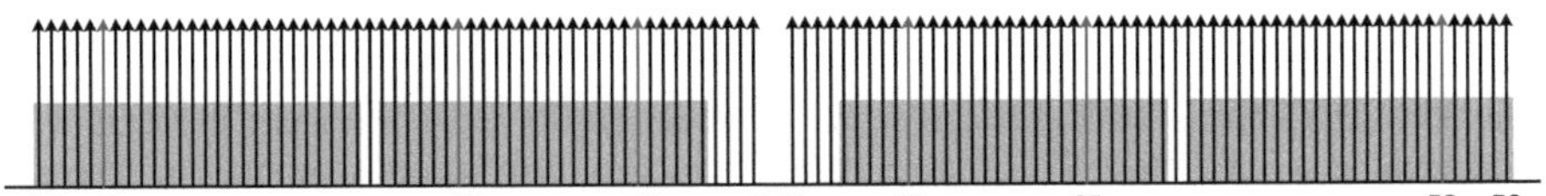

Fig. 8.8 Subcarrier allocations for a 40 MHz channel. Pilots are dark grey. *Light grey blocks* are the subcarriers used for 802.11a, for the legacy part of the mixed-mode preamble, and for the HT-SIG field

The subcarrier mapping for 20 and 40 MHz channels is depicted in Figs. 8.7 and 8.8, respectively. The 20 MHz mode uses 56 subcarriers for the high throughput data symbols, which is four more than the number used by 802.11a. The extra tones increase the throughput at the cost of some extra transmitter complexity to keep the transmitted spectrum within the spectral mask. The 802.11n spectral mask for 20 MHz mode is actually more tight than the 802.11a mask, so the use of more tones does not decrease the adjacent channel performance relative to 802.11a. The legacy part of the mixed-mode preamble and the high throughput signal field use the same subcarriers as 802.11a, which are shown as grey blocks in the figures. The 20 MHz 11n modes use four pilots just like 802.11a, while the 40 MHz 11n mode uses six pilots. A difference with 802.11a is that 802.11n uses a space time mapping for the pilots when transmitting from multiple antennas. In this way some extra transmit diversity is obtained on the pilots, and undesired beamforming effects are prevented.

8.5 LDPC Coding

The mandatory code used in 802.11n is the same binary convolutional code (BCC) that is used by 802.11a/g. An optional LDPC code is specified to get some extra gain over the mandatory BCC. The LDPC code is a systematic block code with

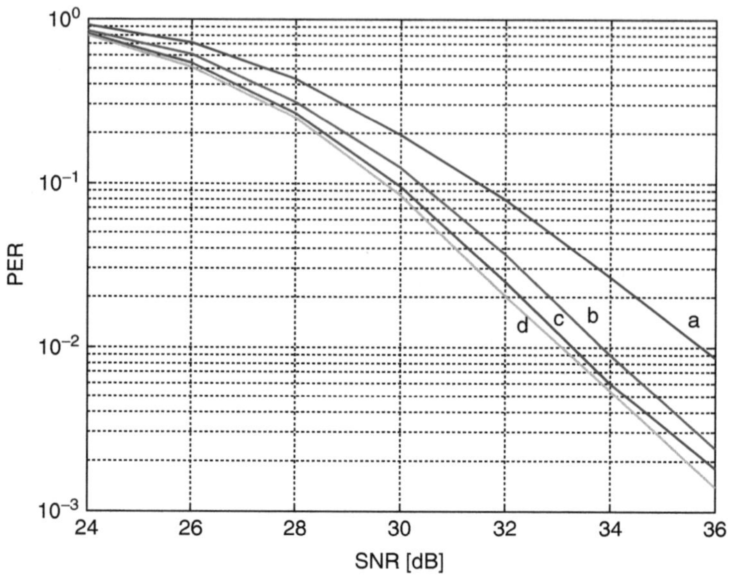

Fig. 8.9 LDPC versus BCC PER curves for 1000B packets in 802.11n channel D NLOS with ideal training, using a rate of 270 Mbps in 40 MHz with two spatial streams, two transmitters, and two receivers. (**a**) BCC, (**b**) LDPC with ten iterations, (**c**) LDPC with 20 iterations, (**d**) LDPC with 50 iterations

possible block lengths of 648, 1,296, and 1,944. The same coding rates as for BCC are provided, which are 1/2, 2/3, 3/4, and 5/6. The parity check matrices are sparse and highly structured, which facilitates both encoding and decoding. Fig. 8.9 shows the gain of LDPC over BCC for the case of two spatial streams using two transmit antennas and two receive antennas, assuming perfect training and an MMSE receiver for MIMO detection. The channel model used in the simulation is the non-line-of-sight channel D, which is a typical indoor wireless channel with a delay spread of 50 ns [9]. The LDPC decoder used layered belief propagation [10] with ten, 20 and 50 iterations. It can be seen that a gain of about 2 dB over BCC can be achieved when doing 50 iterations.

8.6 Space Time Block Coding

Space Time Block Coding (STBC) is an optional feature in the 11n standard to provide extra diversity gain in cases where the number of available transmitters is larger than the number of spatial streams. STBC operates on groups of two symbols, mapping N_{ss} spatial stream inputs onto N_{sts} space time stream outputs. For the case of one spatial stream and two space time streams, for instance, STBC mapping is done as follows: if $\{d_{ke}, d_{ko}\}$ are QAM symbols for an even and odd symbol for subcarrier k, respectively, then STBC encoding maps this single spatial stream input onto two space time stream outputs $\{d_{ke}, d_{ko}\}$ and $\{-d^*_{ko}, d^*_{ke}\}$. Hence, the even symbol contains d_{ke} on the first space time stream and $-d^*_{ko}$ on the second space time stream; the odd symbol contains d_{ko} on the first space time stream and d^*_{ke} on the second space time stream. The benefit of this type of STBC is that it doubles the diversity order of the link. In addition to the single spatial stream STBC mode described above, the 802.11n standard also specifies STBC modes with 2 and 3 spatial streams.

8.7 Beamforming

Beamforming is a way to provide extra performance gain in cases where the number of transmit antennas is larger than the number of spatial streams, and where the transmitter has knowledge about the channel. Beamforming is done by multiplying the N_{ss} spatial stream inputs for every subcarrier by an N_{tx} by N_{ss} beamforming matrix. The 802.11n standard specifies a number of optional methods that can be used to support beamforming. Two categories of beamforming exist, implicit beamforming and explicit beamforming. When using explicit beamforming, the beamformee – i.e. the device that is beamformed to – provides channel information to the beamformer, which is the device that is actually doing the beamforming. It can do this by sending a packet containing the received channel values for all subcarriers, receivers, and spatial streams. It can also calculate the beamforming matrix coefficients and send these to the beamformer. The standard specifies two types of beamforming weight formats, an uncompressed format and a compressed format to limit the amount of overhead. To minimize the amount of overhead, implicit beamforming can be used where the beamformer uses received preambles from the beamformee to calculate the beamforming weights, assuming a reciprocical channel. Implicit beamforming does have more stringent calibration constraints than explicit beamforming.

8.8 MAC Enhancements

The 802.11n standard includes several enhancements to the Medium Access Control (MAC) layer that help to increase the net throughput, especially when using the newly defined high data rates. One important new feature is aggregation; by making the packets as large as possible, the relative throughput impact of preamble overhead is minimized. This is very important as a typical Ethernet packet with a length of 1,500 bytes takes only 40 μs of transmission time for the data part at a rate of 300 Mbps, which is the same duration as the mixed-mode preamble for a two-spatial stream packet. Hence, the net throughput is reduced by 50% just by the preamble overhead. To minimize this throughput hit, the 802.11n standard specifies two type of aggregation. Several MAC Service Data Units (MSDU) can be aggregated into one A-MSDU up to length of 7,935 bytes. It is also possible to aggregate MAC Protocol data Units (MPDU) into one A-MPDU with a maximum aggregated length of 65,535 bytes. A limitation of A-MSDU compared to A-MPDU is that the first needs to be targeted to one destination address.

Another new mechanism that can be used to increase throughput is the Reduced Inter Frame Spacing (RIFS). An 11n device may transmit a burst of packets separated by a RIFS of 2 μs, thereby minimizing the amount of protocol overhead duration.

With the introduction of many new data rates in 802.11n, link adaptation becomes more problematic. Rather than searching through a limited set of data rates, the sender now has to decide how many spatial streams and what channel width to use in order to maximize the link throughput. To facilitate link adaptation, the 802.11n standard introduces a way to provide MCS feedback, whereby a receiver can inform the sender what MCS it could use. The receiver can deduce this MCS recommendation from the received channel estimates and the received signal-to-noise ratio.

8.9 Use of 40 MHz Channels

The 802.11n standard allows the use of 40 MHz channels, while legacy 802.11a/g devices only use 20 MHz channels. In the 5 GHz band, the use of a 40 MHz mixed-mode preamble or the use of duplicate 11a RTS/CTS in both primary and secondary 20 MHz channels ensures that all legacy 802.11a devices can correctly defer for each 40 MHz transmission. At the same time, 40 MHz 802.11n devices correctly defer for legacy devices because the standard requires a 40 MHz device to do a Clear Channel Assessment based on activity in both primary and secondary channel.

In the 2.4 GHz band, the situation for using 40 MHz transmissions is much more complicated than it is for the 5 GHz band. First, there are less channels available, only 320 MHz channels. Second, the channels can be partially overlapping as the center frequencies are specified on a 5 MHz grid rather than on a 20 MHz grid like in the 5 GHz band. Fig. 8.10 shows measured percentages of channel occupancy in the 2.4 GHz band. Channels 1, 6, and 11 are used most frequently, but there is also a significant percentage of other channels. The disadvantage of these channel spacings is that is not possible to transmit a mixed-mode preamble that can be correctly received by all legacy devices. For instance, if an 802.11n device would use primary channel 1 and secondary channel 5 to do a 40 MHz transmission, then only legacy devices centered on channels 1 and 5 could receive the legacy portion of a 40 MHz mixed-mode preamble, while devices on channels {2,3,4,6,7} would not be able to receive the legacy portion, while they would be interfered by a partial overlap with the 40 MHz packet.

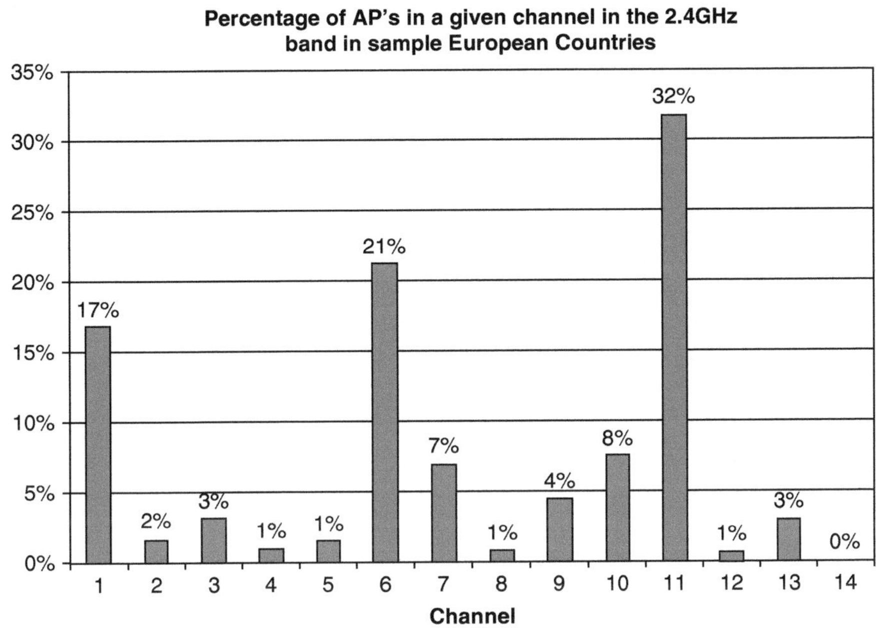

Fig. 8.10 Percentage of Acess points seen in a particular 2.4 GHz channel, based on 1,722 measured Access Points in cities in The Netherlands, Belgium, and Italy

8.10 MIMO-OFDM Performance Results

In 2004, Airgo Networks (acquired by Qualcomm in 2006) launched the first wireless LAN chipset based on MIMO-OFDM. This first generation MIMO-OFDM system uses a 20 MHz channel to transmit at either standard 802.11a/g data rates with a large range increase compared to conventional wireless LAN, or at significantly higher data rates up to 108 Mbps. In 2005, a second generation MIMO-OFDM product was introduced that uses Adaptive Channel Expansion to transmit either in a 20 or 40 MHz channel, increasing the top data rate to 240 Mbps.

Fig. 8.11 demonstrates the performance impact of a wireless LAN using MIMO-OFDM versus a conventional wireless LAN in a 20 MHz channel. The plot shows the cumulative distribution function of the measured TCP/IP throughput, where the client device has been put on a slowly rotating turntable to get throughput results for all possible orientations. From Fig. 8.11 it can be seen that in 10% of all possible orientations, the MIMO wireless LAN has a throughput less than about 33 Mbps, so for 90% of all orientations the throughput exceeds 33 Mbps. For the non-MIMO wireless LAN, this 10% number is only 4 Mbps. Hence, for a 10% outage probability, the MIMO throughput is more than eight times better in this particular case. For the 1% outage probability, the performance difference is even more pronounced.

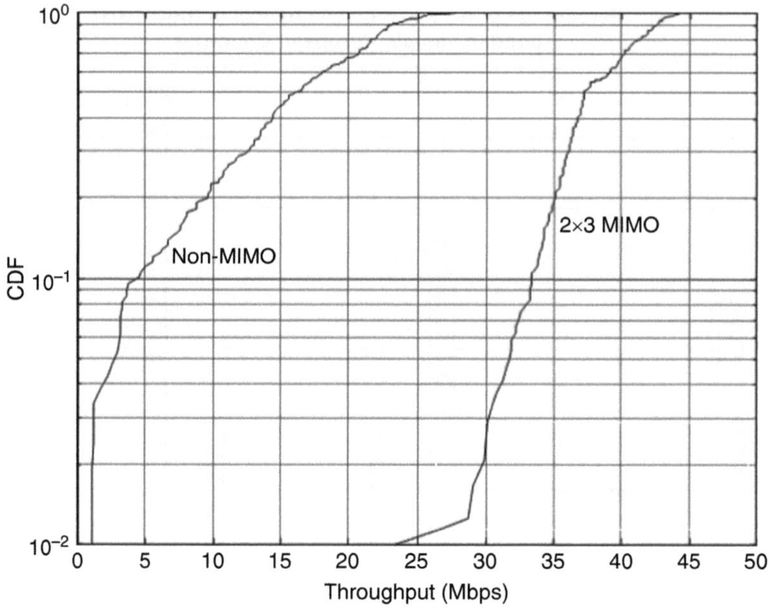

Fig. 8.11 Cumulative distribution of measured throughput

Fig. 8.12 shows measured TCP/IP throughput results for both 20 and 40 MHz channel width. Each throughput curve consists of eight points that correspond to different locations of the client inside a house, with an increased range towards the access point, but also with an increasing number of walls between the client and the access point. For the first test point, the client device is in the same room as the access point at a distance of 17 ft, while at the last test point the distance is 102 ft including five walls in between client and access point. The results show that for any range, the MIMO-OFDM throughput is 2.5–5 times larger than the throughput of non-MIMO products. Notice that several of these other products use channel-bonding to provide a proprietary maximum raw data rate of 108 Mbps in a 40 MHz channel. This explains why these products are able to achieve maximum TCP/IP throughputs over 40 Mbps, while conventional 802.11a/g products have a maximum TCP/IP throughput of about 25 Mbps. The maximum throughput of MIMO-OFDM in a 40 MHz channel exceeds 100 Mbps, which meets the throughput goal set by 802.11n [7].

Another way to look at the results of Fig. 8.12 is in terms of range increase for a given throughput. For instance, for a required throughput of at least 40 Mbps, the best non-MIMO-OFDM product has a maximum range of about 25 ft including one wall. For the same 40 Mbps throughput, MIMO-OFDM has a range of more than 80 ft including four walls. This range increase of more than a factor of

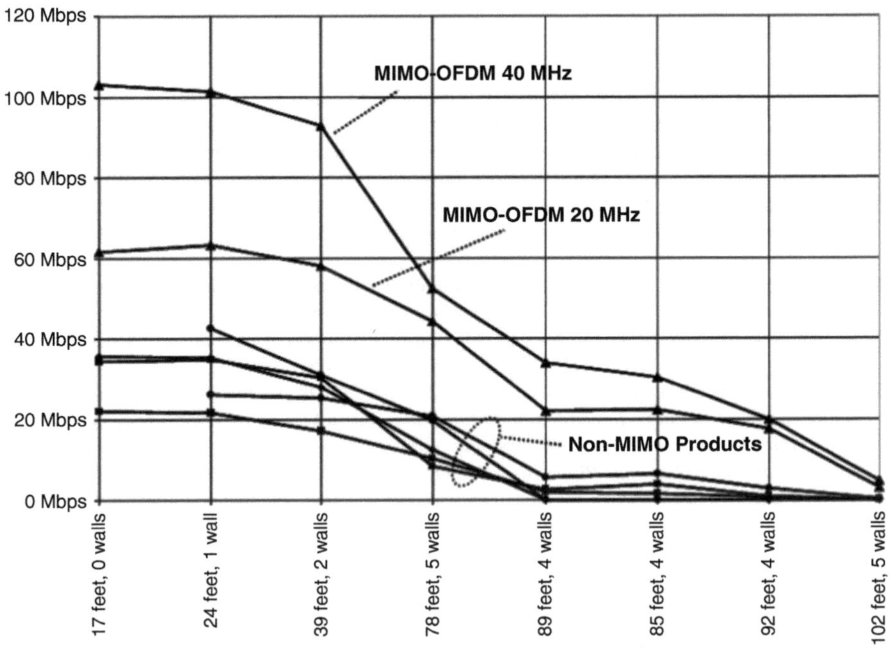

Fig. 8.12 TCP throughput measured at various distances

3 makes it possible to guarantee a high throughput throughout an entire house, which opens the way to new throughput-demanding applications such as wireless video distribution.

8.11 Conclusions

The performance of wireless LAN in terms of range and throughput is increased significantly by the use of MIMO-OFDM which is the basis of the new IEEE 802.11n standard. Performance results show that net user throughputs over 100 Mbps are achievable with just two spatial streams, which is about four times larger than the maximum achievable throughput using IEEE 802.11a/g. For the same throughput, MIMO-OFDM achieves a range increase of about a factor of 3 compared to conventional wireless LAN. This performance boost makes MIMO-OFDM the ideal successor to the current OFDM-only wireless LAN. Also, it enables new throughput-demanding applications such as wireless video distribution. Seeing the effectiveness and superior capability of MIMO-OFDM in enhancing data rate and extending range, other standards organizations have realized that it can do wonders for other technologies, both fixed, mobile and cellular. Standard bodies like 3GPP, WiBro, WiMax and the 4G Mobile Forum have started exploring the use of MIMO-OFDM in their respective technology areas, making it the technology of choice for future wireless networks.

References

1. R. van Nee, G. Awater, M. Morikura, H. Takanashi, M. Webster, K. Halford, New high rate wireless LAN standards. IEEE Commun. Mag. **37**(12), 82–88 (Dec 1999)
2. R. van Nee, OFDM codes for peak-to-average power reduction and error correction, in *IEEE Globecom*, London, 18–22 Nov 1996, pp. 740–744
3. G.G. Raleigh, J.M. Cioffi, Spatio-temporal coding for wireless communications, in *Proceedings of 1996 Global Telecommunications Conference*, pp. 1809–1814, London, Nov 1996
4. G.G. Raleigh, V.K. Jones, Multivariate modulation and coding for wireless communication. IEEE J. Select. Areas Commun. **17**(5), 851–866 (May 1999)
5. R. van Nee, A. van Zelst, G. Awater, Maximum likelihood decoding in a space division multiplexing system, in *IEEE VTC 2000*, Tokyo, Japan, May 2000
6. IEEE 802.11 Working Group, in *IEEE P802.11n/D2.0 Draft Amendment to Standard for Information Technology-Telecommunications and information exchange between systems-Local and Metropolitan networks-Specific requirements-Part 11: Wireless LAN Medium Access Control (MAC) and Physical Layer (PHY) specifications: Enhancements for Higher Throughput*, Jan 2007
7. A. Stephens, 802.11n Functional requirements. IEEE document 802.11-02/813r12 (March 2004)
8. R. van Nee, R. Prasad, *OFDM for mobile multimedia communications* (Artech House, Boston, Dec 1999)
9. V. Erceg et al., TGn channel models. IEEE document 802.11-03/940r4 (May 2004)
10. J. Chen et al., Reduced-complexity decoding of LDPC codes. IEEE Trans. Commun. **53**(8), 1288–1299 (Aug 2005)

Chapter 9
Vision on Radio Resource Management (RRM) and Quality of Service (QoS) for Wireless Communication Systems in Year 2020

Ljupčo Jorgušeski

9.1 Introduction

The historical evolution of major wireless access systems in Europe (similarly world- wide) shows that there is a major introduction of wireless access technology in each decade (Fig. 9.1). In the 1990s we have witnessed worldwide introduction of 2G wireless systems dominated by the GSM standard and voice/SMS applications. In the beginning of the new millennium many operators have extended existing 2G system for supporting data applications (e.g. GPRS and EDGE) and next to that 3G wireless access systems were introduced such as UMTS (later enhanced with advanced packet-data features such as HSPA).

According to the announcements from many wireless operators worldwide from 2010 onwards the existing 3G systems will be upgraded with LTE, which is an intermediate step towards real 4G wireless access systems (according to the ITU categorization) labeled as LTE Advanced. Next to these deployments, 3G wireless access systems are deployed worldwide based on the IEEE 802.16e standard (i.e. Mobile WiMAX) and planned for upgrade towards 4G wireless systems 802.16m. As the wireless systems evolve, the supported range of applications is extended from voice only to voice and data applications and towards full multimedia wireless applications support. Additionally, the supported wireless access speed typically used in these systems increases from few hundreds of Kbps toward few Mbps per user.

At the terminal side we observe a global trend of having smart devices that drastically influence the earlier adoption of wireless broadband access as presented in Fig. 9.2. The most important devices are the so-called smart phones and wireless netbooks. It is important to emphasize that in 2009 more smart-phones are sold worldwide than regular phones and that the netbooks are in the top three consumer electronic products. Newest trend is to decrease price of these devices due to various

L. Jorgušeski (✉)
TNO, The Netherlands
e-mail: ljupco.jorguseski@tno.nl

R. Prasad et al. (eds.), *Globalization of Mobile and Wireless Communications: Today and in 2020*, Signals and Communication Technology,
DOI 10.1007/978-94-007-0107-6_9,

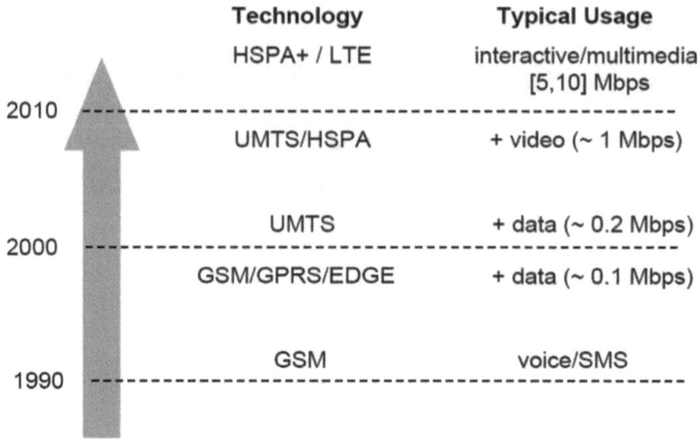

Fig. 9.1 Chronological view of wireless access systems in Europe

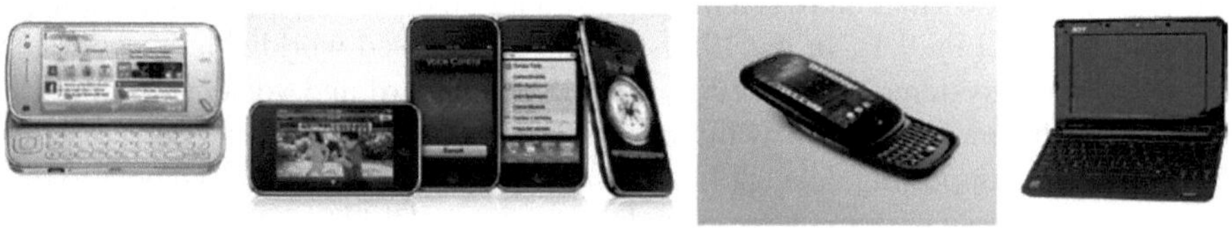

Fig. 9.2 Wireless devices accelerating the wireless broadband: smart phones and netbooks

technology advancements and mass production. Recent announcements indicate that the price of a smart-phone can be decreased below 150 USD.

This chapter presents a vision of the radio resource management (RRM) and QoS support for wireless systems in year 2020. The chapter is organized as follows. Section 9.2 presents an extrapolated view of wireless access systems in year 2020 based on the historical trends and current status worldwide. The vision on RRM in future wireless access systems is presented in Sect. 9.3. The expected QoS support for these systems is presented in Sect. 9.4. Section 9.5 presents conclusions.

9.2 Wireless Access Systems in Year 2020

If we project the chronological trend presented in Fig. 9.1 until the year 2020 then it can be expected that in global perspective the wireless access systems will have architecture as presented in Fig. 9.3.

In the evolution process towards year 2020 the following trends for the wireless access systems can be expected:

- 2G systems (e.g. such as GSM) will be slowly phased out and replaced by either evolved UMTS or LTE systems.

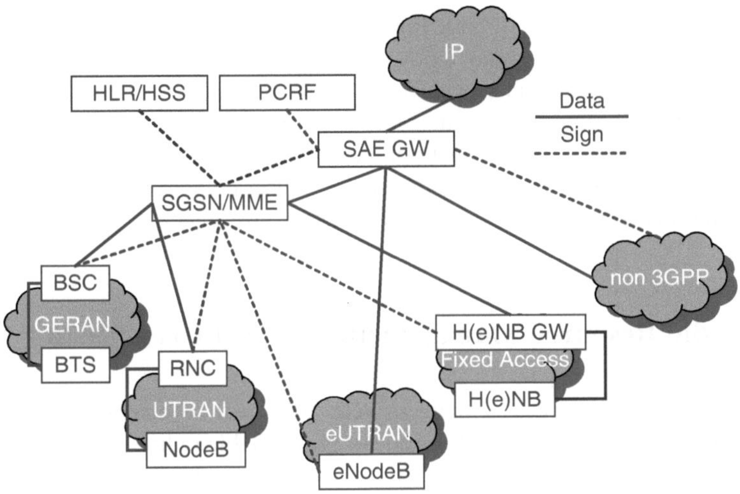

Fig. 9.3 Converged radio access architecture including home wireless access and non-3GPP access

- Advanced 3G systems such as HSPA+, LTE and WiMAX (802.16e) will be at their peak development and deployment, while 4G systems such as LTE Advanced and WiMAX (802.16m) will be in their uptake phase.
- As the physical limitations of the wireless access link is approached, any increase in the wireless access speeds and capacity is foreseen by deployment of smaller cells (including indoor cells at home), and convergence of different wireless access systems.
- At the core part of the wireless access systems, the anchoring of the different access systems will be provided via the System Architecture Evolution (SAE) Gateway as defined in LTE.
- The wireless access architecture will become mor e flat as intelligence and processing is shifted towards the wireless edges of the network reducing the access network latency.
- In the core part of the wireless access network, IP-based transport will be used until the wireless access points and all services will be transported over shared channels making the existence of separate CS and PS domains increasingly obsolete.

In order to provide very high bit-rates (e.g., up to 1 Gbps is the target for LTE Advanced in low mobility scenarios), the deployed spectrum will be at least between 10 and 20 MHz and also using bandwidth aggregation (possibly in different bands) for LTE Advanced and WiMAX (802.16m) systems. The percentage of smart mobile terminals and netbooks is expected to be rather high and the most sophisticated terminals will have high processing power, large and intelligent screens and user interface, multi-band and multi-radio technology support, etc.

These high-end terminals will have high multi-media support and by no means be different in their capabilities than a powerful PC today connected via fixed line and utilizing various multi-media services.

With the proliferation of the smart and powerful wireless end-user devices and the advanced wireless access technologies via small cells the internet usage between wireless/mobile and fixed/nomadic end-user devices will become increasingly similar and approaching true mobile-fixed convergence.

9.3 Radio Resource Management (RRM) in Year 2020

The RRM in wireless access system in year 2020 will be characterized by three major trends: dynamic and effective usage of the available radio spectrum, providing high bit-rates also for users at cell edge, and reduction of the operational cost (OPEX) for the wireless operators.

The dynamic and effective usage of the available radio spectrum will be shaped by the following important evolutionary trends:

(a) The system bandwidth of the advanced HSPA+, LTE and WiMAX systems can be dynamically adjusted based on the availability of wireless spectrum. Note here that HSPA + systems have limited granularity of bandwidth adjustments by 5 MHz chunks.
(b) The newest generation wireless access points will be software reconfigurable in the amount of spectrum that is used for the radio access technologies (RATs) that they support. For example, 3GPP aims to standardize multi-mode base stations that support GSM, UMTS/HSPA + and LTE radio access technologies with adjustable system bandwidths.
(c) As the RF technology becomes advanced and cheaper the end-user terminals can support different RATs with adjustable system bandwidths.

Based on this spectrum usage flexibility the future wireless access systems will abandon the current rather static system bandwidth deployment and adjust the used bandwidth per supported RAT based on current traffic demand. The RRM algorithms will reconfigure the system bandwidth per radio access technology in order to improve its utilization. Currently, this is not the case within the wireless access systems because all RRM algorithms have the given system bandwidth as a fixed starting point. Note here that, based on regulatory limitations, it is expected that in the future, wireless access operators can share spectrum usage among each other in the form of licensing agreements, dynamic spectrum sharing, etc.

As the wireless access technology reaches the physical limits (e.g., Shannon capacity bound) of the radio, the deployment of smaller cells (pico and femto), cooperative transmissions, and relaying concepts [1] are seen as major contributors towards increased data rates for the end user. For example, a wireless operator can deploy pico/femto cells in order to increase the signal level at the end-user and also share the available spectrum with smaller number of users due to the rather limited coverage of the pico/femto cells. An important deployment choice is the spectrum coordination among the different cell layers, i.e., macro layer and the pico/femto layer, as presented in Fig. 9.4.

The overlap of bandwidth between these different coverage layers will be guided by the spectrum availability, quality targets, and capability of the RRM algorithms to coordinate the resource usage at the different coverage layers. From RRM perspective, having no spectrum overlap between the different coverage layers, see Fig. 9.4, is the least challenging deployment scenario. However, for the partial or full overlap the RRM algorithms have to limit the negative interference effects originated from the different coverage layers [2]. Similarly, for relaying concepts the RRM algorithms have to coordinate the usage of the radio resources between the donor Macro cells and the relays/repeaters in order to avoid negative cross-interference effects. In order to facilitate this RRM coordination the necessary measurements and signaling exchange have to be provided between the Macro cells and pico/femto/relay nodes.

Regarding the reduction of OPEX there are two important RRM trends that are expected for future wireless access systems. First, reducing the energy consumption in the wireless access part of the network is considered very important due to the OPEX reduction as well as the environmental targets of operators worldwide to reduce the carbon emissions of their wireless networks and increase their energy efficiency. Next to the enhanced energy efficiency of the wireless access points, the operators consider dynamically switching on/off their wireless access points as the traffic demand varies during the time of the day, as presented in Fig. 9.5. In this example, in the low traffic period during the night-hours sufficient capacity and coverage can be provided by 12 sites while in the busy hour periods the network is reconfigured via switching on additional sites up to, for example, 48 sites. The 3GPP standardization is already introducing this energy saving functionality from Release 9 of UTRAN and e-UTRAN [3].

Second, as the operators have to operate three wireless access networks (2G, 3G, and 4G) and the complexity of these networks increases it is important that the operational costs are decreased by providing intelligent algorithms for automatic configuration and optimization in the network nodes. This concept is illustrated as self-organizing networks. The standardization bodies such as 3GPP have recognized this important property of the future wireless access systems and introduced

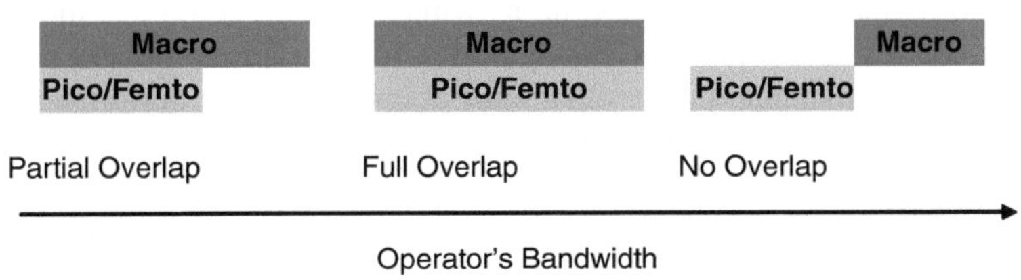

Fig. 9.4 Bandwidth segregation option between Macro-cell (coverage layer) and pico/femto cell (capacity layer)

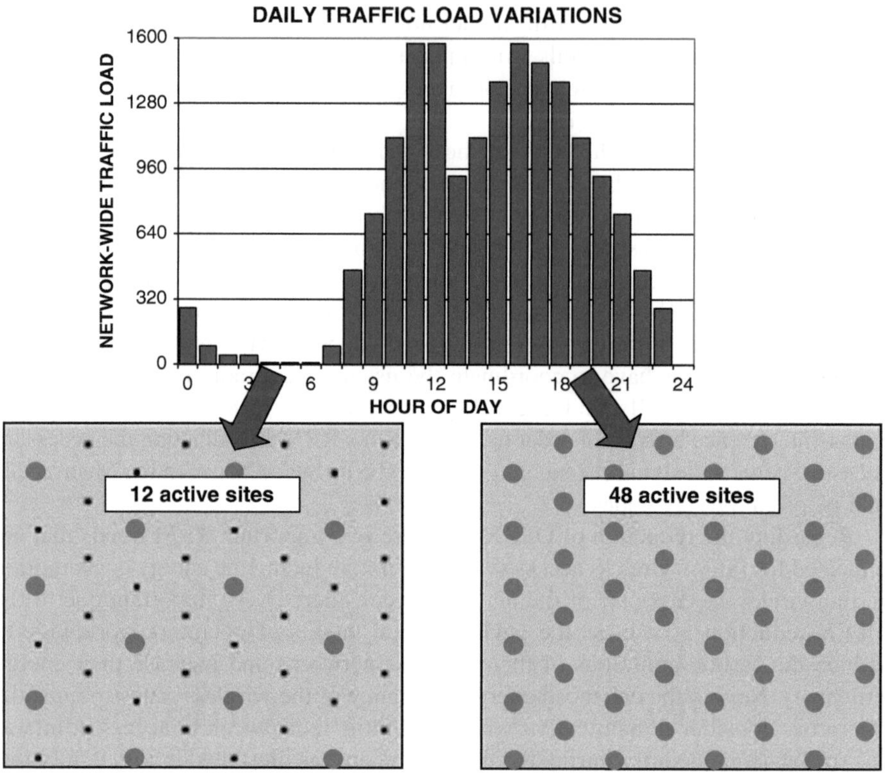

Fig. 9.5 Energy saving via dynamic switching on/off depending on hourly traffic demand

various SON functionalities for the LTE system [4]. The SON concept can be explained via the self-configuration, self-optimization, and self-configuration processes presented in Fig. 9.6.

The self-configuration algorithms take place when new wireless access nodes are introduced in the system or when previously switched off base station is activated. The task of these algorithms is to configure all the necessity parameters in plug-and-play fashion.

The self-optimization algorithms continuously optimize the wireless access network based on measurement data from the wireless access nodes and the end-user terminals. With this concept, the configuration of the wireless access nodes is a continuous and a dynamic process that aims at optimizing the performance of the network as given by the operator's policy.

The self-healing algorithms are activated when a particular network node is having a failure or partially incapable of supporting traffic. The self-healing algorithms aim at mitigating the negative effects of this failure by readjusting the neighboring wireless access points to absorb as much of the traffic, in the coverage area of the failed node, as possible.

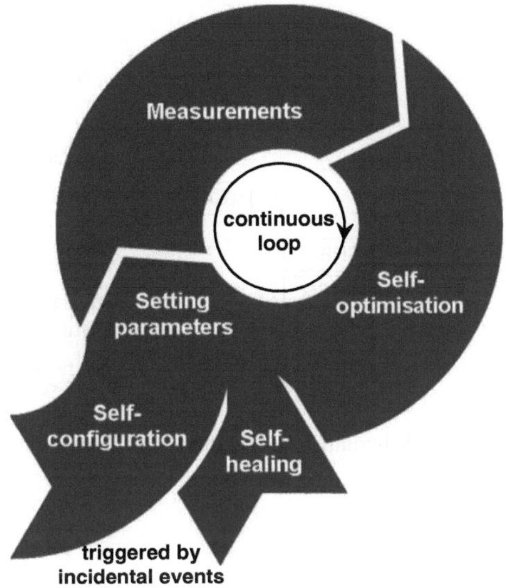

Fig. 9.6 The concept of self-organisation network [5]

9.4 QoS Support in Wireless Access Systems in Year 2020

The QoS support for wireless access system is increasingly important as the range of services supported over wireless systems is extended and becomes more similar with the services provided via fixed lines. The QoS framework for future wireless systems will be also much more dynamic and flexible than today's common practices. Currently, an important trend is to have various types of end-users with regard to their subscriptions. In future, the subscriptions will be much more flexible, open for specification by the end user, and dynamically updated based on the service usage or to the amount of data already used. On the other hand, on the applications side, we are observing a major trend where applications (e.g., application stores) are provided by third parties and enabled for use on the end-user terminals. In future the wireless operators, network vendors and application developers are expected to work jointly via open environments, in contrast to the situation today, resulting in various applications developed for specific end-user requirements and enabled rather smoothly for use on the wireless access system via open service platforms.

The QoS support is provided via the subscription profile in HLR/HSS and PCRF functions and has two major parts, see also Fig. 9.7:

(a) QoS differentiation in the IP based transport towards the wireless access node
(b) QoS differentiation on the radio access part between the wireless access node and the end-user terminal

The QoS differentiation between the wireless access points and the core part will be based on the Deep Packet Inspection (DPI) concept and utilize the DiffServ framework for QoS support in IP networks as presented in Fig. 9.8. The different packet flows, from the different services, will be handled according to their specific

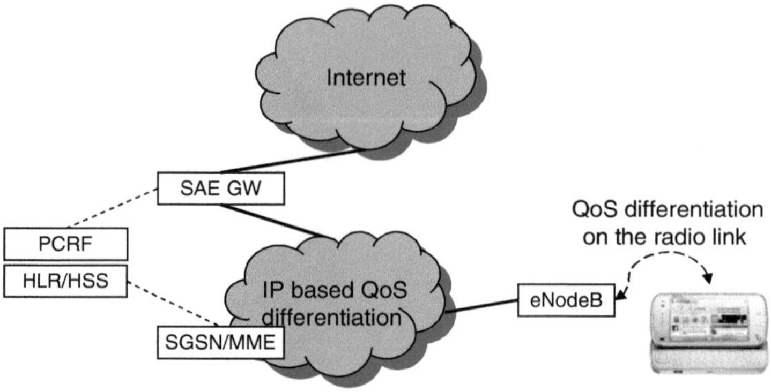

Fig. 9.7 Schematic view on the QoS differentiation in wireless access systems for year 2020

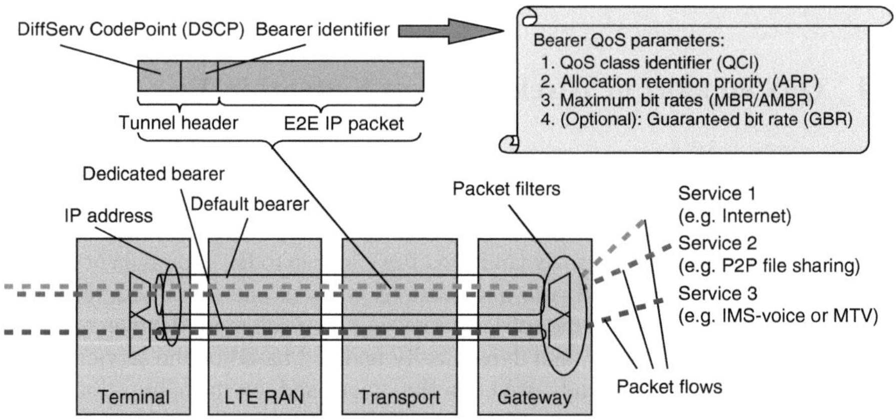

Fig. 9.8 DPI and QoS concept in SAE from [6]

QoS profiles and consequently their quality targets will be satisfied. As the IP traffic transport is established with the wireless access points and there is one shared IP domain for real time and non-real time services, this QoS differentiation framework is necessary to facilitate the different forwarding of the packet streams through the transport links.

Similarly, on the shared radio access link between the wireless access points and the end-user terminal the QoS differentiation is enabled via enhanced and QoS-aware packet scheduling and session control of the radio bearers. If due to mobility reasons the radio access technology towards the end-user terminal is changed, an appropriate QoS mapping will be made between the source and target QoS support functions in order to guarantee seamless QoS experience as the user is served via different RATs.

9.5 Conclusions

The evolution towards wireless access systems that have flexible bandwidth set-up, deployment of small cells and relays/repeaters, flat network architecture with IP based transport, and desired reduction of the OPEX have significant impact on the RRM and QoS functions for the wireless access systems in year 2020.

The most important trends in the RRM functions are the dynamic reconfigurations of the system bandwidth domain of the macro cell layer and between the macro and pico/femto cell layer or the repeater/relaying nodes. Additionally, system bandwidths can be reconfigured among the supported 2G, 3G, and 4G systems at the wireless operators. Due to targets of reducing costs, energy consumption, and carbon emissions, RRM functions will dynamically switch on/off parts of the wireless access systems according to the traffic demand and quality targets. The RRM functions in wireless access points will have self-organization capabilities to continuously optimize network performance based on access point and end-user measurements. This self-organization concept will drastically reduce the amount of operational costs.

Due to the flat architecture, IP based transport to the wireless access node, and the convergence of all traffic flows over a shared radio bearer, the QoS differentiation will be established via DPI functionalities based on the DiffServ concept in the IP transport layer and the QoS aware scheduling of the packets in the shared radio bearer.

Recommendations for further study in RRM and QoS for future radio access networks are in the domain of conflict resolution and policy enforcement considering that self-optimization functions will be active at the same time in the wireless access systems that might have contradictory actions and/or goals. The interaction of the DiffServ IP functions for providing QoS differentiation in the IP based transport and the QoS aware scheduling in the shared radio bearer deserves to be further studied. Algorithms for dynamically adjusting system bandwidths and the activity of the wireless access networks for energy conservation purposes should be investigated as well. Finally, the coordination of the RRM and QoS functions when inter-RAT handovers take place to enable seamless QoS experience should be an important area for further investigation.

References

1. O. Teyeb et al., Dynamic relaying in 3GPP LTE-advanced networks. EURASIP J. Wireless Commun. Netw. **2009**(731317), 11 (2009). doi:10.1155/2009/731317
2. D. Lopez-Perez et al., OFDMA Femtocells: A roadmap on interference avoidance. IEEE Commun. Mag. (Sept 2009)
3. 3GPP TR 32.826, Telecommunication management: Study on Energy Savings Management (ESM)
4. 3GPP TR 36.902, Evolved Universal Terrestrial Radio Access Network (E-UTRAN): Self-configuring and self-optimizing network (SON) use cases and solutions

5. FP7 SOCRATES project, http://www.fp7-socrates.eu/
6. H. Ekstrom, QoS in the 3GPP evolved packet system. IEEE Commun. Mag. (Feb 2009)

Ljupčo Jorgušeski is a senior consultant/innovator in the area of wireless access networks. His current working responsibilities are: optimization and performance estimation of wireless access systems such as HSPA+, WiMAX and LTE, strategic technology assessment and knowledge development of future wireless access systems, and monitoring and active participation in 3GPP standardisation and NGMN. He holds a Dipl. Ing. degree (in 1996) from the Faculty of Electrical Engineering at the university "Sts. Cyril i Metodij", Skopje, Republic of Macedonia and a Ph.D. degree (in 2008) in wireless telecommunication from Aaalbog University, Aalborg, Denmark. From 1997 to 1999 he worked as applied scientist at TU Delft, Delft, The Netherlands on the ETSI project FRAMES that defined the European proposal for the UMTS standardization. From 1999 to 2003 he was with KPN Research, Leidschendam, The Netherlands where he was working on the radio optimization and performance estimation of the GPRS, UMTS and WLAN networks. He has published over 15 conference papers, one journal paper, two book chapters, and co-authored one patent (including six pending patent applications).

Chapter 10
On Designing Future Communication Systems: Some Clean-Slate Perspectives

Petar Popovski

10.1 Introduction: The Constrained Wireless Evolution

Radio or wireless has been the main system of the general communication engineering since the dawn of that discipline, as seen in the efforts of Tesla and Marconi. Nevertheless, during the past 2 decades, we have witnessed an extraordinary growth pace and innovation in the area of wireless technology. During that time, the mobile phone has been promoted from its role of amenity to the role of necessity, while the Wireless LAN (WLAN) has been put forward to be the commonly used gate to access Internet. These two examples illustrate how much the end consumer has been affected by the wireless developments, but the overwhelming wireless growth has featured many engineering milestones and intellectual advances.

The wireless evolution has been subject to many constraints, determined by the spectrum regulation, system architecture, standardization, development stage in the synergistic technologies, the market/social factors, etc. On one hand, those constraints have guided the wireless networking towards proliferation and ubiquity. On the other hand, they have limited, or even sometimes prohibited, innovative ideas and concepts to be implemented in actual wireless networking solutions. The constraints and their impact on the wireless design/operation can be summarized as follows:

- The **regulatory constraints** represent an important factor which imposes operating rules on the devices that communicate by using radio transmission. The regulation segments the spectrum into frequency bands, which is used to provide a framework for managing the interference among different wireless systems. This requires early design decisions, as the wireless device/network must be created to operate in a defined set of frequency bands. Furthermore, the interaction with the other, potentially interfering devices, is restricted by requiring that each device complies with the rules of its current operating band.

P. Popovski (✉)
Aalborg University, Aalborg, Denmark
e-mail: petarp@es.aau.dk

R. Prasad et al. (eds.), *Globalization of Mobile and Wireless Communications: Today and in 2020*, Signals and Communication Technology,
DOI 10.1007/978-94-007-0107-6_10, © Springer Science+Business Media B.V. 2011

- The **architectural constraints** are introduced chiefly by the layered design of the communication protocols. The idea of layering is to decompose the general task of communication over a network into subtasks, such that the coordinated execution of the subtasks can achieve the global task. For example, multi-hop communication is decomposed into a routing subtask and a subtask that consists of data transmission over each of the selected links. Layering defines interfaces by which the subtasks communicate and coordinate and therefore fosters a black-box design, in which the module that executes a given subtask can be implemented (and debugged) in isolation. Such a layered architecture enables proliferation [1], but on the other hand it also features early design decisions which are restricting the freedom for building the most efficient solution for a particular communication scenario/task.
- The **standardization constraints** are represented by the context in which a particular communication system or protocol is developed and operated. Clearly, an evolving wireless system builds on the existing implemented concepts. The requirement for backward compatibility necessarily leads to incrementalism in devising technological improvements or drafting of the new standard versions. Nevertheless, properly supported standard is a *condition sine qua non* for achieving proliferation, as the examples of GSM and WLAN have demonstrated. The standardization constraints are tightly related to the **technological constraints**, which reflect what is implementable and technologically feasible at the time when the system is conceived. These constraints are referring not only to the communication technology per se, but also to the technologies that should operate in synergy with it, such as the hardware.
- The **market/social factors and constraints** are instrumental in giving rise to new applications and requirements from the communication systems, but also in delivering the verdict regarding the ultimate usability and successful presence of the communication system. Although very important, these factors are not the main focus of this text and will be treated only in the context of the other constraints.

These constraints and context–setters have paved the way for evolution of wireless systems. The continuous improvement in those systems comes from the large volume of novel techniques that are adopted in the standards, such as, for example, a particular multi-antenna technique or an efficient ARQ protocol. Using the language of biology to characterize such a wireless evolution, we can say that such techniques represent *mutations* (minor or major) that drive the evolution. Another important mechanism is the *crossover*, which has given rise to several important developments, such as the wireless Internet [2] or the combination of communication and localization in order to provide location-based services [3]. Fig. 10.1 exemplifies some contours of the wireless evolution. For example, the system ws1 and ws2 get combined and the system ws5 emerges. When ws4 is created it is not possible to implement the *concept c2* due to the active constraints, but due the increasing requirements (also by ws3) the constraints are changed and ws7 emerges. By the time some of the systems become obsolete – for the example on the Figure,

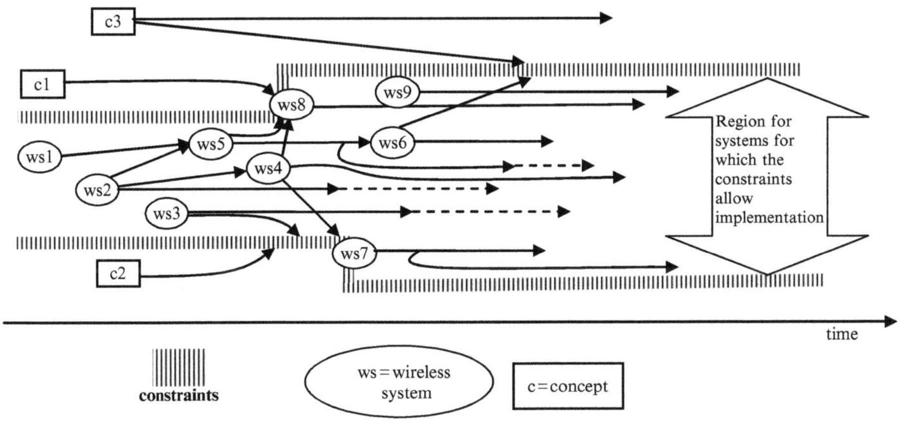

Fig. 10.1 The constrained evolution in the wireless systems. It can be seen that the constraints are changed by the means of the "forces" of the evolving systems and concepts hit a barrier. The *solid line* denotes that the system operates and further evolves. The *dashed line* denoted a system that gets obsolete

this happens the earliest for the system ws2. Note that the constraints do not allow the concept c3 to be implemented during the time frame depicted on the figure.

In such a vibrant technological area it is hard to put the monocular towards the distant future and make a precise account of the attributes featured in the wireless communication technology. Yet, with the present knowledge, we can identify some good candidates that can trace the future evolutionary paths. The goal of this text is to explore the possibilities, but also the reasons for introducing major changes (mutations) in very-long-term wireless evolution.[1] The main hypothesis is that the central element in the very-long-term-evolution is the treatment of interference. Changes in the mechanisms that deal with interference will create new regulatory paradigm. The road to such a paradigm leads through a *clean-slate design* approach in wireless communication, which means to think about wireless systems in terms of attaining the desired behavior and performance, not necessarily following the set of active constraints.

The chapter is organized as follows. In the next section we explore the nature of technological changes in wireless systems, classifying them as linear and non-linear, respectively. In Section 3 we discuss the interference as the central point in the very-long-term evolution of the wireless systems and outline some ideas that can change the paradigms for regulating and dealing with the wireless interference. Finally, Section 4 concludes the chapter.

[1] We use the phrase *very-long-term evolution* in order to avoid confusion with the Long Term Evolution (LTE) used in 3GPP. The time frame behind the phrase very-long-term is more than 10 years.

10.2 The Way Forward for the Wireless Technologies

The central notion of this text is *clean-slate wireless design*, which is achieved by creating a wireless communication system by considering the level of technological sophistication and implementability that is available today, or at least in a very near future. This definition is far from being precise and the constraint removal should be taken with caution, in particular regarding the regulatory constraints, as it will be clarified further in the text.

The need for clean-slate design occurs when the design decisions and constraints made in the past cannot offer the desired performance/applications due to their inherent limits. It is always instructive to recall the definition attributed to J. L. Kelly, that a channel is a part of the communication system that one is "*unwilling or unable to change*" [4]. If we understand the notion of communication broadly, then the constraints discussed above are determining the channel and thus setting its fundamental limits. Therefore, the required improvements necessitate recognizing the inhibiting constraints and changing them appropriately.

Before discussing the possibilities for having a long-term impact on the wireless systems, it is sobering, and at the same time inspiring to reflect on the reasons which motivate the continued research efforts and the innovation in the communication technologies, in particular the wireless ones. This is because the rapid wireless growth in the recent years may be a basis for skepticism towards further major developments in the wireless technology. The large volume and extensive quality of wireless networks that are either operating or being developed in the standards may point towards saturation in the actual need for fundamental wireless innovation and render it as an activity of purely academic interest. In simple words, it might seem that the "*physical wireless bricks*" are already there and any significant innovation effort should combine those bricks in a novel way to give rise to new applications and user scenarios. A strong support for this comes from the emergence of the Apple iPhone or the Android project by Google [5].

Contrary to such a standpoint, the wireless research community sees the existing wireless bricks as insufficient, both in quantitative and qualitative sense, as described in the following subsection.

10.2.1 Linear and Non-linear Wireless Changes

When speaking **quantitatively**, we refer to the improvement of the traditional performance parameters of the communication systems, such as the throughput and delay. For example, it is not uncommon to hear that the objective of the wireless systems is to provide communication speeds in the range of Gigabits per second. Those improvements are expected to be consumed by the increased video communication over the Internet (e.g. YouTube, Web 2.0 applications), but also the future applications that involve telepresence and interactive communication with low latency.

Still, these quantitative improvements are stretching the wireless evolution in a rather *linear* way. The intuition behind such term is that, for example, the changes in the data rate/delay lead to an improvement that can already be anticipated now. On the other hand, the required **qualitative changes** make a much stronger case for driving the wireless evolution forward. It is worth noting that the qualitative changes may also directly yield to increased data rate/performance, which means that they are already containing the quantitative changes. We refer to the qualitative changes as to *non-linear changes* since they require fundamental shifts in the wireless network design, leading to many unforeseen uses, applications, and interactions.

The following list is perhaps not exhaustive, but it tries to capture some of the most important qualitative changes in wireless network evolution.

- *Coping with the crowded spectrum.* The rising volume and variety of the wireless devices brought worries that the spectrum will become very crowded, thus slowing down the growth in the number of wireless devices. Despite the obvious fact that the spectrum is physically limited, the real factor that makes it look scarce is the conservative way in which the spectrum is allocated and used. Hence, in the recent years the myth of "scarce spectrum" has been seriously challenged by the concepts such as *dynamic spectrum access* [6] and *cognitive radio* [7]. These concepts require fundamental changes in spectrum usage and are perhaps the chief drivers of the clean-slate design.
- *Treatment of the interference.* In the majority of the existing communication systems, the approach has been to avoid or eliminate the interference. This is in discrepancy with the information-theoretic approaches that recognize the fact that the interference is not noise and therefore can be useful in devising more efficient communication schemes. To a large extent, such a discrepancy has been present due to the constraints coming from a feasible implementation. However, the reason has also been architectural. Namely, the layered OSI architecture has been created for wired networks, where the interference is a limited and fairly easily controllable phenomenon. On the other hand, interference is the basic phenomenon in a shared wireless medium, which should be reflected in the architecture. As the discussion further in the text will show, the proper treatment of the interference can bring significant values to the wireless network in terms of throughput, security, efficient spectrum sharing, etc.
- *Low power consumption.* This change has already gained momentum with the inception of the wireless ad hoc networks [8], wireless sensor networks [9], personal area networks, etc. The devices in these systems are usually battery-powered, such that every joule of energy should be spared in order to ensure longer operating times. There has been overwhelming evidence that a significant amount of energy is wasted in the non-optimized interactions between the protocol layers. Therefore, the methodology of *cross-layer optimization* has been applied to mend such energy leakages in the communication protocols [10]. This has in turn increased the revisionist feelings towards the existing layered architecture and motivated the research community to think about re-layering concepts.
- *Task-centric protocols.* The wireless sensor/actuator networks have brought scenarios in which the wireless connectivity should be used to somehow coordinate

the sensors and carry out certain global task that should be collectively carried out by the sensors/actuators. Such task can be, for example, control of industrial process or monitoring of the ecosystem in given region. The communication protocols in such networks should be designed to successfully attain the overall objective (task) of the network, rather than being focused on reliably transferring the data from each individual sensor. The value of the data carried by a single sensor is determined by the synergistic effect that emerges at the network level. This again calls for revision of the layered architecture, since the application layer (the task) will be markedly impacted by the decisions done at the protocol entities that transport the wireless data.

- *Alternative media for communications.* In the future communication technologies it is likely that the notion of "wireless" will transcend the radio media and become a more generalized concept to encompass any communication technology which uses unguided, broadcast-based and interference-prone medium. For example, the envisioned nano-networks [11] may use molecular communication and chemical transmitters in order to transfer information. The objectives and the design philosophy in those networks may be completely different from the approaches used today, but considering the rather modest set of existing solutions in that area, these networks are naturally approached via a clean-slate design.

The qualitative categories listed above are not mutually exclusive. For example, to relate the first three categories, we can conjecture that certain future wireless system may operate under regulation that requires treatment of the interference that is different from the current one, and such a regulation paves the way for building very low-power systems.

10.2.2 Clean-Slate Design is a Non-linear Change with a Very-Long-Term Impact

From the discussion above and the illustration on Fig. 10.2, it is clear that the non-linear changes will have tangible impact over a very-long-term, instead of appearing as efficient patches for incremental evolution of the current standards. In particular, the clean-slate approach is the proper response to the **untimeliness effect**, exhibited by the constrained technological evolution. This effect can be explained as follows. At a given time, the efficient mechanism B is not invented or is not feasible and the network system architecture A is built by using a different solution C which is less efficient, but functional. At a later time, after the implemented standard becomes widespread, the studies show that there are plausible scenarios where the mechanism B is feasible and can bring a significant performance benefit compared to C. However, the mechanism B is not compatible with the architecture A, but contrariwise, the ubiquitous usage of the standard renders the change of A infeasible. Does such a situation leave B to be a nice idea that ends up in a drawer?

The obvious answer might be: no, we just need to start creating a new standard and wait for the market/social mechanisms to obsolesce the devices that use the

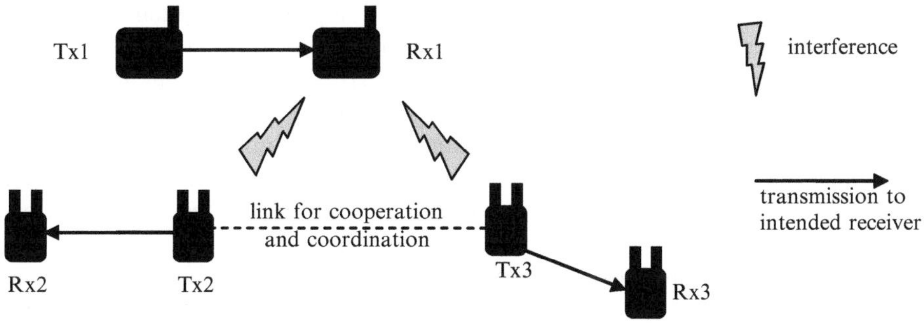

Fig. 10.2 Example of three links under the spectrum commons model. T × i and R × i denote the transmitter and the receiver of the i-th link, respectively. We observe the interference made to the receiver R × 1 by the devices T × 2 and T × 3

current standard. However, that is practically difficult, if possible at all, since by assumption B is incompatible with the (layered) architecture A, which can mean that many parts of the architecture A should be changed in order to accommodate B. Commonly, a given standard is addressing only part of the architecture (e.g. the IEEE 802 standards deal with the two lowest layers), while B can be a technique which spans several protocol layers. The latter may require coordinated change of several standards and it is this change which, on one hand, is an obstacle and, on the other hand, opportunity for major changes over a very-long term.

The coordinated change of standards is an action paradoxical in itself. Namely, the original idea of the standards is to modularize the problems, each module being a different standard, in order to have efficient framework to capture the complexities and the requirements of different systems. The need to coordinately change the modules in order to implement the technique B is a signal that such a modularization (layering) is not appropriate for the systems that feature that technique. However, technological/architectural constraints can be understood more as a recommended practice, which means that, in principle, a re-layered system can be built as a proprietary one, but in that case the ubiquitous usage of the system is at stake.

The described architectural constraints are in stark contrast with the regulatory constraints, which are mandatory and if the mechanism B does not formally satisfy the prescribed regulation for its frequency band, then it cannot be put into operation.

10.3 Interference and the Future Regulations

Within the complex set of parameters and factors for clean-slate design, we consider interference to be the single most important feature. The interference is central both to the regulation and the protocol design in the wireless networks, such that any significant intervention in the methods and rules for treating the interference will have a profound impact on the wireless networks over a very-long-term. In this section we illustrate such a status of the interference by considering its regulatory impact.

As mentioned previously, the spectrum regulation provides a systematic way to manage interference within a wireless system of given type (intra-system) as well as among systems of different types (inter-system). With the existing approaches, the spectrum is divided into *licensed* and *unlicensed* bands. In brief, the devices that use the licensed spectrum pay certain fee for the spectrum utilization, which on the other hand brings a guaranteed level of inter-system interference. Conversely, the unlicensed devices do not pay a spectrum usage fee, but they have to leave with the interference created by the other unlicensed devices.

Such an a priori division of the spectrum into disjunctive bands simplifies the interference management. However, it does not necessarily lead to an efficient spectrum usage and is the main reason behind the perceived spectrum scarcity. On the other hand, any liberalization of the regulatory constraints that will even hint increased interference to the licensed users is deemed unacceptable, as it directly harms their large payments to obtain spectrum licenses. Hence, any radical regulatory change should start by considering the spectrum economy and create convincing arguments that the change of the regulation can over a very-long-term bring economic benefits and new economic opportunities due to the new forms of spectrum ownership.

We take closer look to the models for licensed and unlicensed spectrum usage. A *licensed frequency band* is allocated to/owned by a certain spectrum user. Here the spectrum owner has guaranteed low level of interference from other systems. For example, cellular operators provide high fees to obtain spectrum licenses in the licensed bands in order to provide target level of service to its customers. Still, the actual level (quality) of service depends on how the intra-system interference is handled. Nevertheless, the spectrum license is largely decreasing the interference uncertainty that would have otherwise been induced by wireless systems over which the spectrum owner does not have control. Other examples of licensed bands are the bands used by military, radar, etc. On the other hand, in the *unlicensed bands* the spectrum users can freely access the radio spectrum as long as they obey certain rules/policies that are prescribed for that frequency band. These rules are constraining the temporal, spatial, and radiation patterns used by the users for radio communication in the unlicensed band. These bands are normally used by short-range technologies, such as Wireless LAN, Bluetooth, sensor networks, etc., where the low radiated power makes provisions for better spatial reuse of the frequency resources. The open access policy in the unlicensed bands necessarily leads to larger uncertainty of interference and hints (but does not prove) lower communication performance.

Starting from these two existing models, there is an ongoing debate [12] whether the future spectrum management should be based on property rights (licensed/exclusive) usage model or on the model of open access/unlicensed/spectrum commons [13]. The two models are referred to as *property rights model* and the *commons model*, respectively. Conceptually, the two models are belonging to the same framework, but they have different values for a certain set of discriminating parameters. A good candidate for such a parameter is the ***interference uncertainty*** which is minimal in the property right model and (has a potential to be) maximal in the commons model. By tuning this interference uncertainty, we can span a bridge between the two opposed spectrum usage models and populate the bridge with many innovative spectrum usage models.

The important question is – which are the mechanisms that will give rise to innovative spectrum usage models over a longer term? We provide three observations that may be instrumental in redefining the spectrum policy and the future usage models: (1) introduction of network-centered rules to access the spectrum commons, (2) recognition of the fact the interference is not noise and (3) treatment of the interference as a currency.

10.3.1 Network-Centered Rules to Access the Spectrum Commons

The first observation is that, at present, the policies for using the spectrum commons are **device-centered** and they should evolve towards **network-centered policies and rules**. The existing regulatory rules tend to see the transmitter device as an isolated entity, which sends information towards the intended receiver(s), but its interaction with the other devices is limited to inducing/avoiding mutual interference. We advocate an approach in which the devices, not necessarily belonging to the same type of wireless system, exchange information and cooperate in order to decrease the interference uncertainty and achieve more efficient spectrum usage. The essence of such an approach is described in the following example.

Consider the scenario given on Fig. 10.2, where three links T × 1-R × 1, T × 2-R × 2, and T × 3-R × 3 are sharing common spectrum, but they are belonging to different operators or even different types of wireless systems. To facilitate the discussion, we assume the simplest interference model based on collision, which implies that: if R × i is receiving from T × i but at the same time T × j is transmitting and R × i is in the range of T × j, then R × i does not receive the transmission from T × i successfully. One rule that is often used in the existing policies for unlicensed operation is the *duty cycle limit*, which limits the percentage of time for which given unlicensed device can act as a transmitter. Assume that such a duty cycle is 40% and consider the interference that R × 1 experiences from T × 2 and T × 3. In a simplified analysis, the probability that T × i (i=2, 3) is transmitting is 0.4, such that the probability that R × 1 experience interference is:

$$P_{Rx1}(\text{int}) = 1-(1-0.4)(1-0.4) = 0.64$$

On the other hand, the fraction of time during which T × 2 successfully transmits to R × 2, not being interfered by T × 3, is

$$P_2(\text{succ}) = 0.4 \cdot 0.6 = 0.24$$

Now consider a changed policy, where we allow T × 2 and T × 3 to exchange information and cooperate, and we put duty cycle restriction on the network of transmitting devices (in this case T × 2 and T × 3) to be at most F (100F%). If we select

F=0.64, then the interference experienced by R × 1 is not changed. On the other hand, T × 2 and T × 3 use the cooperative link to exchange non-conflicting transmission schedules, which grant each of the devices T × 2 and T × 3 right to use the medium for 32% of the time. Since the mutual interference between the links 2 and 3 is removed, then the fraction of time in which T × 2 (T × 3) transmits successfully is 0.32, which is a 33% increase compared to the device-centered policy rule. The latter regulation is network-centric because imposes policy rules on set of devices that can be interconnected (though not to communicate useful data, but only network management data).

This simple example clearly carries the message that the network-centered regulation can foster cooperation in order to improve the utility of the selfish spectrum users. The cooperation *reduces the interference uncertainty*, which is one of the main sources of the inefficient spectrum usage. This can have a profound impact on the clean-slate design, since it requires revision of the existing approaches which do not rely on the cooperation between the different spectrum users, even if they belong to the same wireless network.

The opponents of the network-centric regulation may argue that the cooperative links require that the devices must belong to the same type of wireless system (identical air interface) or at least incur additional complexity to some short-range devices whose primary goal is low-power operation and simplicity (e.g. wireless sensors). It should be noted that the cooperative link need not use the same air interface as the system, but it can be a dedicated signaling channel. Clearly, the signaling channel requires additional resources, but with the emerging flexible implementation based on software-defined radios (SDR), it becomes a matter of a well specified channel, which can fit easily in the SDR implementations of various systems. Nevertheless, the network-based regulation can be made to incorporate devices with various capabilities and not necessarily every device should be able/willing to cooperate with the neighboring devices and networks. The network-centered regulation policy should include carefully designed rules that can enable the devices and networks to choose their level of cooperation based on their capabilities and the expected incentives in given scenario. Such rules can emerge from the large volume of ongoing studies that apply game-theoretic methods [14] to study the spectrum access in wireless networking.

Note that a single device is an example of a degenerate network; hence the device-centered rules are only special cases in the network-centered regulation. The latter argument suggests that, despite the clean-slate flavor, the network-centered regulation can be introduced in the current regulations in a backward-compatible manner. This is important because the clean-slate approaches in the regulation should exercise caution not to cause changes that are unfavorable to the legacy spectrum users.

10.3.2 Interference is Not Noise

It is an obvious, but often forgotten fact that the wireless interference is not a Gaussian noise. If we fix the power of the undesired signal at the receiver to be N, then, in the worst case, this undesired signal is a Gaussian signal which entails maximal uncertainty. However, the wireless interference is coming from a source that has finite data rate and has much less uncertainty then a Gaussian noise. The recognition of this fact can be used to devise new rules and operation modes in different spectrum usage models.

We again use an example to illustrate the point. Assume there is a licensed system (primary spectrum user) where a base station (BS) has coverage of 1 km. We define a terminal to be in the coverage of the BS if its signal-to-noise ratio (SNR) is 20 dB or larger. Hence, the terminals that are at the cell edge have, on average, SNR of 20 dB, but the terminals closer to the BS have SNR which is, on average, much larger than 20 dB. Assume that inside the coverage area of this primary spectrum user we want to deploy a secondary spectrum user – devices that can communicate with a high spectral efficiency, without disturbing the primary terminals. The key assumption that we make is that the secondary spectrum users are designed by treating the primary as a legacy system, which in this case means that a secondary spectrum user is capable to decode the signals sent in the primary system.

The question we want to ask is – where is the best place to deploy the secondary link? A seemingly paradoxical answer is that the best place is close to the primary BS and here is the rationale. Refer to Fig. 10.3 and consider a primary terminal (PT) that is close to the BS and receives its signal with an SNR of $\gamma_{PT} = 30$ dB. The secondary receiver (SRX) is assumed to be in the immediate vicinity of the PT, thus receiving the primary signal also at $\gamma_{P-S} = 30$ dB. The secondary transmitter (STX) transmits to SRX, such that the received power at SRX from STX, normalized with the noise power, is $\gamma_{S-S} = 5$ dB. Clearly, this secondary power disturbs

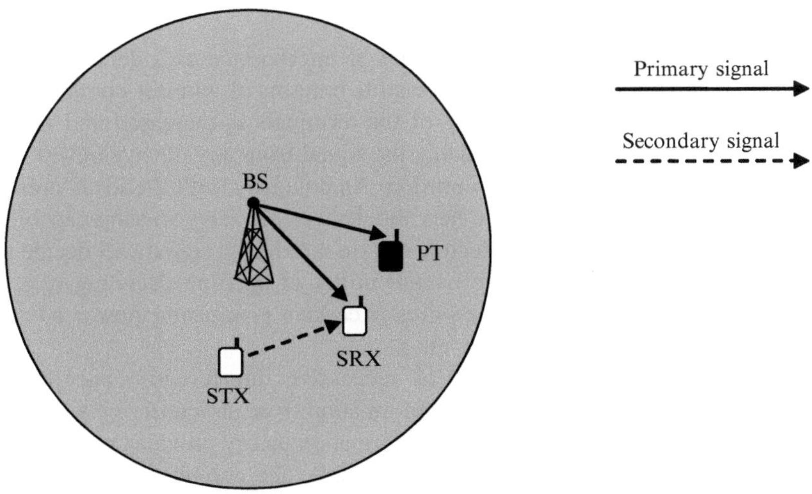

Fig. 10.3 Illustration of a primary system in which the Base Station (BS) serves the primary terminal (PT), while the secondary link (STX-SRX) is deployed in the coverage area of the primary system

the reception of the signal from the primary BS at both PT and SRX. In the worst case, this additional power from STX can be treated as a Gaussian noise by the receivers that intend to decode the primary signal. Thus, the available signal-to-interference-and-noise ratio (SINR) at PT and SRX is:

$$\frac{\gamma_{P-S}}{1+\gamma_{S-S}} = 240 = 23.8 \text{ dB}$$

which is sufficient to decode the signal from the primary BS and thus the communication in the primary system is practically not disturbed by the secondary transmission. Now let us consider the communication in the secondary system. If the secondary signal is decoded by treating the primary as a noise, then the SNR for decoding the secondary signal at the SRX is $\frac{\gamma_{S-S}}{1+\gamma_{P-S}}$, which leads to the normalized capacity of the secondary link:

$$R_{\text{sec}}^{\text{int}} = \log_2\left(1+\frac{\gamma_{S-S}}{1+\gamma_{P-S}}\right) = 0.0046 \ [\text{bits/symbol}] \qquad (10.1)$$

If, as assumed, the secondary system is capable of decoding the primary signal, then it can successfully decode it, since it receives it at SNR of 23.8 dB, cancel out the primary signal from the received signal and decode the desire secondary signal at an SNR of 5 dB, providing a link capacity of:

$$R_{\text{sec}}^{no-\text{int}} = \log_2\left(1+\gamma_{S-S}\right) = 2.057 \ [\text{bits/symbol}] \qquad (10.2)$$

which is significantly larger than (10.1).

This example reveals that the treatment of interference as a decodable signal rather than noise can bring significant benefits in terms of wireless communication performance. Indeed, the complexity of the terminals is increased and not each device can be made capable of decoding the signal from any other wireless device that operates in the same spectrum portion. Analogously with the ideas presented in the network-centered regulation, here the devices can have varying capabilities, while the devices that are able to decode certain interfering signal will decide to do so based on their judgment of the overall utility of such an decoding (e.g. they might not do the decoding if it requires excessive processing power to do the synchronization, channel estimation, etc.).

Finally, the presented technique of successive interference cancellation is perhaps the simplest way of extracting information from an interfering signal when it is not treated as a noise. The network information theory provides many sophisticated techniques to combine interfering signals. At the current (and the expected) development of the technologically feasible solutions, it becomes realistic to start to consider these techniques as part of the regulations.

10.3.3 Interference as a Currency

From the previous discussions, it is clear that the new paradigms of interference management that decrease/eliminate the interference uncertainty are crucial to the very-long-term evolution. However, it is essential to observe that interference uncertainty *is not* the actual utility function of the spectrum users. The users value the spectrum through the level that it fulfills their requirements for reliable/dependable transfer of information over the air. Hence, the spectrum use based on the pure property rights model has questionable foundations, as in that case the spectrum owner is paying to directly decrease the interference uncertainty. It is interesting to look at the interference to as a *currency*, which can be used to trade the spectrum usage rights among the users that have various level of spectrum ownership. For example, there can be a wireless network WN1 that has bought a right to use power P with a certain spatio-temporal transmission pattern. This usage pattern implies certain interference pattern to the devices in the radio proximity and that interference is the resource owned by WN1. Then WN1 can offer part of that resource on the secondary spectrum market to another wireless device/network, such that the resulting interference pattern is not changed.

Another interesting case of trading the interference is related to the critical radio services, such as alarms or radar. These applications are usually the main argument against the spectrum commons model, as the proponents of the property rights model claim that the uncertain interference in the open spectrum usage will significantly harm the proper operation of these critical services. However, one might think that allowing other usages of the same spectrum can be actually beneficial to these services. Namely, the critical services require a high dependability = availability × reliability. Then the critical application can trade off some of its reliability (allow interference) in order to gain availability (the devices creating the interference may appear as helpers/relays for transferring information in the critical system).

10.4 Discussion

In the previous sections we have given reasons why the clean-slate approach can be followed in designing wireless communication systems and we have also provided representative ideas and observations that promote the clean-slate thinking for spectrum regulation. The picture is not complete unless we outline the first step of the roadmap for adoption and implementation of the clean-slate approach.

In architectural sense, the clean-slate approaches can enter by overlaying the older systems and respecting the backward compatibility. That is, the communication protocol in a "new device" is implemented such that it can behave as a legacy system, if required to communicate with legacy devices. As long as the architectural clean-slate ideas are implemented within parts of the system (without mandating global changes in the entire system) and they respect the current regulation, these ideas can be used as performance boosters. The cost is the main obstacle – the performance gains should be sufficiently convincing to warrant proliferation.

The situation is different with the clean-slate ideas in regulatory sense. As mentioned before, the primary purpose of the regulation today is to control the uncertainty of interference in a rather conservative way. The described regulatory ideas markedly increase the spectrum dynamics and no user can implement them in an isolated way as the regulation affects the systems that are outside of the control of that user and there can be unforeseen consequences. Regarding that, the entry point for innovative regulatory changes is a proper parameterization of the interference allowed by the current regulation, as well as the scenarios where it is assessed. This will be used to characterize the interference coming from any device/network with new spectrum behavior and thus provide quantitative basis to judge the acceptability of a certain regulatory change. In principle, if the new spectrum behavior does not induce more interference than what is allowed by the existing spectrum rules, that new behavior can be deemed acceptable. Hence, the parametrization/quantification of the interference allows a kind of backward compatibility approach in the regulatory clean-slate approach, similarly to the architectural approach.

10.5 Conclusions

The evolution of the wireless communication technologies has been guided by several types of constraints. The future evolution over a very-long-term (more than a decade) will largely depend on how these constraints are changed. The constraints that regulate the use of the radio spectrum are mandatory, such that conservative spectrum usage models can hinder the wireless innovation. On the other hand, the regulatory constraints can be changed by assuming that the wireless devices can possess at least the technological sophistication that is available today. This can give rise to a *clean-slate* thinking in designing the wireless systems, where the compatibility with the legacy devices is less important as compared to the potential to foster the technological innovation over a long term. We have identified the interference to be the crux of the clean slate approaches. We have presented three observations that treat the interference in a different way as compared to the mainstream methods applied in today's spectrum regulation policy. The first observation is that the regulation should evolve towards regulation of the behavior of a network of interconnected devices rather than regulating the behavior of each individual device. The second observation is that the interference is not noise and therefore we can use transmission methods that can extract information from the interfering signal and cancel or alleviate the interference. The third observation is that in the new spectrum economy, the interference will be a currency, but not the final utility, such that the networks and devices can trade off the interference for either economic gain or overall improved communication performance.

References

1. V. Kawadia, P.R. Kumar, A cautionary perspective on cross-layer design. IEEE Wireless Commun. Mag. **12**(1), 3–11 (Feb 2005)
2. D.J. Goodman, The wireless Internet: promises and challenges. IEEE Comput. **33**, 36–41 (2000)
3. A.H. Sayed, A. Tarighat, N. Khajehnouri, Network-based wireless location. IEEE Signal Process. Mag. **22**(4), 24–40 (July 2005)
4. J.L. Massey, *Channel Models for Random-Access Systems*. NATO Advances Studies Institutes Series E142 (Kluwer, Dordrecht, 1988), pp. 391–402
5. Android Mobile Platform, http://code.google.com/android/
6. Q. Zhao, B. Sadler, A survey of dynamic spectrum access: signal processing, networking, and regulatory policy. IEEE Signal Process. Mag. Spec. Issue Resour. Constrained Signal Process. Commun. Netw. (May 2007)
7. S. Haykin, Cognitive radio: Brain-empowered wireless communications. IEEE J. Sel. Areas Commun. **23**(2), 201–220 (2005)
8. L. Gavrilovska, R. Prasad, *Ad Hoc Networking Towards Seamless Communications* (Springer, New York, 2007)
9. I.F. Akyildiz, W. Su, Y. Sankarasubramaniam, E. Cayirci, A survey on sensor networks. IEEE Commun. Mag. **40**, 102–114 (8 Aug 2002)
10. M. Conti, S. Giordano, G. Maselli, G. Turi, Cross-layering in mobile ad-hoc network design. IEEE Comput. Spec. Issue Ad Hoc Netw. **37**(2), 48–51 (2004)
11. International Conference on Nano-Networks, http://www.nanonets.org/
12. Spectrum Policy: Property or Commons?, http://cyberlaw.stanford.edu/spectrum/
13. W. Lehr and J. Crowcroft, Managing shared access to a spectrum commons, in *Proceedings of the first IEEE Symposium on New Frontiers in Dynamic Spectrum Access Networks (DySPAN)*, Baltimore, MD, Nov 2005
14. A.B. MacKenzie, S.B. Wicker, Game theory and the design of self-configuring, adaptive wireless networks. IEEE Commun. Mag. **39**, 126–131 (Nov 2001)

Petar Popovski received the Dipl.-Ing. in electrical engineering and M.Sc. in communication engineering from the Faculty of Electrical Engineering, Sts. Cyril and Methodius University, Skopje, Macedonia, in 1997 and 2000, respectively and a Ph.D. degree from Aalborg University, Denmark, in 2004. He worked as Assistant Professor at Aalborg University from 2004 to 2009. From 2008 to 2009 he held part-time position as a wireless architect at Oticon A/S. Since 2009 he is an Associate Professor at Aalborg University. He has more than 90 publications in journals, conference proceedings and books and has more than 20 patents and patent applications. In January 2009 he received the Young Elite Researcher award from the Danish Ministry of Science and Technology. He has received several best paper awards, among which the ones at IEEE Globecom 2008 and 2009. He has served as a technical program committee member in various conferences and workshops, including IEEE SECON, IEEE ICCCN, IEEE PIMRC, IEEE Globecom, etc. Dr. Popovski has been a Guest Editor for special issues in EURASIP Journal on Applied Signal Processing and the Journal of Communication Networks. He serves on the editorial board of IEEE Transactions on Wireless Communications, IEEE Communications Letters, Ad Hoc and Sensor Wireless Networks journal, and International Journal of Communications, Network and System Sciences (IJCNS). His research interests are in the broad area of wireless communication and networking, information theory and protocol design.

References

[illegible]

Chapter 11
Long Term Development of Air Interfaces: Is It CDMA or OFDM?

Ernestina Cianca

11.1 Introduction

Looking into the future to predict the ICT and define the long term research objectives to make it come true is a necessary and also exciting exercise. The *freedom of thought* that is needed to define really advanced and novel concepts is sometimes obscured by the difficulty to get rid of the existing infrastructure. A *tabula rasa* approach is for sure interesting, but it must be based on a clear understanding of the current development path, relieved from myths and misconceptions.

This Chapter aims to review some current trends in air interface development and remove some of the myths, and thus identify more clearly the key elements that must be taken into account in the long term development of air interfaces. The current evolution of air interfaces often appears as a battle between CDMA/single-carrier and OFDM/multi-carrier, a battle that OFDM/multicarrier would probably win. The questions that arise are: What would be the impact on globalization vis-a-vis adoption of the CDMA or OFDM? It the multi-carrier approach really needed? Together with Frequency Domain Equalization (FDE) [1], what are the most efficient choices for air interfaces in future wireless systems? Answering those questions may help to draw a more clear view of the development of air interface towards 4G and beyond in the year 2020 time-frame. The last question that remains is: can this "path path towards 4G" meet the challenges posed by communications beyond 2020? An initial attempt to answer this question is made at the end of this Chapter.

E. Cianca (✉)
University of Rome, Tor Vergata, Rome, Italy
e-mail: cianca@ing.uniroma2.it

R. Prasad et al. (eds.), *Globalization of Mobile and Wireless Communications: Today and in 2020*, Signals and Communication Technology,
DOI 10.1007/978-94-007-0107-6_11,

11.2 CDMA-OFDM Battle: Single Carrier with FDE as a Compromise

Let us have a quick look to the current use of these two key technologies: CDMA is used in the following 2G/3G standards for mobile communications: IS-95, cdma2000, wideband CDMA (W-CDMA), time-division synchronous CDMA (TDSCDMA). Moreover, the principle of spread spectrum multiplexing has found applications in early IEEE 802.11 and 802.11b/g WLAN, Bluetooth, and cordless telephony.

OFDM and the related multiple access scheme, OFDMA, are used in IEEE 802.11a/g/n WLAN, HIPERLAN/2, WiMAX, DVBT, asymmetric digital subscriber line (ADSL), very high rate DSL (VDSL). Moreover, OFDM has also been chosen as the physical layer architecture for 3GPP long-term evolution (LTE) and it is easy to predict that even if with some CDMA/spread spectrum component in the near future OFDM will be used. While in trade magazines one would read sentences like "death of CDMA" or that CDMA will continue to be the leading platform for mobile communications well beyond 2020, in scientific literature there are often statements like that traditional CDMA technologies are suited only for slow-speed continuous transmissions applications such as voice, but not for high-speed all-IP wireless applications [2]. A multicarrier solution such as OFDM appears as the necessary replacement in 4G systems. The main reason is that OFDM offers a lower complexity solution than current single carrier systems to the problem of performance degradation over severely frequency selective channels, such as wireless channels for very high speed communications (several tens of Mbps): with one-tap equalizer and the introduction of the cyclic prefix is able to keep the orthogonality among different users' signals and reduce the intersymbol interference (ISI).

Conventional DS-CDMA receivers consist of a RAKE combiner whose complexity increases with the increase of the number of propagation paths in the channel. This limits its performance over severely frequency selective channels. Moreover, in a frequency-selective channel, multicode operation severely suffers from the loss of orthogonality among the orthogonal spreading codes and the performance with coherent Rake combining severely degrades. Table 11.1 [3] shows a comparison between OFDMA and CDMA systems and outlines the significant advantages of OFDMA over CDMA.

However, it is getting clearer that the advantages of OFDM over traditional CDMA systems are not related to the multicarrier nature of OFDM, but to the use

Table 11.1 OFDMA-CDMA comparison from [3]

	OFDMA	CDMA
Multipath interference	Avoids	Needs equalization, noise enhancing
Frequency selective waterfilling	Yes	No
Frequency selective beamforming/precoding	Yes	Difficult
Multiuser interference on DL	Avoids	Needs equalizer

of the equalization in the frequency domain (FD). It has been recently shown that cyclic-prefixed single-carrier systems (i.e. DS-CDMA, multicode DS-CDMA and also High Speed Downlink Packet Access) with FDE can achieve performance close to the one of a multicarrier-based system with the same receiver complexity [4–6]. It is also worth noting that the best candidate for the uplink of 4G systems is the technique Interleaved FDMA (IFDMA), which is a single carrier scheme with FDE [7]. We can conclude that over frequency selective channel, the winning choice towards 4G is not multicarrier or single carrier but FDE-based: FDE brings the combination Cyclic Prefix + one tap equalizer + FFT implementation, which represents a low complexity and efficient way to remove ISI in frequency selective channels. Whilst this is a pretty straightforward conclusion, in this chapter we wonder if, in the evolution path towards 4G, it is possible to identify other key features that have not been clearly identified so far. We wonder whether once FD equalization is chosen, there are "optimal" choices that must be made to take full advantage of the FDE. To answer this question, it helps to consider the multi-code CDMA scheme with FDE that has been recently proposed [8]. It is presented in the next Section.

11.2.1 Novel Multicode-CDMA Scheme with FDE

Fig. 11.1 shows the transceiver of the multicode scheme with FDE that has been proposed in [8]. The main difference with the DS-CDMA scheme is in the way the transmitted block is built. This is better shown in Fig. 11.2. The steps to build the transmitted block per user are the following: (i) the duration of each of the N/L symbols of each block is extended up to N/L times. We refer these symbols with

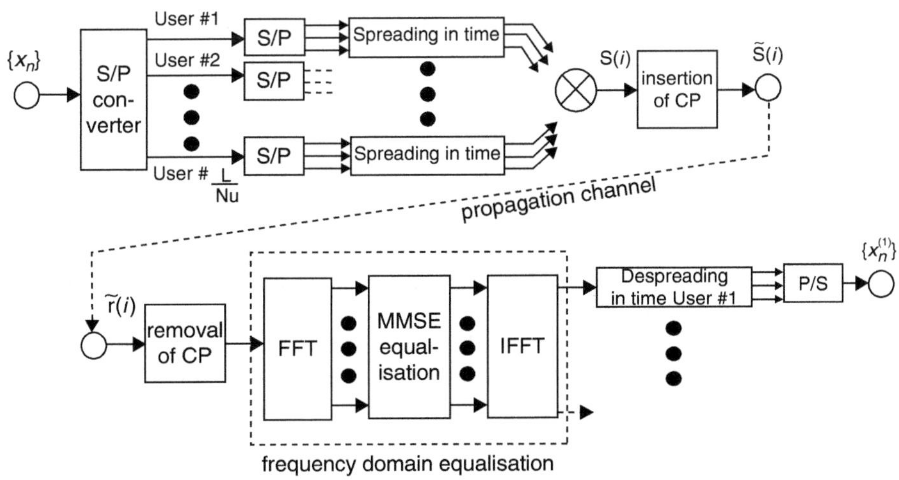

Fig. 11.1 Proposed multicode CDMA scheme with FDE

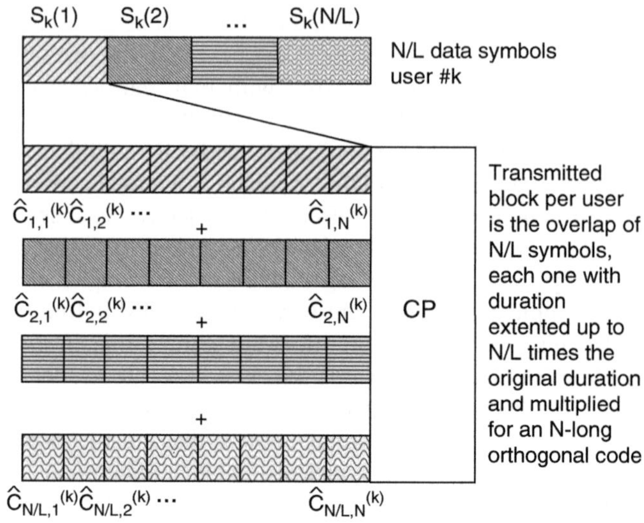

Fig. 11.2 Proposed multicode CDMA scheme with FDE: Transmitted block for the user k

longer duration as extended symbols; (ii) each extended symbol is then multiplied by an N-chip long orthogonal code; the N/L spread symbols are overlapped and the CP is added before transmission. It is worth noting that there is a spreading process since the chip duration is 1/L the original symbol duration. However, even if the orthogonal codes are N-chip long, there is also a symbol duration extension and the actual spreading factor is still L as for an equivalent DS-CDMA with FDE (L=2 in the shown example). In the proposed scheme N/L symbols per user are sent in parallel by using N/L different orthogonal spreading codes and the maximum number of users that can be overlapped is, as in the previous case, N/(N/L)=L. Let us denote with $\left\{\hat{C}_i^{(k)} = \hat{C}_{i,1}^{(k)}, \hat{C}_{i,2}^{(k)}, \ldots\ldots\hat{C}_{i,N}^{(k)}\right\}$ the i-*th* code, with $i=1, 2, \ldots N/L$, of the user *k*, respectively, the lowpass representation of the transmitted signal in the downlink is:

$$s = \sum_{k=0}^{K-1} \sum_{i=0}^{N/L-1} s_k(i)\, \hat{c}_i^{(k)^T} \tag{11.1}$$

It is worth noting that when N/L=1, the proposed approach falls back to a traditional DS- CDMA scheme. However, the case N/L=1 is not efficient in high data rate transmissions: too short blocks should be processed by the frequency domain equalizer. Moreover, the proposed modification of the DS-CDMA scheme can be applied also to a traditional multicode approach. In case of multicode CDMA with FDE, if N_u is the number of codes assigned to each user (for sake of simplicity, we assume the same number of codes per user), the transmitted signal is given by N/L consecutive spread symbols, and each of them is given by overlapping the N_u spread symbols of the user of interest, and the spread symbols belonging to the other users. The maximum number of active users is L/N_u. Therefore, in full load conditions, the transmitted block (to which the CP is added) contains in total N different symbols. In the proposed multicode scheme, these N different symbols are sent in parallel over the entire block duration, by using orthogonal spreading codes.

The intuition behind the proposed approach is that if the CP is introduced after each block to reduce the inter-block interference in frequency selective channels, it does not provide protection to the inter-symbol interference among the spread symbols that are sent in serial in case of DS-CDMA and causes loss of orthogonality). When they are sent in parallel within one block using orthogonal codes, as in the multi-code scheme, the FDE receiver is able to restore the orthogonality of spreading codes over frequency selective channels and the CP (longer than the maximum delay spread of the channel) protects from the inter-block interference.

11.2.2 OFDMA as an Instance of the Multicode CDMA Scheme with FDE

OFDMA, whose transmission scheme is shown in Fig. 11.3, can be seen as an instance of the proposed multicode CDMA with FDE when the following complex orthogonal spreading codes are used:

$$\hat{C}_{i,l}^{(k)} = exp\left\{\frac{j2\pi\, li}{N}\right\} \tag{11.2}$$

where k is the specific user; the index $i=0\sim N_u$ denotes one of the sub-carriers assigned to the user, assuming that N_u is the number of sub-carriers per user; $l=0\sim N-1$ where N is the total number of sub-carriers. In fact, doing the FFT of N symbols (where N is the number of symbols transmitted by each user within the OFDMA frame) is equivalent to taking each of the N symbols, increasing its duration from T_c to $N_{FFT} * T_c$ multiplying for specific spreading codes and summing together the N_{FFT} signals that are $N_{FFT}-$chips long; where N_{FFT} is the number of system sub-carriers and of FFT-points.

Figure 11.4 shows a performance comparison between DS-CDMA with FDE, the proposed multicode scheme and OFDMA, assuming full load conditions and same bandwidth for each scheme. For DS-CDMA and the proposed multicode approach, Walsh-Hadamard (WH) orthogonal codes are assumed. The number of points of the FFT at the equalizer is $N=256$ for all schemes. The spreading factor of the DS-CDMA scheme is $L=16$. Two different channels are considered: (1) a channel with exponential decay power delay profile with M = 8 paths; (2) a channel with uniform power delay profile with M = 8 paths.

The former is less frequency selective than the latter. Therefore, we can observe the following: (1) a noticeable performance gain of the proposed multicode approach with respect to DS-CDMA with FDE in low frequency selective channels; (2) the proposed multicode approach shows lower frequency diversity gain;

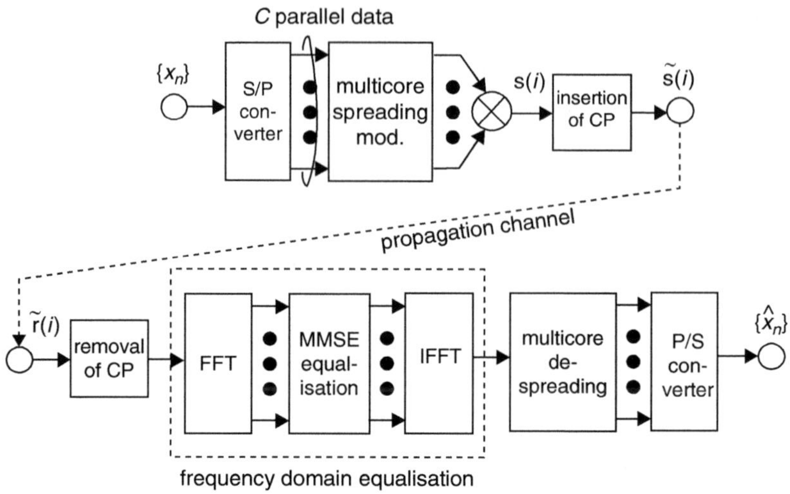

Fig. 11.3 OFDMA transmission scheme

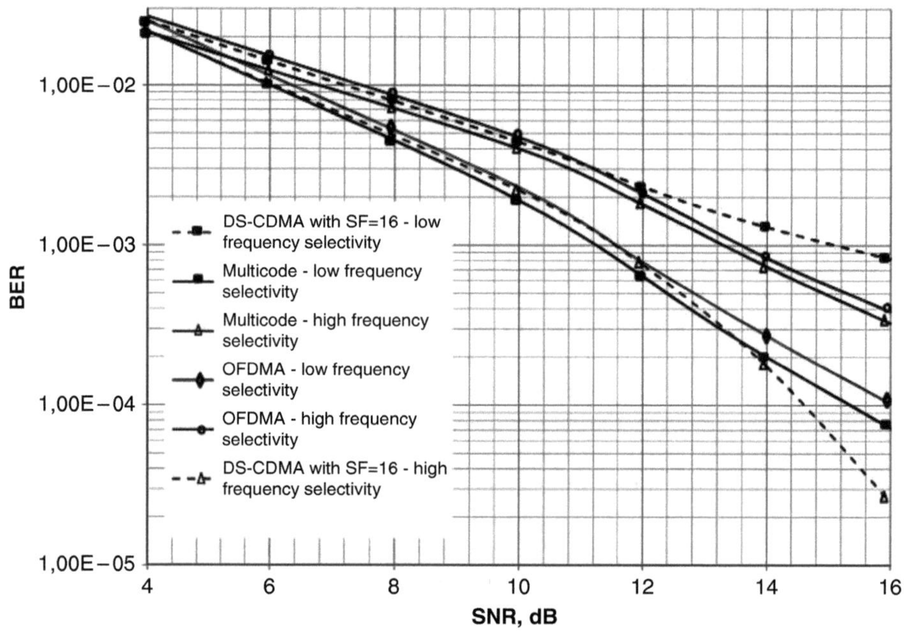

Fig. 11.4 Uncoded BER of DS-CDMA with FDE and L = 16, multicode scheme and OFDMA for: (**a**) low frequency selective channel; (**b**) high frequency selective channel

(3) in both cases, OFDMA shows the same behavior of the proposed multicode approach but with slightly worse performance. It is expected that uncoded DS-CDMA with FDE has higher frequency diversity gain than uncoded OFDMA.

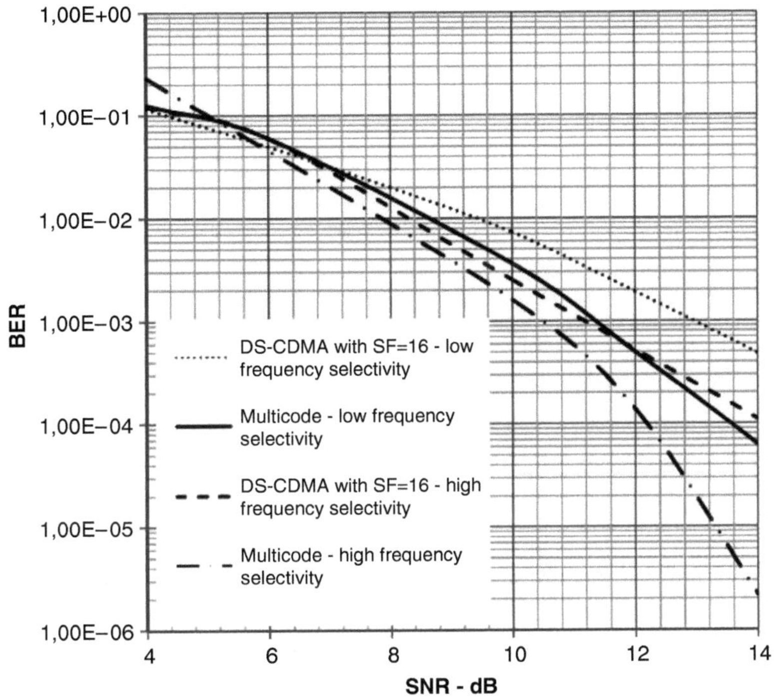

Fig. 11.5 Coded BER of DS-CDMA with FDE and L = 16, multicode scheme for: (**a**) low frequency selective channel; (**b**) high frequency selective channel

Therefore, the proposed multicode approach loses this typical advantage of DS-CDMA system over multicarrier systems and shows a behavior more similar to OFDMA than to DS-CDMA. Fig. 11.5 shows the same comparison including channel coding. A convolutional code of rate $R_c = 1/2$ and constraint length of seven is considered. In the presence of coding, the proposed multicode scheme has always better performance of DS-CDMA with FDE and it shows higher frequency diversity gain than without coding, as OFDMA systems. This further confirms that the proposed scheme is more *similar* to an OFDMA system than to a DS-CDMA system with FDE. The slight difference in performance between OFDMA and multicode can be explained by the fact that in a OFDMA system each symbol is sent over one subcarrier while in the multicode case with WH codes, each symbol is sent over more than one subcarrier (we can see WH codes as a linear combination of the codes in Eq. 2). Therefore, even if less frequency diversity gain with respect to a classical DS-CDMA system with FDE is observed, it still has a higher frequency diversity gain than OFDMA. However, this diversity gain is at the expense of slightly higher receiver complexity compared to the OFDMA case.

Finally, Fig. 11.6 shows that the novel multicode scheme has same Peak-to-Average- Power Ratio (PAPR) of OFDM in the downlink, which is much higher than

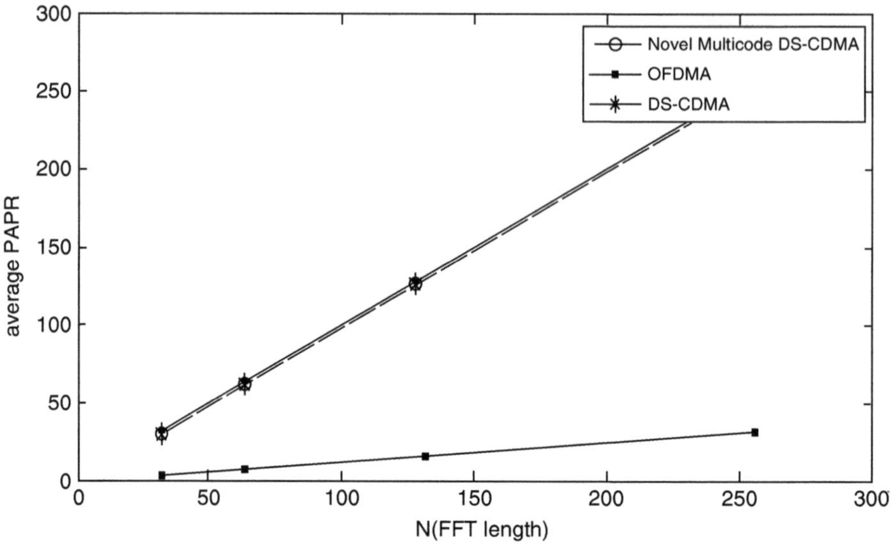

Fig. 11.6 Average PAPR in case of eight symbols per block for the downlink

the PAPR of single carrier systems as DS-CDMA. Therefore, we can state the following:

1. The presented scheme is single carrier but it has high PAPR as in any multicarrier system.
2. The presented scheme has better performance than traditional DS-CDMA with FDE. Performance is similar to an OFDMA system.
3. Like an OFDMA system, it does not offer frequency diversity gain without channel coding.

11.2.3 IFDMA as an Instance of the Multicode CDMA Scheme with FDE

In an attempt to draw the main features of air interfaces towards 4G, we cannot avoid talking also about IFDMA [7], which is the main option foreseen in 3GPP LTE standard for the uplink. The scheme has been described under several names: it has been called IFDMA in [9], OFDM-FDMA in [10], and CDMA using FDOSS in [11]. Even if it is a single carrier system, with many of the advantages of them, it is cyclic prefix-based: information is sent in blocks which are separated by a prefix and then are processed in parallel at the receiver. Therefore, in addition to low envelope fluctuations of the transmit signal and low computational complexity for equalization and user separation, IFDMA provides an efficient implementation at the transmitter that has an even lower complexity than a transmitter for OFDMA. Moreover, due to spreading of the data over the total available bandwidth, IFDMA provides high frequency diversity and thus, good spectral efficiency even without channel state

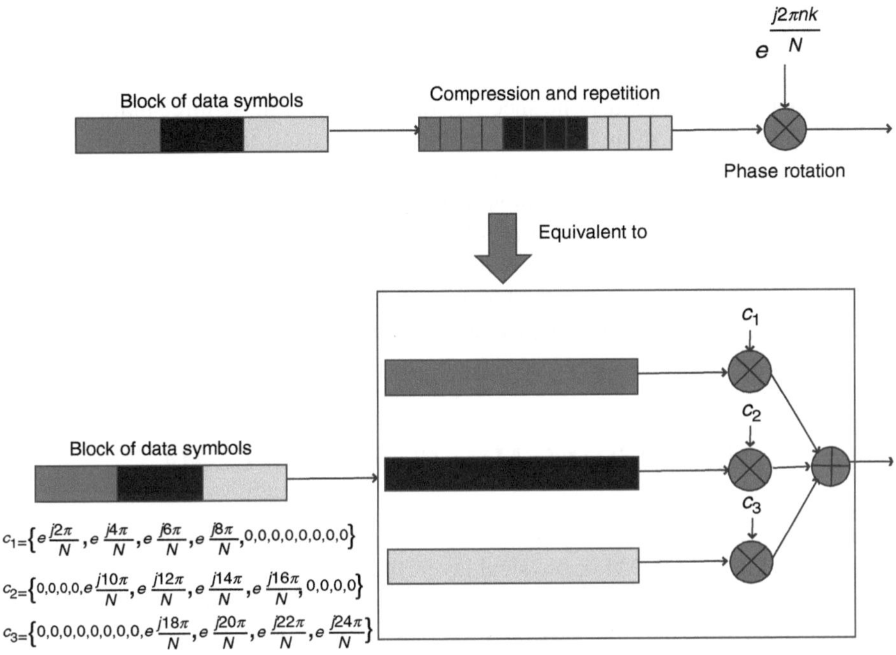

Fig. 11.7 IFDMA signal generation based on multicode-CDMA

information at the transmitter. As for OFDMA, it can be shown that IFDMA is a special case of presented multicode-CDMA with FDE scheme. Fig. 11.7 shows the processing of one block of symbols for the IFDMA scheme and how it can be derived by the multicode scheme previously presented. In both cases one symbol is extended over the block duration and then it is multiplied by one "signature sequence". In case of IFDMA, this "signature sequence" has "zeros" where the symbols of other users are placed (in IFDMA symbols do not overlap in time). Basically, the information of one symbol is time-interleaved with the information carried by other symbols in one "block". The phase shift assigned to each user and the insertion of the CP guarantee that different users are orthogonal in frequency even if the ISI in one received block can destroy the orthogonality among users in the time domain. A cyclic prefix is added after each transmitted block of symbols and equalization is performed after the removal of the cyclic prefix and FFT operation.

11.3 Path Towards 4G and… Beyond?

From previous paragraphs we can conclude that the presented multicode scheme is a more general framework that includes most of the features of the main air interfaces for future wireless communications. Those features are:

1. Orthogonal transmissions

2. Block transmission with cyclic prefix
3. Equalization in the frequency domain

Multicarrier systems are just one specific combination of those three main features. This Chapter aims to show that it is not multicarrier itself that makes it a winning choice. This claim is supported by the previous paragraphs but it is still weak if we do not consider fundamental features of future wireless systems such as adaptive and flexible resource allocation, use of MIMO techniques, cognitive radio technology and use of millimeter waves. A detailed comparison between SC and MC in the presence of the above mentioned features is still lacking, but in the next sections we outline the main conclusions that can be drawn so far.

11.3.1 A Myth: OFDM Is More Flexible than CDMA

Even if SC-FDE has been adopted by IEEE 802.16 standard (WiMAX) as an alternative technique of OFDM in physical layer, the parameters of SC-FDE are not as detailed as in OFDM. One of the factors limiting the application of SCFDE is that Channel State Information (CSI) has not yet been utilized as in OFDM system where it is possible to change the modulation order at subcarrier level based on the gain of the corresponding subchannels [4]. Adaptive OFDM improves spectral efficiency through adopting high level modulations on subchannels whose gains are high. This concept has not been applied widely in SC-FDE system because of its high complexity. In [12] adaptive OFDM and SC-FDE are compared. Whilst system performance of SC-FDE is better than OFDM, adaptive OFDM utilizing CSI outperforms SC-FDE. The combination of MIMO and OFDM has been extensively studied and incorporated in many current standards such as IEEE802.11n that in a 20–40 MHz bandwidth can reach up to 600 Mb/s communication speed (under optimal conditions in a very short range using a frequency band at 2.4 GHz or 5 GHz). Again, one of the key points for the success of the combination MIMO-OFDM is the "natural access" to the frequency domains, also at the transmitter, which gives the flexibility to better use all possible "domains" to achieve diversity and/or multiplexing gains. This is evident when Space Frequency Coding (SFC) schemes are considered [13]. SFC schemes spread the symbols over different subcarriers and different antennas and this can have advantages over Space-Time-Codes (STCs) in very high mobility environments where the channel cannot be considered constant over the OFDM block [14].

However, again we want to point out that all these advantages offered by OFDM are not inherent to multicarrier schemes. Recent work has shown that same or close performance as OFDM can be achieved by using adaptive (also at subcarrier level) techniques and MIMO in SC-FDE systems. For instance, in [15], it is shown that using one bit CSI per-subchannel under IEEE 802.16 scenario, spectral efficiency of a specific kind of adaptive SC-FDE (SC-FDE/FS) is higher than coded OFDM. Moreover [16] shows that in high mobility environments and with a slight increase in the transmitter complexity, SFC SC-FDE schemes can be efficiently implemented and can achieve same performance as OFDM-SFC schemes, even if subcarrier access at the transmitter is not easy as in OFDM systems.

11.3.2 Cognitive Radio and Millimeter Waves Communications

Use of millimeter waves (60 GHz and beyond) and cognitive techniques with multiband terminals are two of the main features of air interfaces under development for future wireless systems as enablers of multigigabit wireless communications. Multigigabit/s wireless will enable many new applications such as high definition multimedia interface (HDMI) cable replacement for uncompressed video or audio streaming and multigigabit file transferring, all of which are intended to provide better quality and user experience. New discussions in the IEEE802.15.3c already try to go beyond the 1 Gb/s and achieve 2–6 Gb/s by using the 60 GHz Band [17]. Partly because of an oxygen absorption peak at 60 GHz, various regulators across Asia, Europe, and the Americas allow for ten's to 100's of Watts of Equivalent Isotropic Radiated Power (EIRP) for wireless transmissions. The wide bandwidth and high allowable transmit power potentially enable multigigabit wireless transmissions, though several key issues must be addressed before low cost gigabit per second links can become a reality.

The Cognitive radio concept is to share spectra by sensing the spectral environment in frequency, time and spatial domains and then to transmit in the unused dimensions. This new radio functionality will involve various new analog and digital signal processing techniques to accomplish the time varying spectral sensing, wideband frequency agility, and spatial discrimination. In both frameworks, such as multi-band cognitive wireless systems and millimeters waves communication, some work has already explained how SC-FDE are less sensitive to RF impairment than OFDM; for this reason, they can be applied to multiband transmissions exploiting cognitive radio principles in a more efficient way [18]. This further explains that even if the wireless world is more and more dominated by OFDM/multicarrier-like schemes, it is of outmost importance to consider SC/CDMA-FD schemes in the development of air interfaces beyond 4G in the year 2020 time-frame.

11.4 Beyond 4G?

Can air interfaces designed according to the previously identified key features meet the challenges of communication beyond 4G? Some of the requirements for beyond 4G air interfaces are:

1. Very low power consumption: no longer slaves of batteries
2. High time variability of the channel (for instance, full-motion TV)

The identified key features of air interfaces towards 4G lead to high PAPR (both in single carrier and multi carrier configuration). High PAPR leads to low power efficiency and hence it is not suitable for power-efficient air interfaces. However, many of the issues related to the power consumption/efficiency will probably be faced at higher layers (MAC and above) by properly managing the idle states and the transitions between different operative modes, collisions, routing (cooperation?) and so on. The second challenging issue is for systems which perform block processing at the receiver, as OFDMA, IFDMA, and CDMA with FDE. The performance of these systems is conditioned on the strong hypothesis that might not hold in highly time variable channels: channel constant over one block. Does this mean that it is necessary to go back to the Time Domain Equalization (TDE)? The answer to this question will be one of the drivers in the definition of the proper air interface. This choice is expected to influence the final definition of the air interface (orthogonal or not, multilevel or not) since the preference so far given to FDE has influenced the choices made in the air interface evolution towards 4G.

References

1. D. Falconer et al., Frequency domain equalization for single-carrier broadband wireless systems. IEEE Commun. Mag. **40** (Apr 2002)
2. Special issue on IEEE Wireless Communications, NEXT-GENERATION CDMA VS. OFDMA FOR 4G WIRELESS A PPLICATIONS, June 2007
3. A. Paulraj, Is OFDMA, MIMO and OS the Right Stu_ for Mobile Broadband? IEEE VTC Fall 2005, Dallas (Sept 2005)
4. D. Garg, F. Adachi, Performance of DS-CDMA with chip interleaving and frequency domain equalisation in a fading channel. IEE Proceedings **152**(6), 753–763 (2002)
5. F. Adachi, D. Garg, S. Takaoka, K. Takeda, Broadband CDMA techniques. IEEE Wireless Commun. (Apr 2005)
6. F. Adachi, T. Sao, T. Itagaki, Performance of multicode DS-CDMA using frequency domain equalisation in frequency selective fading channel. Electron. Lett. **39**(2) (23 Jan 2003)
7. T. Frank, A. Klein, E. Costa, IFDMA: A scheme combining the advantages of OFDMA and CDMA. IEEE Wireless Commun. (June 2007)
8. E. Cianca, D. Valente, R. Prasad, A modi_ed multicode CDMA scheme with FDE and its equivalence with OFDMA, in *IEEE PIMRC 2008*, Cannes, Sept 2008
9. U. Sorger, I. De Broeck, M. Schnell, *IFDMA A New Spread-Spectrum Multiple-Access Scheme, Multi- Carrier Spread-Spectrum* (Kluwer, The Netherlands, 1997), p. 11118
10. D. Galda et al., A low complexity transmitter structure for OFDM-FDMA uplink systems, in *Proceedings IEEE VTC2002*, pp. 173741, Birmingham, UK, May 2002
11. C.-M. Chang, K.-C. Chen, Frequency-domain approach to multiuser detection in DS-CDMA. IEEE Commun. Lett. **4**(Nov 2000)
12. A. Czylwik, Comparison between adaptive OFDM and single carrier modulation with frequency domain equalization. IEEE VTC'97 **2**, 865–869 (1997)
13. H. Bolcskei, A.J. Paulraj, Space-frequency coded broadband OFDM systems. IEEE WCNC 2000 (Sept 2000)
14. T.-W. Yune, C.-H. Choi, G.-H. Im, Single carrier frequency-domain equalization with transmit diversity over mobile multipath channels. IEICE J. Trans. Commun. **89-B**(7), 2050–2060 (July 2006)

15. Jianfei Li, Yan Du, Yang Liu, Comparison of spectral efficiency for OFDM and SC-FDE under IEEE 802.16 scenario, in *IEEE Symposium on Computers and Communications, ISCC 2006*, pp. 467–471, 26–29 June 2006
16. N. Marchetti, E. Cianca, R. Prasad, Linear dispersion codes in space-frequency domain for SCFDE: A general framework. IEEE PIMRC 2007 (Sept 2007)
17. Chia-Chin Chong, Kiyoshi Hamaguchi, Peter F.M. Smulders, Su-Khiong Yong, Millimeter-wavewireless communication systems: Theory and applications. EURASIP J. Wireless Commun. Netw. **2007**(53203), 5 (2007)
18. D. Falconer, M. Muck, Feasibility of multi-bandwidth transmission in future broadband wireless systems, in *Proceedings of Information Society Technologies (IST) Summit 2005*, pp. 1923, Dresden, Germany, June 2005

Ernestina Cianca received the Laurea degree in Electronic Engineering "cum laude" at the University of L'Aquila in 1997. She got the Ph.D. degree at the University of Rome Tor Vergata in 2001. She concluded her Ph.D. at Aalborg University where she has been employed in the Wireless Networking Groups (WING), as Research engineer (2000–2001) and as Assistant Professor (2001–2003). Since Nov. 2003 she is Assistant Professor in Telecommunications at the URTV (Dpt. of Electronics Engineering), teaching DSP, Information and Coding Theory and Advanced Transmission Techniques. She has been the principal investigator of the WAVE-A2 mission, funded by the Italian Space Agency and aiming to design payloads in W-band for scientific experimental studies of the W-Band channel propagation phenomena and channel quality. She has been coordinator of the scientific activities of the Electronic Engineering Department on the following projects: ESA project European Data Relay System (EDRS); feasibility study for the scientific small mission FLORAD (Micro-satellite FLOwer Constellation of millimeter-wave RADiometers for the Earth and space Observation at regional scale); TRANSPONDER2, funded by ASI, about the design of a payload in Q-band for communications; educational project funded by ASI EduSAT. She has worked on several European and National projects. Main research interests – Novel Air Interfaces for future wireless systems, in particular MIMO in Single carrier systems with Frequency Domain Equalization, Multicarrier Systems, communications is EHF bands (in particular, W-band), energy efficient wireless systems, power control and resource management, heterogeneous networks, UWB for biomedical applications. She has been TPC Co-Chair of the conference European Wireless Technology 2009 (EuWIT2009), joint event with the European Microwave Week 2009; TPC Co-Chair in the conference Wireless Vitae 2009 (http://www.wirelessvitae2009.org). She is Guests Editors of some Special Issues in journals such as Wireless Personal Communications (Wiley) and Journal of Communications (JCM, ISSN 1796–2021). She is author of about 50 papers, on international journals/transactions and proceedings of international conferences.

Chapter 12
Flexible and Dynamic Use of Spectrum: The Cognitive Radio Approach

Enrico Del Re, Gherardo Gorni, Luca Simone Ronga, and Rosalba Suffritti

12.1 Introduction

In the future, the increasing and continuous demand of services *anytime* and *anywhere* and, therefore, the mandatory globalization of mobile and wireless communications for next generation networks services will require a more and more efficient use of the scarce radio spectrum resources. The traditional communication systems which imply an a priori association of the frequency band, the service assigned to it and the technology used, need to become much more flexible, efficient and easy-to-use dynamic systems able to cope with the requirements and constraints of the environment and the users. A *Cognitive Radio* (CR) approach can be considered as a promising and suitable solution to solve this problem. The CR technologies are able to modify this current communication paradigm because they define a system able to sense the electromagnetic environment (spectrum sensing), detect the spectral resources actually occupied in a given temporal interval and in a given location, and use the free bands (holes) for its own communication.

In this work the cognitive radio idea, named nowadays also as the last frontier of the technology, will be presented in this chapter as one of the most innovative technology which can help in the novel development of new communication systems. It will be analyzed with two viewpoints: heuristic methods and Game Theory approach. The former approach deals with the coexistence problem of two systems, one licensed and one named "cognitive" terminal, which can share the radio spectrum resources if the cognitive device is able to use free spectrum holes efficiently, without interfering with the licensed communication services. In the latter one, instead, the Game Theory is considered as suitable mathematical instrument able to analyze, model and manage the contention for radio resources.

E.D. Re (✉), G. Gorni, L.S. Ronga, and R. Suffritti
University of Florence, Italy
e-mail: enrico.delre@unifi.it

R. Prasad et al. (eds.), *Globalization of Mobile and Wireless Communications: Today and in 2020*, Signals and Communication Technology,
DOI 10.1007/978-94-007-0107-6_12,

The chapter is organized as follows. In Sect. 12.2 the new challenges which the next generation of wireless systems will have to face, will be introduced. In Sect. 12.3 the cognitive radio concept will be described in more detail while in Sect. 12.4 the open issue in this field will be presented. In Sect. 12.5, the adopted cognitive strategy, based on a heuristic method, and the considered application scenario will be described and some results presented. Section 12.6 will include discussions on the Game Theory approach and on its potential application in telecommunications. Some concluding remarks are made in Sect. 12.7.

12.2 Challenges in Next Generation Networks

The increasing needs for a global and interactive environment, wherein users can access more easily services and applications and the continuous demand for bandwidth in wireless communications generates serious shortage of spectrum both in licensed and unlicensed bands. But is the scarceness of spectrum resources a real issue? Traditional spectrum policy is driven by the general concept of protection of relevant services from interference. The effect is that large portions of spectrum are locked on vast geographical areas for a very long time, [1].

Although this approach has led to the development of many successful services (i.e., TV broadcasting, cellular telephony, wireless access to Internet), it also introduced a diffused inefficiency in large portions of the radio spectrum. As cited in [1], a recent report presenting statistics regarding spectrum utilization shows that even during the high demand period of a political convention such as the one held between August 31 and September 1, 2004 in New York City, only about 13% of the spectrum opportunities were utilized [2]. The real issue in the spectrum debate appears to be focused more on efficiency rather than exploration of new frequencies. This motivates the increasing interest in capacity exploiting technologies like MIMO, Ultra Wideband or Cognitive Radio. These technologies could enable for next generation networks which aim at the convergence of digital contents (video, data, voice) towards IP-based networks and the diffusion of high-speed broadband access [3].

The Next Generation Networks will enable a number of key features that can be particularly beneficial to a wide array of potential services. While some of these services can be offered on existing platforms, others benefit from the advanced control, management, and signaling capabilities of NGNs. The most relevant services we are likely to use in the near future are strictly related to the features offered by NGNs. They are:

- **Voice Telephony**: NGNs will likely need to support various existing voice telephony services (e.g., Call Waiting, Call Forwarding, Three-Way Calling).
- **Connectivity Services**: Allows for the real-time establishment of connectivity between endpoints, along with various value-added features (e.g., bandwidth-on-demand, connection reliable/resilient Switched Virtual Connections (SVCs), and bandwidth management/call admission control).

- **Multimedia Services**: Allows multiple parties to interact using voice, video, and/or data. This allows customers to converse with each other while displaying visual information. It also allows for collaborative computing and groupware.
- **Virtual Private Networks (VPNs)**: Voice VPNs improve the interlocation networking capabilities of businesses by allowing large, geographically dispersed organizations to combine their existing private networks with portions of the PSTN, thus providing subscribers with uniform dialing capabilities. Data VPNs provide added security and networking features that allow customers to use a shared IP network as a VPN.
- **Public Network Computing (PNC)**: Provides public network-based computing services for businesses and consumers.
- **Unified Messaging**: Supports the delivery of voice mail, email, fax mail, and pages through common interfaces. Through such interfaces, users will access, as well as be notified of various message types (voice mail, email, fax mail, etc.), independent of the means of access (i.e., wireline or mobile phone, computer, or wireless data device).
- **Information Brokering**: Involves advertising, finding, and providing information to match consumers with providers.
- **E-Commerce**: Allows consumers to purchase goods and services electronically over the network.
- **Call Center Services**: A subscriber could place a call to a call center agent by clicking on a Web page. The call could be routed to an appropriate agent, who could be located anywhere, even at home (i.e., virtual call centers).
- **Interactive Gaming**: Offers consumers a way to meet online and establish interactive gaming sessions (e.g., video games).
- **Distributed Virtual Reality**: Refers to technologically generated representations of real world events, people, places, experiences, etc., in which the participants and providers of the virtual experience are physically distributed. These services require sophisticated coordination of multiple, diverse resources.
- **Home Manager**: With the advent of in-home networking and intelligent appliances, these services could monitor and control home security systems, energy systems, home entertainment systems, and other home appliances.

All these services will be accessed both from fixed and mobile locations. In the latter case, innovations in wireless communication are essential. The path to next generation wireless communication is characterized by a few but challenging milestones. One of them is the full IP convergence of all services. Several driving protocols and technologies are ready for the complete abstraction of transport to IP (i.e. VoIP, IP QoS, broadcasting and conferencing services over IP) but several issues are still open in their integration. Another key element of future generation systems is the context awareness giving access to communications services which depend on current user location and activity. VICOM project [4] collects many technologies in this area. The third milestone is the ubiquitous availability of large communications resources. This last step is obtained through consistent improvements in transmission and reception techniques, the use of Shannon-limit achieving channel coding, cooperative techniques, spectrum sharing and dynamic spectrum access (DSA) through the adoption of the Cognitive Radio approach. Finally, a fourth challenging milestone for NGNs can be identified in the huge number of wireless devices that will be active in next decades. This aspect becomes critical considering that, in the 2020 time-frame, the estimated global population will be around seven billion people communicating through more than seven trillion wireless devices, [5]. In this futuristic, but likely, scenario it is absolutely necessary to think of a flexible technology able to guarantee the communication among this huge number of devices. The Cognitive Radio technology thanks to its flexibility and dynamic nature is one of the best candidate to solve this problem, allowing the wireless NGNs to exploit efficiently the scarce radio resources.

12.3 Cognitive Radio: The Last Frontier of Spectrum Exploitation

The term *Cognitive Radio* (CR) was introduced in [6] with reference to a communication system able to observe and learn from the surrounding environment as well as to implement and adapt its own transmission modalities also to user requirements. The concept of CR originated because of an increasing demand of broadband services and the scarcity of radio resources. Recent studies of the FCC Spectrum Policy Task Force demonstrate that a large amount of licensed bands are under-utilized [7], i.e., a lot of spectral resources are reserved for specific services, but, actually, they remain unused for most of the time or unused in several locations. From these studies, the possibility of a CR is envisaged, i.e., a system able to sense the electro magnetic environment (spectrum sensing), detect the spectral resources actually occupied in a given temporal interval and in a given location, and use the free bands (holes) for its own communication. The search for available resources is not limited to spectrum portions dedicated to unlicensed communications, but is also extended to licensed bands. In this case, a CR system, called the *secondary system*, must coexist with a *primary system*, i.e., the license owner, without producing harmful interference. Both earth and satellite systems can be considered for the role of primary and secondary users.

A CR system assumes that there is an underlying hardware and software system infrastructure that is able to support the flexibility demanded by the cognitive algorithms [8]. In this case, the abstraction of hardware capabilities for radio software architecture is a primary design issue because it is advisable to isolate the cognitive engine from the underlying hardware. In this context, the Software Defined Radio (SDR) represents the essential enabling technology for its characteristics. SDR is a methodology for the development of applications in a consistent and modular fashion such that both software and hardware components can be promptly reused for different implementations. SDR technology can also provide an advanced management of the available resources and facilitate the definition of interfaces for the coexistence of different communication services.

Considering what has been said before, a CR approach turns out to be one of the most promising technologies which can help to overcome the constraints of the current wireless systems in terms of efficient use of the spectrum resources and, therefore, it can play a key role in the development of a global access environment, where there are increasing number of user devices and decreasing free spectrum available.

12.4 Open Issues in CR Technologies

The *Cognitive Radio* idea of exploiting the unused spectrum is extremely powerful and efficient of flexibility in spectrum management. The use of licensed frequencies by secondary users in a transparent way to primary users is quite complex. With transparent it means that cognitive radios terminals do not impair licensed users communications. There are three key issues in particular that make this problem particularly complex.

Firstly, there is the matter of sensing and detecting primary licensed users. This functionality is mandatory for CR terminals since through this process let them determine the presence or absence of licensed users. Technically speaking this can be done by estimating the noise temperature over a certain range of frequencies, and then determining eventual variations due to primary users. The issue is that not all primary users can be detected, but however they can be affected by interference due to transmission by secondary users.

Second issue is how the CR terminals access the spectral holes. Indeed, if a frequency band is found unoccupied after sensing, there will be strong interference if all secondary users try to access to the same free resource without any control. Suited multiple access criteria must be defined in order to have the network working. Two different approaches can be pointed out to solve this issue: the first is given by heuristic methods which ca nbe quickly applied even if not optimal, whereas in the second approach game theory, which is a mathematical tool able to manage complex communications contexts, can be applied. Both methods will be presented in later sections.

Assuming that a channel access policy is able to resolve the second issue, the third issue is about the amount of information required to realize coordination between Cognitive Radios. Here a trade off has to be found between complete information, describing opportunely the environment of each CR terminal, and partial information to CR terminals which provides a less accurate description of the operating environment but it produces less overhead.

12.5 Heuristic Methods for Cognitive Radio

This section describes the potential benefits of the adoption of cognitive radio strategy, based on an heuristic method, to the coexistence problem of different services and systems on the same radio bands. The basic model of a Cognitive Radio (CR) system will be introduced and described as well as the assumptions made for the overall considered system. In this context, a CR system is defined by a *secondary system* which has to coexist with a *primary system*, which is the license owner, without producing harmful interference. In this study the problem is formalized as a constrained maximum search, where the objective function is the rate of the secondary system subject to the preservation of the primary system quality requirements.

12.5.1 Cognitive Radio Model

In the following, the model of the overall system, composed of a primary and of a secondary component, and the coexistence problem, i.e., the exploitation of radio resources unused by the primary system, are investigated.

All the considered systems are OFDM-based. Several reasons motivate to consider OFDM as the principal candidate for implementing the physical layer of a CR system, ranging from its ability to combat multipath fading to a relatively simple way to manage spectrum occupancy.

Let $P_{Rx}(k)$ and $P_{Tx}(k)$ be the received and the transmitted power on the k th carrier of the system, $k = 0,1,\ldots,K-1$, with K total number of carriers, $C(k)$ the instantaneous channel matrix relative to the k th channel and N_x the noise power. The index x stands for either 1 (denoting the primary system) or 2 (the secondary system). The solution space is represented by the $P_{T2}(k)$ power vector, obtained by the cognitive radio strategy. Fig. 12.1 depicts schematically the CR system.

The adopted cognitive strategy is derived for a secondary system which is able to collect all the relevant propagation information of both systems. The power vector transmitted by the primary system, the channel impulse response and the statistical measures of the thermal noise are assumed to be known by the secondary system. The considered system is regulated by (12.1) and (12.2), where k is the subcarrier index.

$$P_{R1}(k) = C_{1,1}(k)P_{T1}(k) + C_{2,1}(k)P_{T2}(k) + N_1(k) \tag{12.1}$$

$$P_{R2}(k) = C_{2,2}(k)P_{T2}(k) + C_{1,2}(k)P_{T1}(k) + N_2(k) \tag{12.2}$$

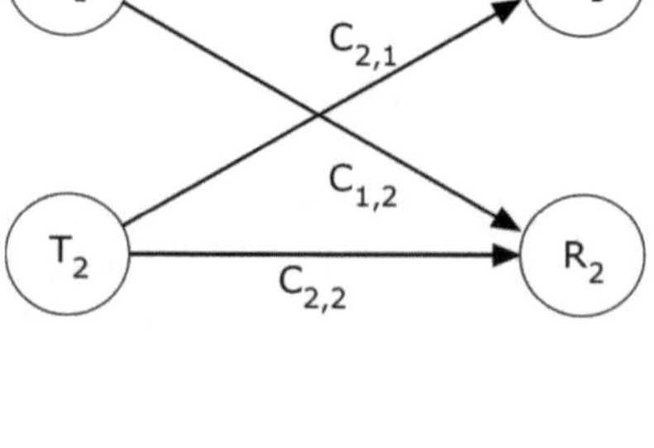

Fig. 12.1 Cognitive scenario

The problem can be formulated as a constrained multi-variable maximum search, where the objective function, namely R, is the secondary system bit-rate and the constraints are given by the minimum quality, in terms of maximum tolerable BER, that must be assured for the primary and the secondary system. The objective function and the constraints are given by:

$$\max_{b_2(k)} R = \sum_{k=0}^{K-1} b_2(k) \tag{12.3}$$

$$\frac{P_{T2}(k)C_{2,2}(k)}{N_2(k)+P_{T1}(k)C_{1,2}(k)} \geq SINR_{R2,min}(b_2(k)) \tag{12.4}$$

$$\frac{P_{T1}(k)C_{1,1}(k)}{N_1(k)+P_{T2}(k)C_{2,1}(k)} \geq SINR_{R1,min} \tag{12.5}$$

$$\sum_{k=0}^{K-1} P_{T2}(k) \leq P_{T2,tot} \tag{12.6}$$

In the (12.3), $b_2(k)$ indicates the value of the secondary bit-rate for the k th carrier. This is equal to 0 if the considered carrier is not used by the secondary system. Otherwise, it assumes a value depending on the modulation used. Constraints (12.4) and (12.5), instead, are derived according to the signal-to-noise-and-interference ratio ($SINR$) at the secondary and primary receiver, respectively. Let $SINR_{R1,min}$ and $SINR_{R2,min}(b_2(k))$ be the minimum SINR values required for a reliable communication, i.e., values that allow a given BER to be achieved, on a given carrier of the primary and secondary system, respectively. We assume that an adaptive modulation is used for the secondary system, so that the $SINR_{R2,min}(b_2(k))$ is a function of the number of bits delivered by the k th carrier, i.e., the modulation scheme used on the carrier (to be chosen from a given set). The primary receiver target quality is expressed by a given value of $SINR_{R1,min}$, which results independent of $b_2(k)$ (a QPSK modulation is assumed to be used by the primary system). During the iterative allocation process the simpler modulation is considered first (e.g., QPSK) to compute $SINR_{R2,min}$, and then, whenever possible, a modulation upgrade is performed recursively (see the power allocation algorithm described below). According to the constraints (12.5) and (12.4), the power on the k th carrier of the secondary system is upper and lower bounded by the following expressions:

$$P_{T2}(k) \leq \frac{P_{T1}(k)C_{1,1}(k) - N_1(k)SINR_{R1,min}}{SINR_{R1,min}C_{2,1}(k)} \tag{12.7}$$

$$P_{T2}(k) \geq \frac{SINR_{R2,min}(P_{T1}(k)C_{1,2}(k)+N_2(k))}{C_{2,2}(k)} \tag{12.8}$$

The constraint in (12.6) is the total amount of power transmitted by the secondary device. This is ruled by (12.6) and derived by regulatory limitations on the secondary transmission power.

The described constraints can sometime be contradictory: in this case the applied rule is to protect the primary service.

Regarding the quality constraint of the primary system, it is worth noting that for each subcarrier k two possible cases may occur at the primary receiver. Let $SINR_{R1,2off}$ be the signal-to-noise ratio experienced at the primary receiver before the secondary begins its transmission. The two possible cases are:

1. $SINR_{R1,2off}(k) < SINR_{R1,min}$
2. $SINR_{R1,2off}(k) \geq SINR_{R1,min}$

In the former case, the primary system does not reach the desired quality level on the k th carrier, so the primary does not use the carrier whereas the secondary system can fully exploit it. In the latter case, the primary system reaches the desired minimum $SINR$ and eventually provides a useful margin to be exploited by the secondary.

The underlying idea of the proposed resource allocation strategy is to create a power allocation vector for the secondary allocation which is, then, iteratively updated until a stop condition is reached (i.e., when the total amount of secondary power is reached or the constraint is violated).

The allocation procedure is an iterative process where the instantaneous secondary resources at iteration n are represented by two vectors, an integer vector $b^{(n)}$ and a real valued vector $P_{T2}^{(n)}$, representing respectively the modulation and the allocated power for each subcarrier. The possible values of $b(k)^{(n)}$ are $\{0,2,4,6\}$ which identify an unallocated carrier, a QPSK, a 16-QAM and a 64-QAM symbol on the k th carrier, in that order. An auxiliary vector $d^{(n)}$ represents the amount of power, at iteration n, necessary to increase the rate of the secondary system on each carrier. The rate increments are those that, if applied, change the corresponding value of $b(k)$ from 0 to 2 (activation of the carrier with a QPSK symbol), from 2 to 4 (modulation upgrade from QPSK to 16-QAM), from 4 to 6 (modulation upgrade from 16-QAM to 64-QAM). The algorithm is as follows:

Step 0: $n = 0$, $b^{(0)}$ and $P_{T2}^{(0)}$ are initialized to all-zero vectors of length N (no carrier allocated to the secondary).

Step 1: $d^{(n)}$ is computed taking into account the constraints in (12.8) and (12.7), as well as the current allocation $b^{(n)}$ and $P_{T2}^{(n)}$. The carriers where the lower power constraint is greater than the upper one are marked as "unusable".

Step 2: find the carrier for which a rate increment is possible with the lowest amount of secondary power, i.e. $\hat{k} = \arg\min_{k} d(k)^{(n)}$ (excluding the unusable carriers and those already allocated with the highest modulation order).

Step 3: if a valid $\hat{k}$ has been found, allocate the carrier by defining the modulation:

$$b^{(n+1)}(k) = \begin{cases} b^{(n)}(k) + 2 & \text{if k} = \hat{k}, \\ b^{(n)}(k) & \text{otherwise.} \end{cases}$$

and the corresponding power:

$$P_{T2}^{(n+1)}(k) = \begin{cases} P_{T2}^{(n)}(k) + d(k)^{(n)} & \text{if k} = \hat{k}, \\ P_{T2}^{(n)}(k) & \text{otherwise.} \end{cases}$$

Step 4: increase the iteration index n

Step 5: compute the total amount of power allocated to the secondary; if $\sum_k P_{T2}^{(n)}(k) \le P_{T2,tot}$ than go back to Step 1, otherwise use the allocation vectors $b^{(n)}$ and $P_{T2}^{(n)}$.

It is worth noting that at each iteration there are two possible choices: allocate a new carrier to the secondary with a QPSK modulation or perform a modulation upgrade (e.g., QPSK to 16-QAM or 16-QAM to 64-QAM) on a carrier already allocated to the secondary service. The choices are equivalent in terms of rate increment and it is selected the one with the minimum required power.

12.5.2 Application Scenarios

Cognitive Radio strategies are strictly related to their operating environment. In this section, two scenarios are described: a fully terrestrial and a mixed satellite/terrestrial scenario. Fig. 12.2 depicts schematically the mixed terrestrial-satellite scenario. The considered scenarios have the objectives to investigate the coexistence problem, following the model illustrated previously. A terrestrial infrastructure or a satellite system, considered as primary systems, have to coexist with a mesh-based terrestrial telecommunication service. All the considered systems are OFDM based, but they exhibit different propagation conditions.

12.5.2.1 Fully Terrestrial Context

The first scenario is composed of a licensed primary system and a "cognitive" secondary system. Both are terrestrial systems and are characterized by a single transmitter and a single receiver. The two systems are not independent since they share the same radio resource. The signal from each transmitter represents an interfering component to the other system receiver. The secondary system operates with OFDM with the same carrier spacing as the primary one. It has, however, a more flexible power allocation scheme. Being *cognitive*, the operating parameters like frequency, modulation and power are modified by its software radio implementation. The possible

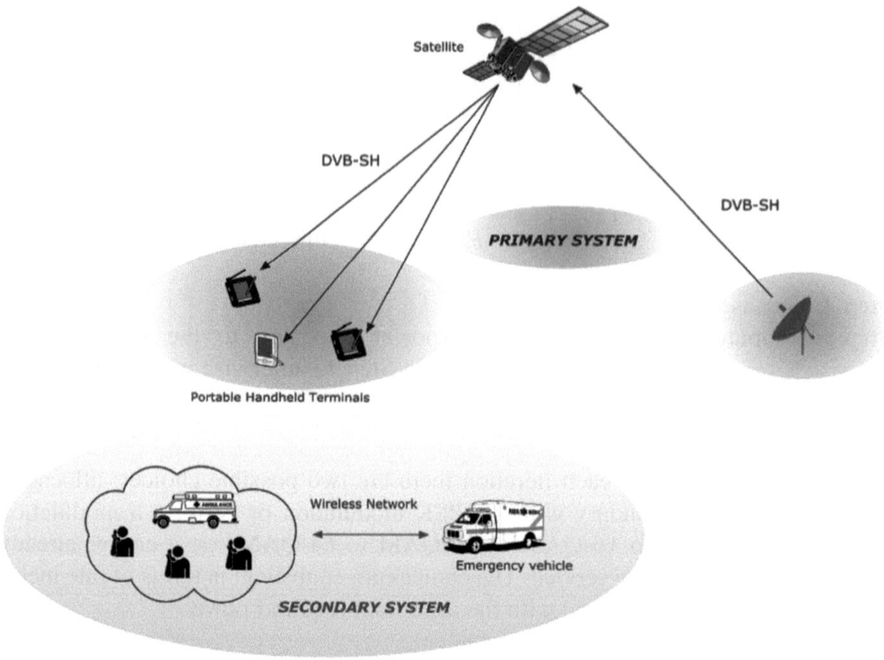

Fig. 12.2 A cognitive scenario for mixed terrestrial-satellite emergency communications

modulations for the considered secondary system are QPSK, 16-QAM and 64-QAM. With a constant bit error probability, the minimum required signal-to-noise ratio depends on the modulation order. The actual transmission mode is automatically selected by the secondary device based on the available sensed information.

12.5.2.2 Channel Model

The received signal is corrupted by different phenomena: a *path loss* term due to the transmitted distance calculated at the middle-band frequency, and *multi-path* fading due to the propagation environment and terminal motion. The path loss is modeled by:

$$L = 10log_{10}\left(\frac{4\pi d}{\lambda_0}\right)^{\alpha} dB \tag{12.9}$$

where the exponent α models the attenuation dependence from the distance, λ_0 is the central frequency wavelength and d is the transmitter distance from the receiver. In the time-domain the channel exposes a finite impulse response of L

samples, resulting in a frequency selective channel response. A tapped-delay line model with Rayleigh distributed coefficients has been adopted. The power delay profile is exponential as follows:

$$\sigma_n^2 = e^{-\beta n} \tag{12.10}$$

where σ_n^2 is the variance of the n-th coefficient and β is computed for a normalized mean power response.

12.5.2.3 Mixed Terrestrial/Satellite Context

Differently from the case previously described, the second scenario is mixed, i.e. it is characterized by a licensed primary satellite system and a "cognitive" secondary terrestrial system. The primary system is a mobile satellite system based on DVB-SH standard [9], while the secondary system is a terrestrial wireless meshed network that can be used in emergency situations. The two considered systems work in L band (0.39–1.55 GHz) and exploit a context of coexistence in which the secondary one is allowed to take resources from another system without interfering with its normal operations.

12.5.2.4 Channel Model

The propagation channel for the considered mobile satellite channel at L-band is the Lutz (et al.) model [10]. It is based on a two-state (GOOD-BAD) Markov-chain for the fading process. According to this class of models, the amplitude of the fading envelope is divided into fast and slow fading. Slow fading events, normally due to large obstacles, are modelled as a finite state machine. Fast fading events, due to the irregularity of the obstacles (e.g., vegetative shadowing) and the multipath propagation phenomenon can be additionally represented as superimposed random variations that follow a given probability density function (PDF) for each state. This channel, differently from the previous one, has a flat frequency response. Also in this case, the *path loss* term due to the transmitted distance calculated at the middle-band frequency, has to be considered in the channel model.

12.5.3 Performance

In the following, the proposed CR strategy is validated through computer simulations for both considered scenarios. The terrestrial propagation exponent (α) is considered equal to 3.0, valid for a medium density urban scenario while for the satellite scenario the value of α is assumed to be 2.0. It is assumed that the number of carriers is $K = 128$ for the fully terrestrial scenario and $K = 853$ for the mixed terrestrial/satellite case.

12.5.3.1 Fully Terrestrial Context

The performance of the fully terrestrial scenario has been evaluated considering the achieved rate of the secondary system as the main performance index. Simulations have been conducted for six target BER values of the secondary system, from 10^{-3} to 10^{-8} and for three E_b / N_0 values for the primary receiver equal to 10, 15 and 20 dB.

The primary achieved rate depends only on the primary received E_b / N_0, since the cognitive radio strategy always preserves the primary rate.

The considered context has terminal displacements as shown in Fig. 12.3. Units are normalized to the primary distances.

The performance of the secondary system heavily depends on its target BER and on the primary working point. In Figs. 12.4 and 12.5, for both considered displacements and each E_b / N_0 value of the primary, we show the BER values in log-scale function of the achieved secondary rate. In the first case (Case A), i.e. where the secondary terminals are further from the primary system, it is worth noting that, for a fixed BER, the achieved rate increases as the primary SNR increases. This is a well known limitation of cognitive radio systems, where low $SNRs$ at the primary heavily impair the secondary achievable performances.

In the second case (Case B), where the two systems are closer, it can be noted that the outcomes are different in comparison to the Case A. In fact, for a fixed BER, the achieved rate increases as the primary SNR decreases. This is due to the closeness between the two systems. In this case, the interference of the secondary is higher than in Case A. The secondary can use only the carriers that the primary discards because of the fading and these increase as the primary SNR decreases.

Another interesting element is represented by the modification of the experimented $SINRs$ at the primary and secondary receivers before and after secondary

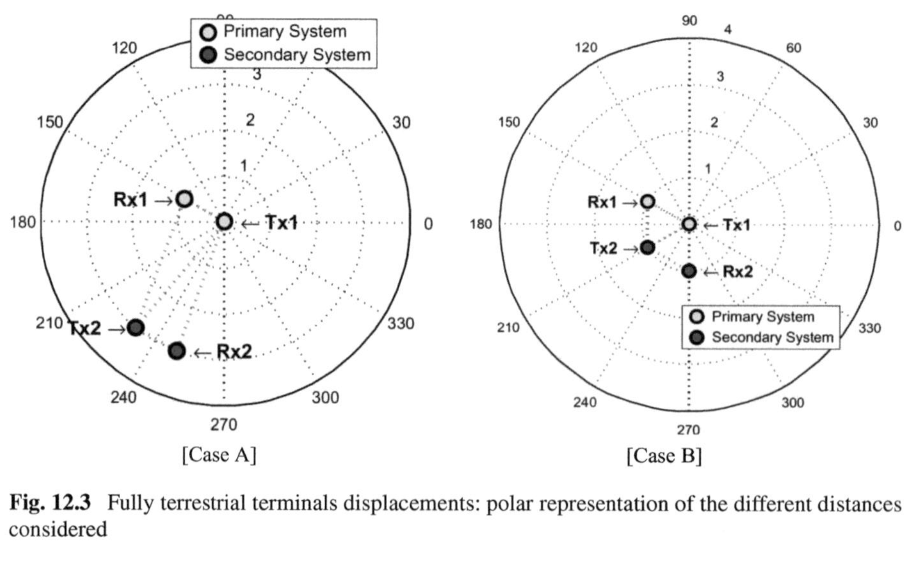

[Case A] [Case B]

Fig. 12.3 Fully terrestrial terminals displacements: polar representation of the different distances considered

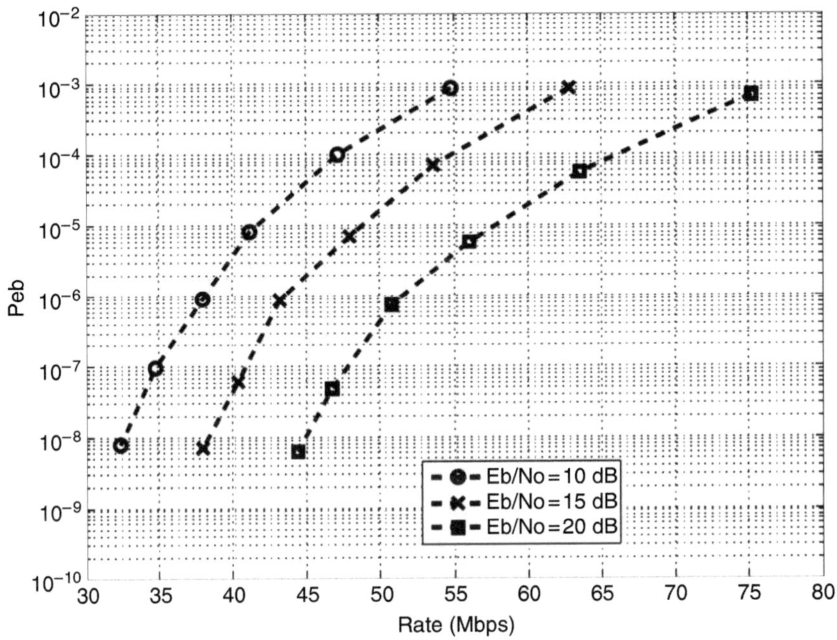

Fig. 12.4 BER versus Rate for the secondary system: Case A

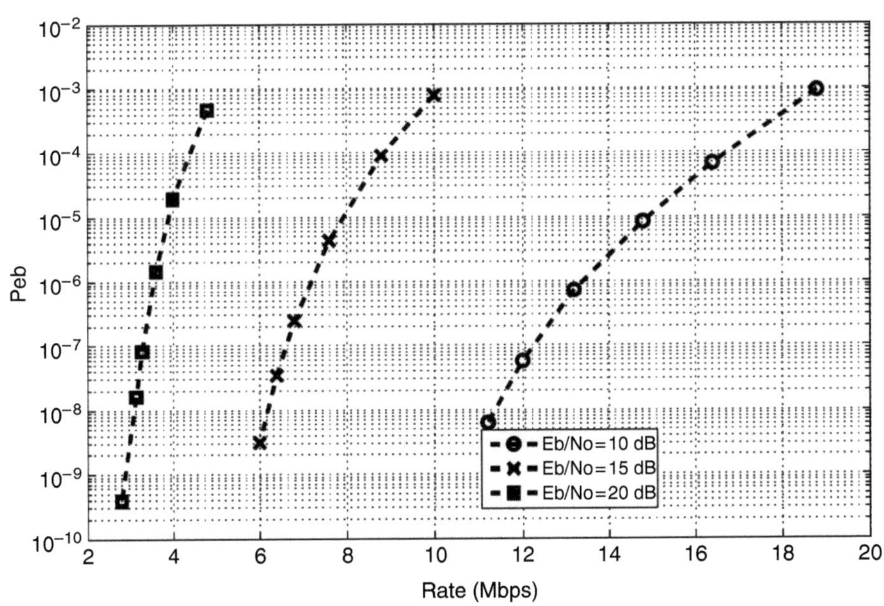

Fig. 12.5 BER versus Rate for the secondary system: Case B

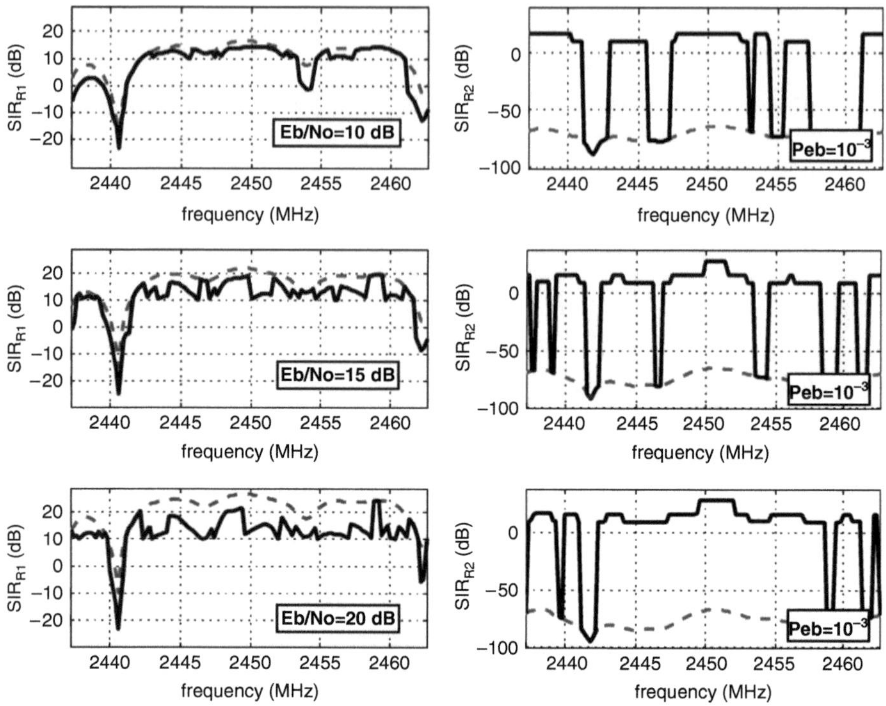

Fig. 12.6 SINR versus Frequency for the primary (*left*) and secondary (*right*) systems: Case A

transmission. Fig. 12.6 shows for the Case A and for each subcarrier, the primary (left) and secondary (right) $SINR$ (named SIR in the Figure) before and after secondary activation, for a fixed secondary BER. The considered primary E_b / N_0 values are 10, 15 and 20 dB, whereas the secondary BER is fixed to 10^{-3}. The dashed line represents the $SINR$ before the secondary activation. The channel frequency selectivity provides a variable response for both systems. The cognitive process places power where the secondary channel is in a good state, and after its activation, the corresponding primary $SINR$ lowers (i.e. the interference from the secondary increases). As can be seen, the larger is the E_b / N_0, the larger is the number of carriers where the secondary is allowed to allocate power. This explains the operating point dependence in Fig. 12.4.

12.5.3.2 Mixed Terrestrial/Satellite Context

Also for the mixed terrestrial/satellite scenario, as the main performance index, the achieved rate of the secondary terrestrial system has been considered. In this case, the six target BER values of the secondary system are different from the fully terrestrial scenario and they are: $10^{-2}, 5\cdot 10^{-3}, 10^{-3}, 5\cdot 10^{-4}, 10^{-4}, 5\cdot 10^{-5}$; the three E_b / N_0 values for the primary receiver are, instead, the same as the previous scenario: 10, 15 and 20 dB.

For this context, two different environments, CITY and HIGHWAY, have been considered. The terminal displacements are shown in Fig. 12.7. As it can be seen, the secondary terminals, Tx_2 and Rx_2, are fixed and their distance is set to 50.0 m while the receiver of the primary system, Rx_1, approaches the secondary cluster at different distances. The performance has been evaluated for three distance values among the three terminals: $100-100-50$ m, $70-70-50$ m and $25-25-50$ m.

In Fig. 12.8, for each distance and for each E_b / N_0 value of the primary, we show the secondary rate value in the CITY environment with the target BER equal

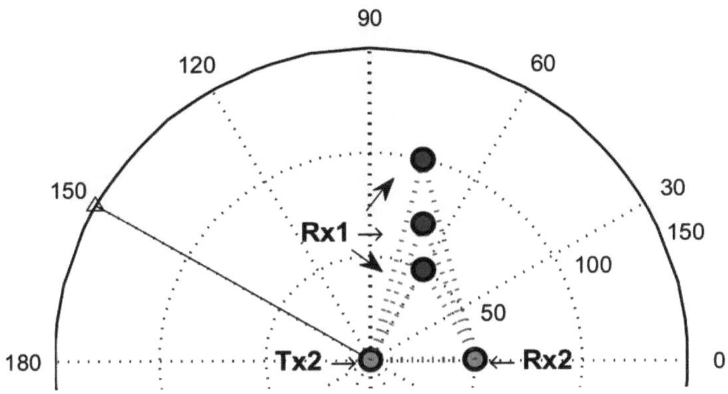

Fig. 12.7 Terminals displacements: polar representation of the different distances considered

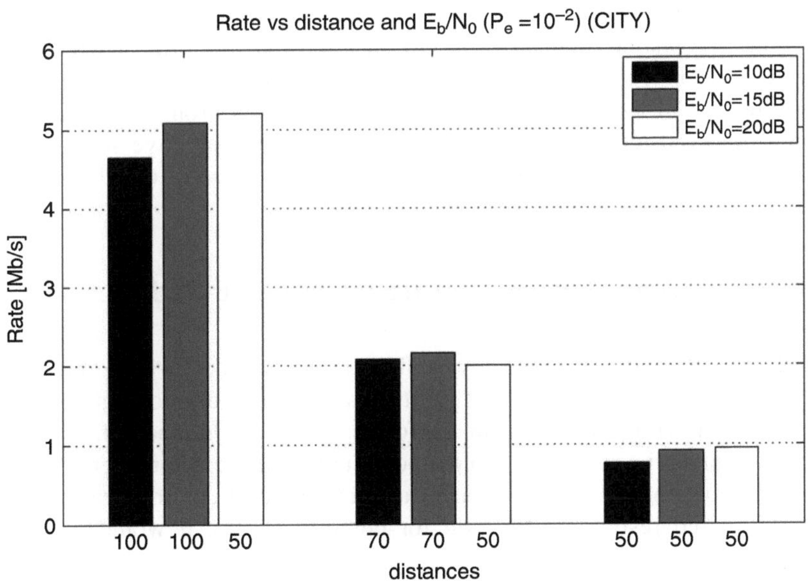

Fig. 12.8 Secondary rate versus Distances (CITY environment)

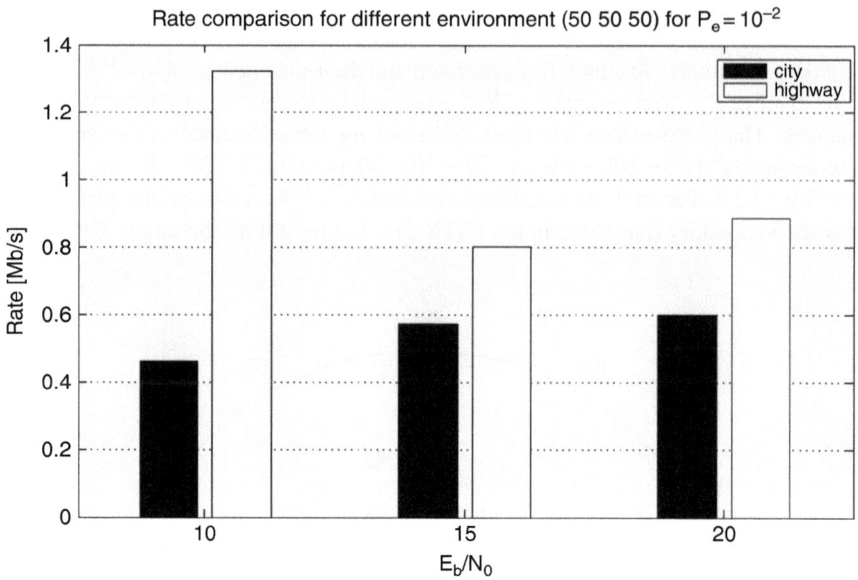

Fig. 12.9 Rate versus E_b / N_0 for different environments

to 10^{-2}. It is important to note that as the distances decreases, the secondary rate decreases as well. As already discussed, this is due to the increase of the interference between the two systems and the lower power that the secondary can utilize when it is approaching the primary in order to protect the licensed service.

Another interesting element is to consider the rate performance for the two analysed environments. In Fig. 12.9, it can be seen that the achieved secondary rate is lower in the CITY case. This is due to the presence of a strong shadowing resulting in a impaired reception of the primary signal. The cognitive strategy has to guarantee the primary rate so the secondary has to minimize its transmission.

12.6 How Game Theory Can Help Cognitive Radio

Very often mathematics provide powerful instruments to deal with complex problems that appear unsolvable at first sight. In a cognitive radio operative environment, the following situation can be found: a certain number of unoccupied frequency bands, i.e. spectrum holes, potentially available for cognitive terminals. Assuming the existence of an infrastructure based communication system with centralized resource allocation for cognitive users, it would solve most problem related to the occupation of spectrum holes. Actually, the former solution is not so convenient since a high number of CR terminals would imply a huge amount of data exchange with the system base station, resulting in corresponding decrease of data throughput. Furthermore each cognitive terminal senses the operating environment which differs from the other terminals one, e.g., a cognitive radio could detect several spectrum holes while a CR terminal nearby could detect only one available hole because it is surrounded by a number of licensed users.

It is clear that in such a case, the problem must be approached differently. Instead of an infrastructure based system managing the radio resources as well, it could be better to distribute the occupation of the spectrum holes. Therefore, there is a need for a tool able to analyze and model the contention of radio resources. Mathematics can help us with the former issue providing a really suitable instrument: the *Game Theory*.

The Game Theory is a mathematical tool proposed by John von Neumann in the last century. It rose to the rank of a discipline after the von Neumann and Morgenstern's book [11]. At the beginning it was created mainly to analyze economic situations in order to study what would be the result of interaction between many users, e.g. a bid for a good, or the battle of prices between two or more industries. Morgenstern and von Neumann dealt with only cooperative games. Afterwards, Game Theory developed through the concept of Nash Equilibrium as solution for non-cooperative games defined by the mathematician J.F. Nash. The Nash Equilibrium idea made the whole theory interesting for various applications: mainly economics but also social sciences, politics, evolutionary biology, and in the last decade also wireless communications.

The idea of a distributed management of spectrum holes makes the Game Theory particularly suitable for its analysis. The attractiveness of the Game Theory is because, differently from the decision theory, it deals with several agent interactions instead of considering only one. In particular, the main difference with the decision theory is the dependence of the outcome on the interaction under study. In fact, while in decision theory a situation where the choice of a unique agent directly affects its outcome, the Game Theory analyzes situations with multiple agents where the outcome achieved by one agent (also referred to as player) strictly depends on both its own choices and the other players actions. The game theory finds a natural application in cognitive radio as a modeling tool to analyze how the resources and the channel access are shared between the terminals. Considering Mitola's original cognitive radio tasks, the game theory role is to translate the observations, orientations and preferences of the cognitive terminals into decisions. These decisions can be taken by cognitive radios opportunistically or in a cooperative way. This consideration marks the difference between application of non-cooperative or cooperative game theory approach in a cognitive radio based network.

12.6.1 Basic Elements and Different Types of Games

The application of game theory implies the formulation of a mathematical framework since now referred to as game. The main elements of a game, independent of its cooperative or non-cooperative nature, are:

- The number of active players looking for radio resources
- A set of strategies players can choose from
- A set of payoff functions mapping the vector of strategies chosen by all the players into a preference parameter or number

In the cognitive radio context the players are the cognitive terminals operating within a certain area and in a given frequency band as secondary users. The set of strategies available to each player is settled by all actions a player can take. In a telecommunications context, some possible actions are the choices of transmission parameters e.g. modulation, coding rate, transmission power, time slot to occupy, amount of bandwidth to be used. Of course the action to be undertaken can imply the choice of just one or more parameters. How many quantities are involved in the decision process is up to the cognitive radio network designer; it has to be remembered however that more parameters are considered in the game framework, more complex it becomes to find a solution. In fact, keeping into account more options to choose from, results in a huge increase in the total number of strategy combinations and in solution computation times, as well. So a proper game theory based decision framework for cognitive radio must consider a tradeoff between potential high performances achieved with an accurate model and the implementation feasibility of the same model. As far as the payoff functions are concerned, they translate the vector of the adopted strategies by all players, i.e. a vector of real numbers, into a real value representing the level of satisfaction for the outcome achieved. The key element of the game theory emerges right here: a change of strategy by one player affects the real value given by payoff functions of the other players. This aspect makes the payoff functions the decisions engine of the whole framework; indeed for a player, through them, it is possible to analyze the structure of the game evaluating the most convenient strategies given the opponent strategies.

It is possible to point out several typologies of games, each one with well defined features, [12 and 13]. In the following some of them are described:

- Non-cooperative games
- Potential games
- Repeated games
- Cooperative games
- Hierarchical games
- Stochastic games

Non-cooperative games they are mostly used in applications. They are suitable to model interactions having the following features: firstly, each player behaves in a selfish way, assuming the context under study is non-cooperative. Secondly, no agreements or coalitions are possible; in third place each player acts independently, i.e. it is not conditioned or piloted by other players in its decision; the interaction between players happens only once. The solution for non-cooperative games is better known as Nash Equilibrium point [14]. Nash Equilibrium can be described as the point, in terms of vector of strategies, at which no user achieves gain deviating from it. Formally Nash Equilibrium, in terms of payoff value for the i-th player, is defined in the following way:

$$u_i(s_i^*, \mathbf{s}_{-\mathbf{i}}^*) \geq u_i(s_i, \mathbf{s}_{-\mathbf{i}}^*) \tag{12.11}$$

where $u_i : R^N \rightarrow R$ is the payoff function for the i-th user, s_i is the strategy adopted by i-th user, and $\mathbf{s}_{-\mathbf{i}}^*$ is the vector of best strategies available for the opponents. Nash Equilibrium requires existence conditions in terms of payoff functions and strategies.

Potential games they are a different approach to games in strategic form, [13], which is one possible representation for game theory framework. Potential games are so called because it is possible to define a potential function, valid for all players, mapping a given strategies profile into a unique real number. Furthermore it has been demonstrated the strategies profile maximizing the value given by potential function corresponds exactly with Nash Equilibrium. Then it is opportune to threat a game as a potential game when the formulation allows it, since the advantages can derive from the potential function: it is more practical to deal with a unique function to maximize instead of researching for an equilibrium point satisfying (12.11) into the strategy space.

Repeated games they are suitable to model situations where the players interact more than once, [12]. Repeated games can treat both finite and infinite number of iterations, with differences in terms of the equilibrium point that is reached. A repeated game is created from a game in strategic form, keeping constant the number of players and the available strategies. As far as it concerns the utility functions since the game is structured in stages there are two possibilities: keeping into account in payoff functions the result achieved in the former stages, or forgetting completely what happened before; the latter one is the subclass of myopic repeated games, where players concentrate only on the present situation not considering neither the future nor the past outcomes.

Cooperative games they are proper to model interactions where the scopes of participants are not the same but they are neither opposed, [14]. Differently from non-cooperative ones, in cooperative games players are able to reach a constraining agreement and respect it. Players can coordinate on the agreement exchanging reciprocal informations. This class of games do not have the Nash Equilibrium as solution, but a point with different characteristics known as Nash Bargaining Point. Differently from the Nash Equilibrium, this new solution point is Pareto-optimal and, with specific payoff functions, it guarantees fairness for the players.

Hierarchical games they have been thought to analyze situations where there are two classes of players organized hierarchically: manager users and common users, [13]. A common user is typically located in a group of players administrated by a manager. Common users within a group interact between them but not with manager players. Managers can determine the game parameters and rules in their own group and interact with the other managers; moreover common user payoffs are involved in the payoff function of their manager. Then the equilibrium concerns both master and common users, and actually it is not trivial to find a solution for this class of games.

Stochastic games they represent a more complex game theory framework which further encompasses the concept of probability, [16]. That is possible extending the basic elements of a game with two additions: a set of states, which is the result of the product of states available for each player, and a set of transition function which expresses the probability to pass from one of the available states to another when a certain profile of strategies is played by the players of the game. Stochastic games are actually a generalization of a Markov decision process that can be seen as a stochastic game with only one player.

12.6.2 Example of Application to Telecommunications

It is interesting to point out a possible application of the game theory in a cognitive radio context. In the following a simple and intuitive example is provided. An FDMA based cognitive radio network where two cognitive terminals are competing for three FDMA slots is considered in Figs. 12.10 and 12.11. The figures represent two possible situations in terms of strategic choices performed by cognitive radio terminals: in the first one, Fig. 12.10, terminal B occupies two FDMA slots while terminal A occupies just one slot; in the second figure, Fig. 12.11, both terminals

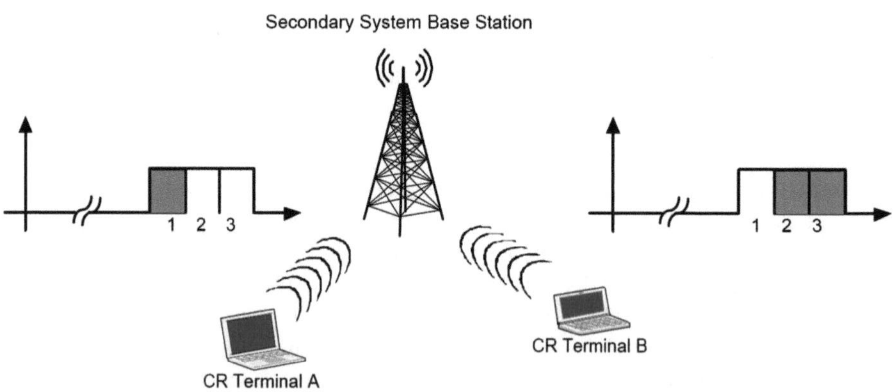

Fig. 12.10 Game theory application example: Case A

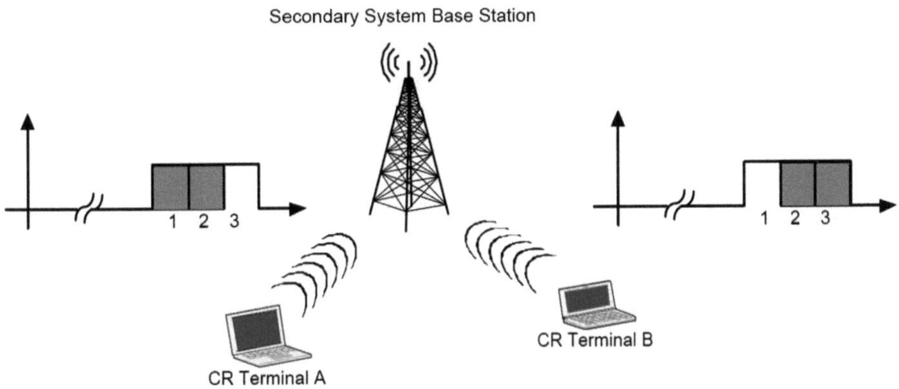

Fig. 12.11 Game theory application example: Case B

simultaneously access two FDMA slots, impairing the transmission on the common slot. It is worth noting that it has been assumed the terminal A starts the slot occupation from slot number 1 and afterwards it can further occupy the slot number 2; differently the terminal B starts the slot occupation from slot number 3 and afterwards it can further occupy slot number 2. This assumption avoids to complicate the model and it allows to study a reduced number of cases for the game outcome.

It is considered that the Cognitive terminals are the players of the game and the number of FDMA slots eligible for occupation are the object of the strategies available to players. It is assumed that there are three adjacent FDMA slots available. The game is simplified assuming the players have only two strategies: to occupy one slot or to occupy two slots. Furthermore it is assumed that the game is played only once and no communication is present. As for the payoff functions, the following three rules to assign revenues to the two players are imposed:

1. A player has a positive gain equal to 3 for each FDMA slot occupied.
2. A player has a penalization equal to 2 for each slot which is also occupied by the other terminal, i.e. the gain is negative.
3. Every player has a positive gain equal to 2 for each slot left free.

The fixed payoff values represent actually the logic of the game. In fact, the gain of 3 for each slot achieved gives convenience to players to occupy as many slots as possible, and this aspect points out that the created game is actually non-cooperative. By the way, in order to compensate the impairing tendency to occupy all the slots a penalization value of 2 has been introduced, in case of interference, in order to dissuade this kind of behavior. Furthermore, in this model particular importance is given to those outcomes of the game that leave one slot free for other users, and actually this leaves a possible cooperative approach to a game originally thought as non-cooperative. This will be more evident when the possible solutions will be

Table 12.1 Matrix representation of the game for FDMA slots

	Terminal B occupies 1 slot	Terminal B occupies 2 slots
Terminal A occupies 1 slot	(5, 5)	(3, 6)
Terminal A occupies 2 slots	(6, 3)	(4, 4)

described. In Table 12.1, the game of allocating FDMA slots is illustrated, given the considered rules.

The rows of the table are the strategies for cognitive radio terminal A, while the columns are available strategies for CR terminal B. The pair of values within brackets are payoff values achieved by players: on the left-side is relative to terminal A, the right-side to terminal B. This game is known in literature as prisoner's dilemma and it is has been exhaustively analyzed. According to the Nash Equilibrium definition, given by equation (12.11), and payoffs of Table 12.1, the equilibrium point of the game is achieved when both users occupy two FDMA slots. This equilibrium point is evidently the sub-optimal one, since the optimal equilibrium point in this case is achieved when both users choose to occupy only one slot, e.g. payoff achieved by both users equal to 5. Allowing communication and coordination between players, and considering as solution of the game, instead of Nash Equilibrium definition (12.11), the Pareto-optimal point, i.e. the Nash Bargaining point, this can be solved with a cooperative approach. By a practical point of view, this last approach can be addressed as cooperative because it avoids interference between the two players and, more fundamentally, it leaves free one FDMA slot perhaps for another incoming terminal.

12.7 Conclusions and Trends

The ever increasing demand for new and improved services is deeply changing the telecommunication industry. Radio access is no longer considered as the bottleneck of the information delivery chain, but a new opportunity for mobile broadband services. In this framework, the efficient use of spectrum resources plays a fundamental role in the deployment of future generation networks and in the fulfillment of the globalization process. A new gold rush has begun. But flexibility in spectrum usage is not only an "engineering" issue: it also represents a great social and economic opportunity for emerging countries. Universal radio devices have the potentials of the global market, with a compact (but smart) production load. This aspect can be successfully exploited by emerging economies willing to gain a position in the global communication industry.

The dynamic and flexible use of the radio spectrum is the key enabling technology for the universal abstraction of the network access. The connection between communication services and their transport mechanism is finally broken, allowing users to use their services without concerns about geographical position, motion, situation and device capability. The "perfect" communication service is only a few years away.

Cognitive radio is the most effective solution to the widespread waste of radio resources due to the poor utilization of large licensed bands. Digital television, video compression and dynamic resource management tend to compact licensed spectrum creating the basis for this new paradigm of radio access. Cognitive radio has two main approaches: heuristic and gaming. The heuristic method allows devices in a clear and defined context to efficiently use free spectrum holes, without interfering with the licensed communication services. Examples in both satellite and terrestrial scenarios have been provided. Game theory gives another perspective to CR systems. In games, players subject to a set of rules can take decisions even if they are in complex situations. When applied to CR, game theory can solve complex coexistence scenarios with distributed decisions, actually inserting some intelligent functions into devices. A game theoretic approach to CR is described, showing basic principles and specific proposals from the scientific community.

In conclusion, a new telecommunications era is approaching, being characterized by a large platform of new services and a powerful convergence of access technologies into a unified universal network. A flexible use of spectrum represents the required milestone in this evolution and the CR can be the key concept to carry it out.

References

1. O. Ileri, N.B. Mandayam, Dynamic spectrum access models: toward an engineering perspective in the spectrum debate. IEEE Commun. Mag. **46**(1), 153–160 (January 2008)
2. M. McHenry, D. McCloskey, *New York City Spectrum Occupancy Measurements, September 2004* (December 2004)
3. Convergence and Next Generation Networks, OECD Ministerial meeting (Seoul, Korea, 17–18 June 2008), http://www.oecd.org/dataoecd/25/11/40761101.pdf
4. VICOM (Virtual Immersive COMmunications) project, http://www.vicom-project.it
5. www.wireless-world-research.org
6. J. Mitola, *Cognitive Radio: Making Software Radios More Personal* (IEEE Personal Communication, August 1999)
7. FCC Spectrum Policy Task Force, *FCC Report of the Spectrum Efficiency Working Group* (2002, Nov.), http://www.fcc.gov/sptf/files/SEWGFinalReport1.pdf/
8. B. Fette (ed.), *Cognitive Radio Technology* (Newnes, 2006)
9. ETSI EN 302 583, v. 1.1.1, *Digital Video Broadcasting (DVB); Framing Structure, Channel Coding and Modulation for Satellite Services to Handheld Devices (SH) Below 3Ghz* (2008–03)
10. E. Lutz, D. Cygan, M. Dippold, F. Dolainsky, W. Papke, *The Land Mobile Satellite Communication Channel-Recording, Statistics, and Channel Model*. IEEE Trans. Vehicul. Technol, **40**(2) (May 1991)
11. J. von Neumann, O. Morgenstern, *Theory of Games and Economic Behavior* (Princeton University Press, Princeton, 1947)
12. D. Fudenberg, J. Tirole, *Game Theory* (The MIT Press, Cambridge, 2001)
13. T. Basar, G.J. Olsder, *Dynamic Noncooperative Game Theory*, 2nd edn. (SIAM, Philadelphia, 1999)
14. J.F. Nash, Non-cooperative games. Ann. Math. **54**(2), 286–295 (1951)
15. J.F. Nash, Two person cooperative games. Econometrica **21**(1), 128–140 (1953)

16. D. Monderer, L.S. Shapley, Potential games. Games Econ. Behav. **14**(1), 124–143 (1996)
17. L.S. Shapley, Stochastic games. Proc. Natl. Acad. Sci. **39**, 1095–1100 (1953)

Enrico Del Re was born in Florence, Italy. He received his Ing. degree in electronics engineering from the University of Pisa, Pisa, Italy, in 1971. Until 1975 he was engaged in public administration and private firms, involved in the analysis and design of the telecommunication and air traffic control equipment and space systems. Since 1975 he has been with the Department of Electronics Engineering of the University of Florence, Florence, Italy, first as a Research Assistant, then as an Associate Professor, and since 1986 as Professor. During the academic year 1987–1988 he was on leave from the University of Florence for a 9-month period of research at the European Space Research and Technology Centre of the European Space Agency, The Netherlands. His main research interest are digital signal processing, mobile and satellite communications, on which he has published more than 300 papers, in international journals and conferences. He is the Co-editor of the book Satellite Integrated Communications Networks (North-Holland, 1988), one of the authors of the book Data Compression and Error Control Techniques with Applications (Academic, 1985) and the editor of the books Mobile and Personal Communications (Elsevier, 1995), Software Radio Technologies and Services (Springer, 2001), Satellite Personal Communications for Future-Generation Systems (Springer, 2002), Mobile and Personal Satellite Communications 5-EMPS2002 (IIC, Italy, 2002) and Satellite Communications and Navigation Systems (Springer, 2008). He has been the Chairman of the European Project COST 227 "Integrated Space/ Terrestrial Mobile Networks" (1992–1995) and the EU COST Action 252 "Evolution of satellite personal communications from second to future generation systems" (1996–2000). He has been the Chairman of the Software Radio Technologies and Services Workshop (2000), the Fifth European Workshop on Mobile/Personal Satcoms (2002) and the Satellite Navigation and Communications Systems (2006). He received the 1988/1989 premium from the IEE (UK) for the paper "Multicarrier demodulator for digital satellite communication systems". He is the head of the Digital Signal Processing and Telematics Laboratory of the Department of Electronics and Telecommunications of the University of Florence. He has been Director of the Italian Interuniversity Consortium for Telecommunications (CNIT) and is now Vice-President. Professor Del Re is a Senior Member of the IEEE and a member of the European Association for Signal Processing (EURASIP).

Gherardo Gorni was born in Firenze, Italy, the 26th September 1982. He achieved the Bachelor Degree and the Master Degree with Honours in Electronics Engineering from University of Florence respectively in the 2005 and the 2007. He spent the fall semester of 2004 at Tampere University of Technology, Tampere, Finland, as Socrates/Erasmus visiting student. Between July 2006 and March 2007 he has been a visiting student at Universitat Autonoma de Barcelona, Barcelona, Spain, in the context of SatNEx project funded by European Commission. Since March 2008 he has a grant as researcher at the Department of Electronics and Telecommunications of University of Florence. His main research interests concern terrestrial and satellite wireless communications, with particular emphasis on Software Defined Radio and Cognitive Radio approaches and technologies.

Luca Simone Ronga (IEEE S89-M94-SM04) received his M.S. degree in electronic engineering in 1994 and his Ph.D. degree in telecommunications in 1998 from the University of Florence, Italy. In 1997 joined the International Computer Science Institute of Berkeley, California, as a visiting scientist. In 1998 obtained a post-doc position in the engineering faculty of the University of Florence. In 1999 he joined Italian National Consortium for Telecommunications, where he is currently head of research. He conducts research activity and project management in various telecommunications areas, mainly in the satellite and terrestrial wireless fields. He has been leader of national and international research groups. He authored over 50 papers published in international journals and conference proceedings. He has been editor of EURASIP Newsletter for 4 years. His interests range from satellite communications to Software Defined Radio and Cognitive Radio techniques.

Rosalba Suffritti received her M.S. degree in Telecommunications Engineering from the University of Florence, Italy, in 2006, and, her Ph.D. from the University of Florence, Italy, in 2010. Her research activity focuses on advanced topics in satellite communications: cooperative strategies in mobile environment, satellite emergency communications and cognitive radio technologies. She is a Student IEEE member and member of ETSI SES – SatEC (Satellite Earth Stations and Systems – Satellite Emergency Communications) working group which carries out the activity of standardization for satellite emergency communications. She is author of about 20 papers, on international journals and proceedings of international conferences.

Chapter 13
Future Perspectives for Cognitive Radio Technology

Nicola Marchetti and Muhammad Imadur Rahman

13.1 Introduction

The number of mobile subscribers is expected to exceed 5 billion by the end of 2015: a key to attract such a huge number of users is to make them experience new multimedia services with a very high bit rate. In fact, the currently available bit rates do not encourage usage of today's services like video downloading, as the video downloads can take tens of minutes, if not hours. The foreseen mobile market growth is shown in Fig. 13.1 [1]. Furthermore, the number of wireless devices is expected to be around 7 trillion by 2017 [2].

To achieve the increase in data rates necessary to enlarge the mobile market and to make possible the coexistence of such a large number of devices, a much better exploitation of the common frequency spectrum than the one employed today, will be necessary for the future generations of wireless communications.

So far a large part of the spectrum has been allocated according to the first-come, first-served principle with broadcast radio and TV getting the most desirable bands. Likewise frequency band has been allocated for military and medical uses. For the future cellular technologies and for the wireless technologies in general -WiMAX/ Wi- Fi/ZigBee/Bluetooth – communication quantity is expected to increase dramatically in relation to the future new wireless applications. Despite the radio and wireless technology having developed significantly in the last decades, the problem of the scarce spectrum for general civilian purposes has increased. It is more than ever important – *and in the long run necessary* – to get a much better frequency access and frequency spectrum usage [3].

Around the year 2000 a new idea was launched concerning improvements on exploiting the frequency spectrum: Cognitive Radio (CR) technology. Cognitive

N. Marchetti (✉)
Center for TeleInFrastruktur, Aalborg University, Denmark
e-mail: nm@es.aau.dk

M.I. Rahman
Ericsson Research, Sweden

R. Prasad et al. (eds.), *Globalization of Mobile and Wireless Communications: Today and in 2020*, Signals and Communication Technology,
DOI 10.1007/978-94-007-0107-6_13, © Springer Science+Business Media B.V. 2011

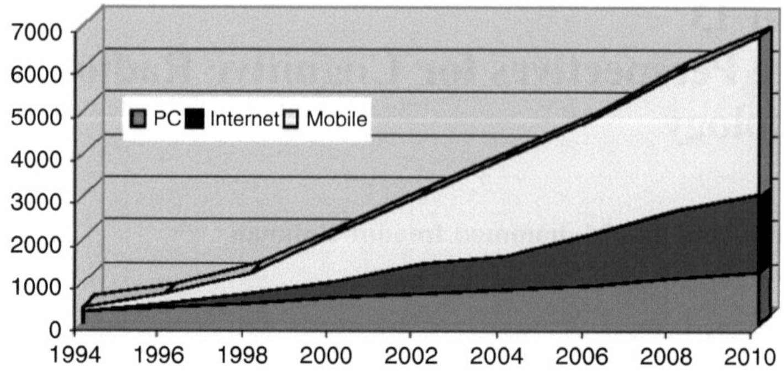

Fig. 13.1 Mobile market growth [1]

radio is the integration of software radio and machine intelligence. The term, coined by Joe Mitola, describes a radio communications paradigm where a network or individual nodes change communications parameters to adapt to a changing environment that includes both external factors like spectrum usage, and internal factors like user behavior. It blends the promises of artificial intelligence and of Software Defined Radio (SDR). Science often advances through the exploration of inspired metaphors like this one [4].

CR is built on top of a SDR [5], and can be defined as an intelligent wireless communication system that is aware of its environment and uses the methodology of understanding-by-building to learn from the environment and adapt to statistical variations in the input stimuli, with two primary objectives in mind [6]:

- Highly reliable communication whenever and wherever needed
- Efficient utilization of radio spectrum

A CR is looking for virtual unlicensed spectrum bands, i.e., bands to be shared between primary licensed users (Primary Users (PU)) and secondary unlicensed users (Secondary Users (SU)), on a non-interfering basis, i.e., without any activity to or from PUs at present [3]. CR techniques provide the capability to use or share the spectrum in an opportunistic manner. Dynamic spectrum access techniques allow the CR to operate in the best available channel [7].

A CR must [8]:

- Recognize the needs of the user.
- Understand the limitations imposed on the radio operation by the channel and external environment.
- Realize its own limitations in flexibility and power.
- Conform to local regulations and policy.

The remainder of the chapter is organized as follows: in Section 13.2 a general overview of CR technology from several points of is be given. In Section 13.3 the concepts of Spectrum Sharing (SpS) and Flexible Spectrum Usage (FSU) will be introduced, and then we discuss these issues to configure concrete study case. A few challenges we face when dealing with a very large spectrum are pointed out in Section 13.4. Some interesting issues to be investigated are introduced as examples of the challenges and benefits related to CR in Section 13.5. Considerations about possible future scenarios of applications of CR and related challenges are provided in Section 13.6. Finally, while conclude the chapter in Section 13.7.

13.2 Cognitive Radios from Different Perspectives

13.2.1 Towards the "Real" 4G: IMT-Advanced

IMT – Advanced (IMT-A) is a concept from International Telecommunication Union (ITU) for mobile communication systems with capabilities which go further than those of IMT-2000. IMT-A was previously known as "systems beyond IMT-2000". It is foreseen that the development of IMT-2000 will reach a limit of around 30 Mbps.

In the vision of the ITU, there may be a need for a new wireless access technology to be developed around the year 2010 capable of supporting even higher data rates than 30 Mbps with high mobility, which could be widely deployed around the year 2015 in some countries. The new capabilities of these IMT-A systems are envisaged to handle a wide range of supported data rates according to economic and service demands in multi-user environments, with target peak data rates of up to approximately 100 Mbps for high mobility such as mobile access, and up to approximately 1 Gbps for low mobility such as nomadic/local wireless access [9].

To support this wide variety of services, it may be necessary for IMT-A to have different radio interfaces and frequency bands for mobile access and for new nomadic/local area wireless access. The band which is envisaged to be needed to provide the afore-mentioned very high rates, is in the order of 100 MHz, and likely such a high band will not be available for each operator of a country, which means that some kind of flexible and smart spectrum usage among different operators and networks will be needed.

A key component enabling this "peaceful" coexistence between systems and networks in the same bandwidth is the so called Spectrum Sharing (SpS), which could be broadly classified as follows:

- Inter-system SpS, allowing the coexistence of different Radio Access Technology (RAT)s
- Inter-network SpS, allowing the coexistence of different operators/networks
- Intra-network SpS, allowing the coexistence of different cells owned by the same network, and operating with the same RAT (we could identify this kind of sharing as multi-cell Radio Resource Management (RRM))

In Section 11.3, deeper and further considerations on spectrum sharing will be provided.

13.2.2 CR from PHY Point of View

Orthogonal Frequency Division Multiplexing (OFDM) is probably the best Physical Layer (PHY) candidate for wideband CR systems because it allows achieving the Shannon channel capacity in a fragmented spectrum. A CR identifies unused spectrum segments in a target primary user sub-band through spectrum sensing.

CRs are designed to be aware of and sensitive to the changes in their surroundings. The spectrum sensing function enables the cognitive radio to adapt to its environment by detecting spectrum holes [8, 10]. The most efficient way to detect spectrum holes is to detect primary users receiving data within the communication range of a CR user. In reality, however, it is difficult for a cognitive radio to have a direct measurement of a channel between a primary receiver and a transmitter. That is why most current work focuses on primary transmitter detection based on local observations by secondary users. Generally, the spectrum sensing techniques can be classified as [8]:

- Transmitter detection: matched filter detection, energy detection, cyclo-stationary feature detection
- Cooperative detection
- Inteference-based detection

Another PHY aspect related to CR is Adjacent Channel Leakage (ACL): to ensure successful operation, a CR must minimize the interference to the primary user bands by reducing the ACL. In [10], two ACL mechanisms have been identified: signal time-domain truncation and transmission nonlinearity. ACL due to signal time-domain truncation is unavoidable but it can be controlled through proper system parameter choices.

For time-domain signals, the parameters affecting ACL include: guard interval size, pulse shaping filter roll-off, truncation window width, and truncation window roll-off. For OFDM, the parameters affecting ACL include: guard band size, subchannel spacing, OFDM symbol window width, and OFDM symbol window roll-off [10].

13.2.3 CR from RRM and MAC Point of View

To discuss the Medium Access Control (MAC) issues related to CR, let us consider the new IEEE 802.22 standard. The 802.22 MAC is very flexible, and can provide an adequate service to secondary users while enforcing the necessary incumbent protection mechanisms [11].

Effective coexistence is one of the key responsibilities of a CR. Within the context of 802.22, coexistence has two facets: coexistence with incumbents and self-coexistence. Coexistence with incumbents deals with Dynamic Spectrum Allocation (DSA) mechanisms for a reliable, efficient, and timely detection of primary services or white spaces, followed by a network recovery procedure once these incumbents are detected. In contrast, self-coexistence addresses Dynamic Spectrum Sharing (DSS) amongst collocated 802.22 cells. Self-coexistence in 802.22 is a critical issue given its unlicensed operation and its very large coverage range.

Regardless of which coexistence method is used, a solid and flexible measurement and spectrum management component is of great importance. This involves mechanisms by which nodes cooperate in performing channel measurements (spectrum sensing), measurement reports are generated and communicated (spectrum analysis), decisions are made as to which channels to use, when, and for how long (spectrum management or decision).

13.2.4 Software Defined Radios

An SDR system is a radio communication system which can tune to any frequency band and receive any modulation across a large frequency spectrum by means of programmable hardware which is controlled by software.

The goal of research on SDRs is to produce a radio that can receive and transmit when there is a change in the radio protocol, simply by running new software. In the future advanced SDR the antenna will remain the only band specific item.

SDRs can currently be used to implement simple radio modem technologies, but in the long run, software-defined radio is expected by its proponents to become the dominant technology in radio communications: SDRs will likely be the enabler of CRs.

13.3 Spectrum Sharing and Flexible Spectrum Usage

The functionalities to enable spectrum-aware communication protocols are [7]:

- Spectrum sensing: aims at detecting unused spectrum and sharing the spectrum with other operators and networks, without harmful interference
- Spectrum management: aims at capturing the best available spectrum to meet user communication requirements
- Spectrum mobility: aims at maintaining seamless communication requirements during the transition to better spectrum
- Spectrum sharing: aims at providing a fair and efficient spectrum scheduling method among coexisting operators and networks

13.3.1 Classification

We will adopt the terminology used in WINNER project [12]. Inter-system Spectrum Sharing (SpS) refers to the peaceful use of the same frequency band by multiple radio access systems using different RATs. FSU refers to the use of the same frequency band by radio access systems using the same RAT. It can further be classified into:

- Intra-system FSU, referring to the use of the same frequency band by multiple Radio Access Networks (RANs), operated by different operators;
- Intra-RAN FSU, referring to methods increasing the spectrum use efficiency and flexibility within a single RAN, e.g. a spectrum re-allocation among neighboring and overlapping cells.

If two or more radio systems (using the same or different RAT) are to be operated in the same frequency band, the basic possibilities for peaceful coexistence are

- Time separation: systems transmit at different times.
- Frequency separation: systems transmit at different frequencies.
- Space separation: path loss from the transmit antennas of one system to the receive antennas of another system is sufficiently high to attenuate the interfering signal. This can be achieved with e.g. with directional antennas.

FSU and SpS can be considered from different viewpoints. From the number and type of involved systems viewpoint (by expanding the set of involved systems, the possible benefits from FSU and SpS increase, but at the price of added complexity), we can have the following cases:

- Several RANs with same RAT of one operator
- Several RANs with same RAT of several operators
- All kind of Beyond 3rd Generation (B3G) systems
- All systems operating in a certain frequency band

One can have several degrees of coordination:

- None: no exchange of information is possible among the involved systems
- One-way: one system might obtain information about the spectrum use of the other, but not vice-versa
- Two-ways: all systems are informed about the spectrum use of all other systems. The assumption in this case is that signaling between systems is possible, enabling the possibility of negotiations
- Joint Radio Resource Management (JRRM): it refers to the presence of a central entity managing the radio resources of all involved systems

Several hierarchical structures can be identified, referring to priorities and privileges in spectrum use:

- Exclusive access: one and only one system has exclusive access to the spectrum.
- Horizontal sharing (equal right access): all systems have the same regulatory status and may access the spectrum on equal footing. As an example, consider the usage of Industrial, Scientific and Medical (ISM) bands by WLAN and Bluetooth. No system can claim any priority over the others.

- Vertical sharing (priority-based access): a certain frequency band is dedicated to a primary system.

Secondary (lower priority) systems may only use this band as long as they do not harmfully interfere with the primary system.

The case of horizontal sharing can be distinguished based on the absence or presence of coordination:

- Without coordination: systems access the spectrum without coordination (e.g. WLAN and Bluetooth in 2.4 GHz bands). The main drawback of this approach is that, since the interference from other systems is unpredictable, no Quality of Service (QoS) can be guaranteed.
- With coordination: systems coordinate their spectrum access based on a set of rules (spectrum etiquette) that all systems adhere to.

SpS and FSU techniques can be classified according to the architecture assumption:

- Centralized: a centralized entity controls the spectrum allocation and access procedures [13]. With aid to these procedures, generally a distributed sensing procedure is proposed such that each entity in the network forwards its measurements about the spectrum allocation to the central entity and this entity constructs a spectrum allocation map.
- Distributed: these solutions are usually proposed when an infrastructure is not preferable [14]; each node is responsible for spectrum allocation and access is based on local (or possibly global) policies.

Another classification can be based on access behavior:

- Cooperative: these solutions (sometimes also called collaborative) consider the effect of the node's communication on other nodes [13], i.e., the interference measurements of each node are shared with other nodes, and the spectrum allocation algorithms consider this information. All centralized solutions can be regarded as cooperative.
- Non-cooperative: these solutions (sometimes also called non-collaborative or selfish) consider only one node at hand [15], and may result in reduced spectrum utilization, but the minimal communication requirements among the nodes could guarantee a suitable trade-off for practical solutions.

Sharing techniques can be distinguished based on the access technology:

- Overlay: a node accesses the network using a portion of the spectrum that is not used [13]: as a result, interference to other nodes is minimized.
- Underlay: exploits the spread spectrum techniques developed for cellular networks [16]. Once a spectrum allocation map has been acquired, a node begins transmission such that its transmit power at a certain portion of the spectrum is regarded as noise by the other nodes. This technique requires sophisticated spread spectrum techniques and can utilize increased bandwidth compared to overlay techniques.

13.3.2 FSU

DSA studied in DRiVE [17] and OverDRiVE projects [18], provides a flexible alternative to dedicated "brick-wall" spectrum allocation. Main goal of DSA is to allow a controlled amount of spectrum sharing within a defined set of radio systems, such that these systems obtain the optimum amount of spectrum available to them, either at a particular time, or in a particular location, depending on their load characteristics. FSU among RANs working with the same RAT can be seen as a meaningful application of DSA.

Significant benefits can be expected from intra-system spectrum sharing, mainly increased spectral efficiency and increased spectral scalability of the Long Term Evolution – Advanced (LTE-A) system. In particular, concerning spectral scalability, this allows for a gradual long term spectral deployment of the system: individual bands do not need to be available for all RANs at the launch of the new system, instead additional bands can be made available to the RANs following the growth in time of subscribers and traffic.

Compared to SpS, the coordination mechanisms can be more advanced, since all systems use the same RAT, allowing therefore a tighter cooperation.

An agile PHY and frame structure is required for FSU: the air interface should be able to dynamically adjust its transmission in frequency and space, while still maintaining the desired QoS requirements, and to minimize inter-RAN and inter-operator interference, as well as guard bands (with a positive effect on spectral efficiency).

The ITU is currently working on specifying the system requirements towards next generation mobile communication systems called IMT-A [19, 20]. 3G mobile communication systems including their evolution are in this respect part of the ITU International Mobile Telecommunications (IMT)-2000 systems. The deployment of IMT-A is believed to take place around year 2015 at mass market level and will facilitate what has been a buzzword for a almost a decade, namely "4G". IMT-A systems are expected to provide peak-data-rates in the order of 1 Gbps in local areas. Such high data rates require both usage of advanced Multiple Input Multiple Output (MIMO) technology to achieve high spectral efficiency and also extreme high spectrum allocation in the range of more than 100 MHz.

Despite that new frequency bands are expected to be allocated for IMT-A, the high system bandwidth requirements will likely demand spectrum sharing among operators. This is a very different situation from today's IMT-2000 systems (e.g., GSM/UMTS), where each network operator has a dedicated licensed band for its own network. In fact, traditionally cellular mobile network operators (e.g. GSM/UMTS) have obtained dedicated spectrum resources from their national telecommunication regulators.

The dedicated radio resources can be utilized independently by each operator without any interference coordination.

On the other hand, the obvious drawback of dedicated spectrum allocation is reduced spectrum trunking efficiency and reduced peak data-rates by restricting the

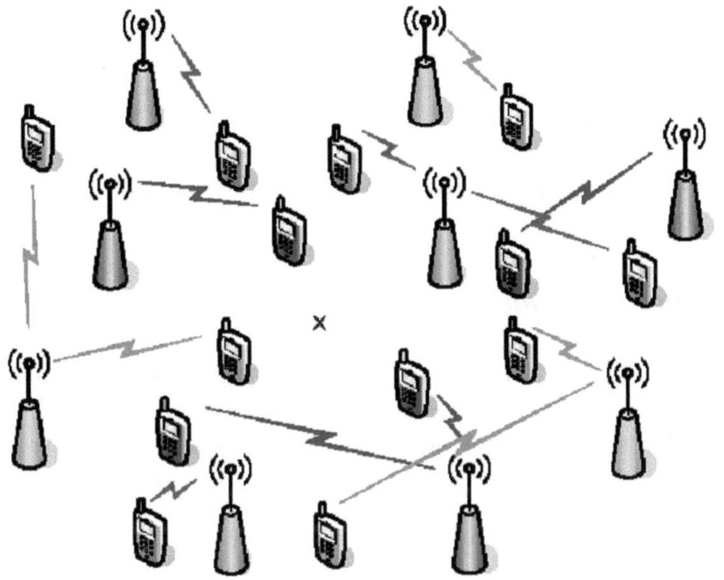

Fig. 13.2 Cellular multi-operator environment [21]

system bandwidth to each operator. This has however not been a severe problem for IMT-2000 cellular systems since the system bandwidth has been relatively small (lower than 5 MHz).

Next generation IMT – Advanced (IMT-A) systems are predicted to facilitate flexible bandwidths of up to 100 MHz. Even with new spectrum allocation to IMT-A by the World Radio Congress in 2007, it is impossible to allocate such high bandwidth to several operators operating in the same geographical area (see Fig. 13.2). Hence new approaches for spectrum sharing among operators are required.

13.4 Dealing with a Very Large Spectrum: The Challenges

13.4.1 Dynamic Radio Range

Radio range changes with operating frequency due to attenuation variation. In many solutions, a fixed range is assumed to be independent of the operating spectrum [14]. However, in SpS/FSU, where a large portion of the wireless spectrum is considered, the neighbors of a node may change as the operating frequency changes. This affects the interference profile as well as routing decisions. Moreover, due to this property, the choice of a control channel needs to be carefully decided: it would be more efficient to select control channels in the lower portions of the spectrum where the transmission range is higher, and to select data channels in the higher portions of the spectrum where a localized operation can be utilized with minimized interference [7].

Therefore, due to direct interdependency between interference and radio range, spectrum sharing techniques should be aware of the operating frequency.

13.4.2 Spectrum Unit

A channel is the basic spectrum unit for operation. In [22], the necessity of a spectrum space is advocated. The possible dimensions of the spectrum space are classified as: power, frequency, time, space, and signal. Although not orthogonal, these dimensions can be used to distinguish signals. In [23], the resource space is defined as a cube with time, rate, and power/code dimensions. The time dimension models the time required to transfer information, the rate dimension models the data rate of the network, and the power/code dimension models the energy consumed for transmitting information through the network. The resource in terms of available bandwidth can be modeled using the time and rate dimensions.

Determining a common spectrum unit is crucial for the efficient utilization of the wireless spectrum and the seamless operability among operators and among networks of a single operator.

13.5 Some Interesting Issues

13.5.1 Application of Machine Learning

CR offers the promise of intelligent radios that can learn from and adapt to their environment. So far, most CR research has focused on policy-based radios that are hard-coded with a list of rules of how the radio should behave in certain scenarios.

A concrete model for a generic CR to utilize a learning engine is described in [24]. The goal is to incorporate the results of the learning engine into a predicate calculus-based reasoning engine so that radios can remember lessons learned in the past and act quickly in the future. Based on a Knowledge Base (KB), a reasoning engine and a learning engine work together to maximize an objective function (see Fig. 13.3 [24]).

At any given time, the cognitive engine generates conclusions based on information defined in the KB, the radio's long term memory. These conclusions are the result of extrapolations of this information based on reasoning or learning. The reasoning engine is often referred in Artificial Intelligence (AI) literature as an expert system. The learning engine is responsible for manipulating the KB from

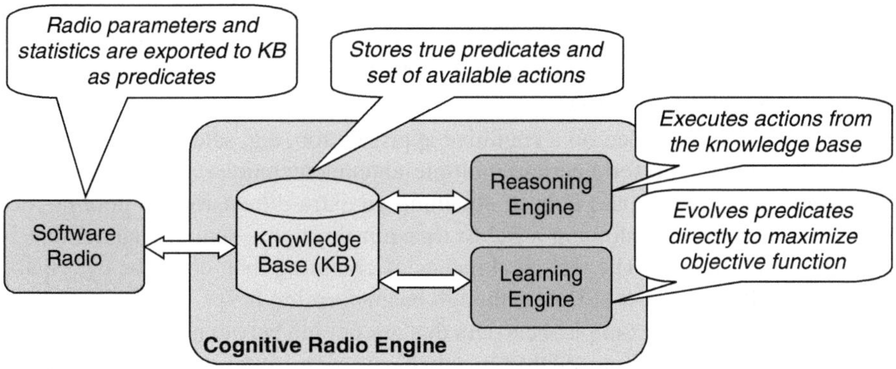

Fig. 13.3 Cognitive Radio architecture including a learning engine [24]

experience. As lessons are learned, the learning engine stores them in the KB for future reference by the reasoning engine. The cognitive engine is responsible for augmenting the list of actions available to the radio that allow it to adapt to a changing environment [24].

A learning technique involves the use of an objective function to determine the value of the learned data: in CR this function will reflect the overall application goal, such as channel capacity maximization. The goal of the learning engine is to determine which input state will optimize the objective function, but while for the reasoning engine there is a simple mathematical relationship between the system inputs and the objective function, this does not hold for the learning engine.

13.5.2 Combination of MIMO and Spectrum Pooling

One possibility recently introduced [25, 26] to realize SpS is the so called spectrum pooling. The spectrum pooling technique allows the coexistence of two mobile radio systems within the same frequency range (e.g. PUs and SUs), by enabling the secondary utilization of already licensed frequency bands, obtaining thus an increase in the spectral efficiency (measured in bps/Hz). OFDM is a candidate for such a flexible system as it allows to leave a set of subcarriers unused, i.e. by feeding certain subcarriers with zeros resulting in no emission of radio power on the carriers that are occupied by the licensed users.

It is well known that MIMO techniques increase the spectral efficiency without increasing the bandwidth. Most of the available MIMO techniques are effective in frequency-flat scenarios. OFDM converts a wideband frequency selective channel into a number of narrowband subcarriers. Therefore, in wideband scenarios, OFDM can be combined with MIMO systems, by applying the considered MIMO technique on each frequency-flat narrowband subcarrier.

It is therefore expected to obtain a very efficient spectrum utilization by combining MIMO techniques with spectrum pooling in OFDM-based systems. The cognitive approach can help concerning decisions on which chunks of the band owned by PU are to be shared with SU. The decisions about what MIMO technique to utilize on each chunk could be based on a cognitive approach too, e.g. selecting in what scenarios and bands to go for a certain multiple-antenna technique.

MIMO technology could help in obtaining an extra-dimension for pooling, i.e. the spatial dimension, allowing a SU to transmit through a spatial channel that is not utilized by the PU. The technical means of obtaining that, could be e.g. Space Division Multiple Access (SDMA) though beam-forming [27].

Let us consider for example 4 carriers that are pooled between two operators $O1$ and $O2$, and that through an SDMA mechanism, each of the two operators' Base Station (BS)s can form a main transmission lobe pointing in a certain direction, thus creating in total two distinguished spatial channels. One operator's transmitter can transmit zeros in the frequencies where the other operator is transmitting, as in normal pooling, but in this case also the space domain is part of the game: let us assume that if for frequency $f1$ operator $O1$ is transmitting on spatial channel 1, and $O2$ on spatial channel 2, then $O1$ will transmit zero on the second spatial channel on $f1$, and the same will do $O2$ on the first spatial channel. The situation is resumed in Table 13.1 and depicted in Fig. 13.4 (for frequency $f1$ and down-link case). Analogously, one can proceed in the same way for all carriers.

13.5.3 Policies

Policy enabled spectrum usage is one of the key features of CRs. Policies have their origin in spectrum usage restrictions imposed by a regulating authority. Further policies may come from other policy makers to reflect for instance preferences of the user or operators. The specification of algorithms for enabling SpS is another important aspect for using policies. Policies might have a limited validity which depends on multiple factors as for instance the local time, the particular geographical location of the radio or the country where it is operating. Thus, CRs have to use policies in an adaptive way. A well defined policy framework is required to enable such a CR capable of updating policies. This framework implies language constructs for specifying a policy, a machine- understandable representation of these policies and a reasoning instance, in [28] called spectrum navigator, which decides

Table 13.1 Space-frequency pooling example

	Operator	f_1	f_2	f_3	f_4
Spatial Channel no. 1	O_1	1	0	0	1
	O_2	0	1	1	0
Spatial Channel no. 2	O_1	0	1	1	0
	O_2	1	0	0	1

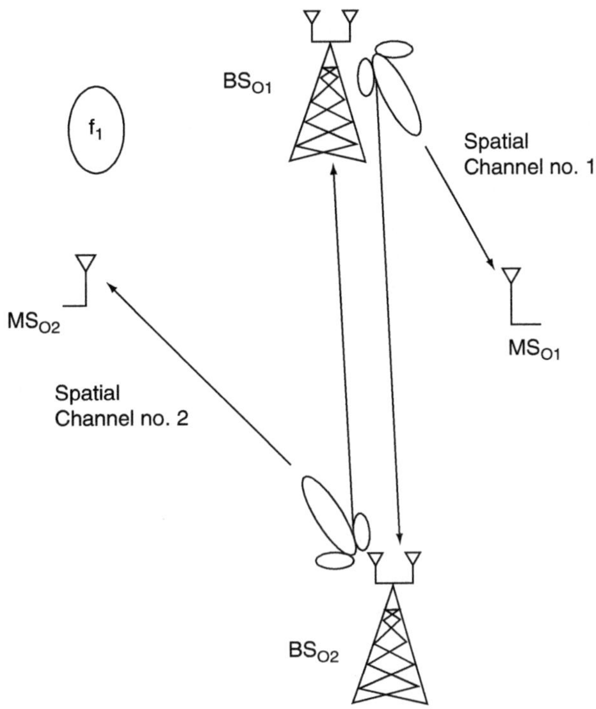

Fig. 13.4 Spatial Pooling concept

about spectrum usage. The policy conformance validation is responsible for downloading, updating and validating policies, and afterwards the syntactical correctness of a policy that has been downloaded to the CR is verified. After conformance validation, the CR translates the policies to a machine-understandable language to enable computation through the spectrum navigator [28].

13.6 Cognitive Radios in Beyond 2020

Several reasons can be outlined for the need of dynamic spectrum and cognitive radio techniques, and among others that the continuously increasing density of wireless devices (especially in terms of home gadgets and sen-sors/pervasive computing) cannot be supported by current static and inefficient spectrum allocation. Indeed wireless is overtaking wired as the primary mode of connectivity to the Internet; nowadays there are 500 million Internet- connected servers/PCs vs. two billion cellular phones (of which 400 million are internet capable). Concerning sensors, their deployment is just beginning and around 10 billion of them are expected to be active by year 2020 [29].

WWRF predicts that in 2017 seven trillion wireless devices will be serving the 7 billion Earth inhabitants, which means 1,000 wireless devices per person will be active [2]. These devices will be belonging not only to traditional mobile telephony, but also to health and medical, smart homes, intelligent cars and Radio Frequency Identification (RFID) applications.

Besides the already mentioned spectrum efficiency, this poses challenges for the network architecture as well. Therefore, CR should not be seen just as a new generation of software radio, aimed at sensing the spectrum, but it shall also be able to sense the communication and networking environments: thus there is a need for integration of CR and existing infrastructure to optimize communications and networks. Future terminal devices shall then form cognitive networks: seeing things from this perspective there is no point to make a distinction between infrastructure and ad-hoc networks. The main challenge is to do the proper networking of potentially trillion wireless devices within a few GHz bandwidth [30].

As a matter of fact, the ubiquitous and pervasive computing and networking is becoming a clear trend in wireless communications: this is confirmed by the fact that computer processors are increasingly becoming part of more and more everyday items, which may form a wireless network. This trend corresponds to a shift from the "very large" to the "very small", leading to *disappearing electronics* with the following characteristics: low-cost, miniature size and self-contained from energy perspective [2]. The ultimate goal is to achieve reliable universal coverage at all times and this brings up several challenges, some of which we have already mentioned [31]:

- Seven trillion devices will very rapidly run out of spectrum.
- Most devices will have limitations in energy consumption.
- Wireless might be unreliable (bigger challenges due to the media than in wired counterpart).
- From the very high heterogeneity of devices it might come incompatibility.

This scenario is related with the so-called Self-Organizing Networks (SONs), which are characterized by self-configuration, self-service, self-knowledge, self-awareness and self-maintenance capabilities. Examples of networks with self-organization capabilities include: ad-hoc networks, wireless sensor networks, and mesh wireless access networks. Concrete examples of SONs are [32]:

- Vehicular networking (among cars, e.g., accidents warnings, and within the car, e.g. network among personal devices)
- Wireless sensor networks (applied, e.g., to monitoring of humans, in the context of assisted living and medical engineering)
- Wearable computing

SONs are in nature decentralized and this implies a limited amount of information available compared to centralized networks. One key issue is to quantify and minimize the information required to be exchanged by the nodes of the a SON such that performance will still be acceptable.

SONs can also be seen as a means to address the growing complexity of networks. The above mentioned trend towards ubiquitous computing adds to the spatial-temporal complexity and dynamics of wireless networks. The solution to increasing complexity, without users to become experts in the field and without network operators having to spend much time and resources in configuring networks, lies in increasing the level of self-organization, i.e., to design networks that minimize human intervention by organizing themselves as much as possible [32].

13.7 Conclusions

In this chapter, we first introduced the CR terminology and the problem this technology tries to address and then provided a general overview of CR State-of-the-Art from several viewpoints. Further, we motivated the use of cognitive radio techniques for better spectrum utilization in future high data rate and high density deployment wireless networks. For extremely high spectral efficiency such as up to 10 bps/Hz, a combination of several component technologies is envisioned, such as multi-antenna techniques, OFDM transmission methods, and cognitive mechanisms based on intelligent learning process.

A glance on the role of CR in wireless systems beyond 2020 was also given: future wireless networks will be pervasive, formed by 1,000 times more wireless devices than in use today, sharing the same limited spectrum resource. The need of CR is felt already today, mainly to provide very high data-rate services, but this technology will become even more vital in 10 years from now in order to have intelligent wireless devices with built-in learning capabilities, i.e., cognitive devices that, in a self-organized way, will form heterogeneous networks. Such networks will be composed of a very large amount of nodes that will be able to adapt to changing conditions, while simultaneously optimizing their functioning in real-time.

References

1 R. Syputa Intel against Qualcomm: Clash of the Titans (2005), http://maravedis-bwa.com/FreeArticles-Clashofthetitans.html. Dec 2005
2. J.M. Rabaey, D.O. Pederson, A brand new wireless day – What does it mean for design technology? in *Ambient Society Symposium* (*ASPDAC*), Seoul, 2008
3. F.B. Frederiksen, R. Prasad, Strategies for improvement of spectrum capacity for WiMAX cellular systems by Cognitive Radio Technology supported by Relay Stations, in *Proceedings 12th International OFDM-Workshop*, Hamburg, pp. 336–340, Aug 2007
4. P. de Vries, Cognitive radio as metaphor (2006), http://deepfreeze9.blogspot.com/2006/10/cognitive-radio-as-metaphor.html. Oct 2006
5. F.K. Jondral, Software-defined radio – Basics and evolution to cognitive radio. Eurasip J. Wireless Commun. Netw. **3**, 275–283 (Apr 2005)
6. S. Haykin, Cognitive radio: brain-empowered wireless communications. IEEE JSAC. **23**(2), 201–220 (Feb 2005)

7. I.F. Akyildiz et al., NeXt generation/dynamic spectrum access/cognitive radio wireless networks: A survey. Comput. Netw. **50**, 2127–2159 (May 2006)
8. F.B. Frederiksen, Improving spectral capacity and wireless network coverage by cognitive radio technology and relay nodes in cellular systems. Wireless Pers. Commun. **45**(3), 355–368 (Feb 2008)
9. H.W. Lee, 3G LTE & IMT-advanced service (2006), http://mmlab.snu.ac.kr/links/hsn/workshop/hsn2006/document/2.24.Fri/8–2.pdf. Feb 2006
10. H. Tang, Some physical layer issues of wide-band cognitive radio systems, in *IEEE International Symposium on New Frontiers in Dynamic Spectrum Access Networks (DySPAN)*, pp. 151–159, Nov 2005
11. C. Cordeiro et al., Cognitive PHY and MAC layers for dynamic spectrum access and sharing of TV bands, in *Proceedings of the First International Workshop on Technology and Policy for Accessing Spectrum (TAPAS)*, vol. 3, Boston, MA, Aug 2006
12. K. Hooli et al., WINNER Spectrum Aspects: Assessment report, 58, p. 1. IST-2003-507581 WINNER D6.3 v1.0, Dec 2005
13. V. Brik et al., DSAP: a protocol for coordinated spectrum access, in *IEEE DySPAN*, 2005, pp. 611–614
14. L. Cao, H. Zheng, Distributed spectrum allocation via local bargaining, in *IEEE Sensor and Ad Hoc Communications and Networks (SECON)*, pp. 475–486, Sept 2005
15. S. Samkaranarayanan et al., A bandwidth sharing approach to improve licensed spectrum utilization, in *IEEE International Symposium on New Frontiers in Dynamic Spectrum Access Networks (DySPAN)*, pp. 279–288, November 2005
16. J. Huang et al., Spectrum sharing with distributed interference compensation, in *IEEE International Symposium on New Frontiers in Dynamic Spectrum Access Networks (DySPAN)*, pp. 88–93, Nov 2005
17. P. Leaves et al., A summary of dynamic spectrum allocation results from DRiVE, in *IST Mobile Summit*, Chalkidiki, June 2002
18. D. Grandblaise et al., Dynamic Spectrum Allocation (DSA) and reconfigurability, in *SDR Technical Conference and Product Exposition*, San Diego, Nov 2002
19. Guidelines for Evaluation of Radio Transmission Technologies for IMT-Advanced, Tech. Rep. Document 8F/TEMP/495-E, Jan 2007
20. Draft Requirements Related to Technical System Performance for IMT-Advanced Radio Interfaces, Tech. Rep. Document 8F/TEMP/496-E, Jan 2007
21. S. Kumar, G. Costa, S. Kant, F.B. Frederiksen, N. Marchetti, et al., Spectrum sharing for next generation wireless communication networks, in *First International Workshop on Cognitive Radio and Advanced Spectrum Management (CogART)*, Aalborg, Denmark, Feb 2008, pp. 1–5
22. W.D. Horne, Adaptive spectrum access: using the full spectrum space, in *Telecommunications Policy Research Conference (TPRC)*, Arlington, Sept 2003
23. M.C. Vuran, I.F. Akyildiz, AMAC: adaptive medium access control for next generation wireless terminals. IEEE/ACM Trans. Netw. **15**(3), pp. 574–587 (June 2007)
24. C. Clancy et al., Applications of machine learning to cognitive radio networks (2006), http://users.ece.gatech.edu/ eps/publications/wcm-crnet06.pdf
25. T. Weiss, F. Jondral, Spectrum pooling: An innovative strategy for the enhancement of spectrum efficiency. IEEE Commun. Mag. **42**(8–14) (Mar 2004)
26. D. Cabric et al., A cognitive radio approach for usage of virtual unlicensed spectrum (2004), http://www.eecs.berkeley.edu/~smm/IST_paper.pdf
27. D. Trajkov et al., Spectrum sharing for directional systems, in *Proceedings of the First International Workshop on Technology and Policy for Accessing Spectrum (TAPAS)*, 2006
28. L. Berlemann et al., Policy-based reasoning for spectrum sharing in cognitive radio networks, *IEEE International Symposium on New Frontiers in Dynamic Spectrum Access Networks (DySPAN)*, pp. 1–10, Nov 2005
29. N.B. Mandayam, Cognitive Radio Networks & The

Nicola Marchetti is currently an assistant professor at the Networking and Security Group, Center for TeleInFrastruktur (CTIF) at Aalborg University. In 2003 and 2007 he received the M.Sc. degree in Electronic Engineering and the Ph.D. degree in Wireless Communications from University of Ferrara, Italy, and Aalborg University, Denmark, respectively. From July 2003 to April 2004 he was a research assistant at the University of Ferrara and he worked as a PhD fellow from May 2004 to May 2007 at CTIF, Aalborg University, Denmark. From June 2007 to December 2008, he has been leading and coordinating a project on spectrum sharing and cognitive radios in collaboration with the Danish Research Council and Nokia Siemens Networks. From January 2009 he is with CTIF at Aalborg University. He is the co-founder of the International Workshop on Cognitive Radio and Advanced Spectrum Management (CogART) and of the International Symposium on Applied Sciences in Biomedical and Communication Technologies (ISABEL). His research interests include: multiple antenna technologies, single- and multi-carrier modulations, radio resource management, advanced spectrum management techniques, cognitive radios and networks, self-organizing networks, tele-healthcare, mathematics applied to wireless communication.

Muhammad Imadur Rahman was born in Sylhet, Bangladesh on 30 January 1976. He is currently with Radio Access Technology Department of Ericsson Research, Sweden, as a Research Engineer. He obtained his Ph.D. degree in Wireless Communications from Aalborg University, Denmark in 2007. Prior to that, he received his M.Sc. degree in Radio Communications from Helsinki University of Technology (HUT), Finland in 2003 and his B.Sc. (Hons) in Electronics Engineering from Multimedia University, Malaysia in 2000. In 2007, he was with Radio Access Technology Section, Aalborg University, Denmark as an Assistant Research Professor, where he taught a number of M.Sc. level courses in mobile communications, supervised 10 M.Sc. students and 3 Ph.D. students. Previously, he was with Communications Laboratory of HUT in 2001–2002 as a research assistant. He joined Ericsson Research in 2008, where he is involved in research related to future radio access technologies and cognitive radios. His current research interests are signal processing, multiple access techniques, multi-antenna issues, interference analysis and inter-system coexistence studies. He has co-authored two books, a few book chapters and around 40 publications in international journals and conferences. He holds around 10 patent grants and applications. He has been involved in a number of conferences and workshops as organizing and technical committee member. He is an active IEEE volunteer; currently he is the vice-chairman of IEEE Sweden Section for 2010 and also the membership development officer for IEEE Sweden VT/COM chapter for 2010. He is also involved in a number of socio-cultural community welfare activities in Stockholm, Sweden.

Chapter 14
Multi-users Participation in Bidding Process in a Congested Cellular Network

K. Sridhara, P.S.M. Tripathi, Ambuj Kumar, Ashok Chandra, and Ramjee Prasad

14.1 Introduction

As we are aware that mobile phones played a vital role in communications. Mobiles proved essential in keeping people in touch with each other during any event whether is manmade or natural. The key advantages of mobile are as follows, these have been emerged from a range of experiences [1]:

- The speed of recovery of mobile networks after physical damage, significantly less than for other communications media such as fixed line networks, TV and radio
- The ability of operators to optimize the use of the network under strain in order to minimize congestion and call failure rates
- People's ability to send and receive SMS messages even when the networks are too busy for additional voice calls
- Mobile's greater accessibility than many other sources of information and communication; and
- Mobile's unique capacity to distribute information directly amongst the people affected by the emergency

As we are aware that traffic pattern is non-uniform throughout the city or service area. As a result, the telecom service area can be classified into various types of region generating different levels of traffic such as areas generating very heavy traffic,

K. Sridhara (✉) and A. Kumar
Global ICT Standardization Forum of India (GIFSI)
e-mail: ksridhara9@hotmail.com

P.S.M. Tripathi and A. Chandra
Department of Telecommunications, Ministry of Communications and IT, Government of India

R. Prasad
Aalborg University, Denmark

R. Prasad et al. (eds.), *Globalization of Mobile and Wireless Communications: Today and in 2020*, Signals and Communication Technology,
DOI 10.1007/978-94-007-0107-6_14, © Springer Science+Business Media B.V. 2011

areas generating medium traffic and areas generating low traffic, etc. Very heavy traffic may be generated due to commercial activity or important event in a particular area. On the basis of traffic pattern, cities can be divided into three layer structure:

(i) Most Dense traffic Area (MDA) like commercial business area – MDA is the area covered by sites having cells with more than 90% peak traffic and traffic density
(ii) Dense Traffic Area (DTA) like industrial area, government offices etc. DTA is the area covered by sites having cells with peak traffic and traffic density between 70 and 90; and
(iii) Normal Traffic Area (NTA) like residential area – Rest of the area, which is not covered as MDA and DTA (as defined above) is categorised as NTA

In a city, system dynamics is continuously changing in terms of traffic patterns and loads, user locations, network congestion, radio channel conditions etc. In MDA, during the busy hours, which can be defined as the period during which maximum traffic load occurs, normally it happens during morning 9–11 a.m. and evening 7–11 p.m. Service operators designed their network in such a way that their network can cope up with traffic load during busy hours. Normally, in India, it has been observed that call drop during busy hours is less than 3%, which can be considered as satisfactory. But still there is a chance that customer face several problems during busy hours like call drop, congestion in network cross talk, etc., due to that he is unable to make calls which thereby affect business of the customer. It has been seen that the congestion in busy hour is location based as congestion in network is more severe in MDA as compared to DTA and NTA. Therefore, major problem of congestion in network is appears in 'Metro' and 'A' class service areas, from where major chunk of revenue is generated.

In addition to above, during some special occasion like new year eve, any major festival, major accidents, natural calamity, etc., call demand as well as call duration increased sharply. The traffic pattern shows that there is a sharp increase in call volume in the given location during the period. The increase in demand, led to a sharp increase in call congestion due to which number of calls drop cases increases. It has also been seen that during any events like cricket match, any live show of entertainment or political party meets etc., spikes in call volumes appears in the cell of that particular location.

These situation can be attributed by three components of a triangle i.e. location, time duration and event. When these three meets, call volume increased abruptly very high, network gets exhausted and unable to handle high volume of traffic and service providers are unable to handle such demand, though it appears for a short duration.

When such phenomena occur, congestion in the network gets increased and many callers could not be able to make any call, which may be very important for them. In view of the above, we need different approach towards our planning to overcome with network congestion during such situation. One method is that we should keep some of the lines free and offer these lines to user during these situations/busy hours for their use 'ON Demand' and 'ON Payment' (Fig. 14.1).

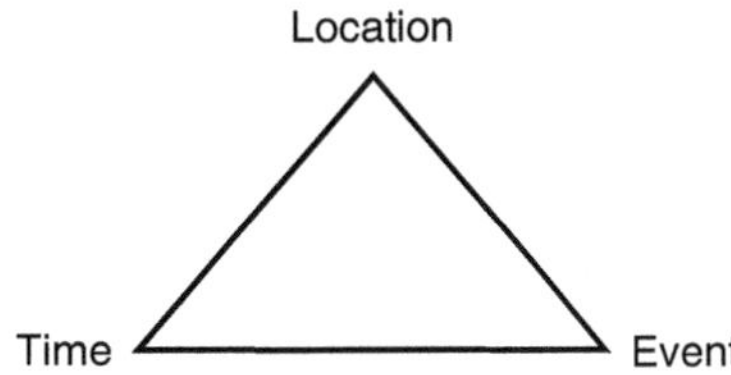

Fig. 14.1 Triangle of three components

The telecom operators all over the world charges fixed amount for a call. This model is known as the 'Fixed Price Model (FPM)'. In a fixed Price model, a user can get access the network only if there are free resources in the cell and user pay a fixed price for using the resource irrespective of demands in the cell. In busy hours or in case of any events, for a certain time, in a certain area, few cells are located and the Radio Resource of these cells will be in highly demanded. Due to this, user could not get preference though he is willing to pay more. In FPM, preference to those users who are willing to pay more cannot be given and at the same time, operators miss the chance to earn more money from the limited RR available in a cell [2].

To overcome this problem, it is suggested that some of the slots in such area can be kept open for those users who are ready to pay more. The allotment of these slots to users can be done through specialized periodically auction. Such type of auction is known as Dynamic Spectrum Auction(DSA) in which a small bunch of spectrum auctioned for short period. DSA increase the efficiency of spectrum by exploiting temporal and spatial variation according to demand pattern [3]. Auctions allow the user to incorporate their needs and demand instantaneously according to market situation and it also provide operators to earn more revenue by announcing reserve price according to market situation. It is pertinent to mention that no any other method of allocation can give appropriate value of spectrum as well as ensures efficient utilization of spectrum.

In this chapter we have discussed pricing of hotlines through dynamic spectrum auction. The chapter has been organized as follows: GSM Telecom scenario in India is given in Section 14.2. We provide a brief of various methods of auctions prevailing at present in Section 14.3. In Section 14.4, Prasad Algorithm will be discussed. In Section 14.5, methodology of auction of slots in cellular network based on the Prasad algorithm will be explained. Implementation issues and Extensions and future actions will be discussed in Sections 14.6 and 14.7, respectively. Conclusion will be discussed in Section 14.8.

14.2 Telecom Scenario of India

Telecom is one of the major contributors of global economy. The global telecom services market is poised to generate US$ 1.4 trillion in 2009 despite the economic slowdown. Worldwide particularly India has registered an impressive growth in the

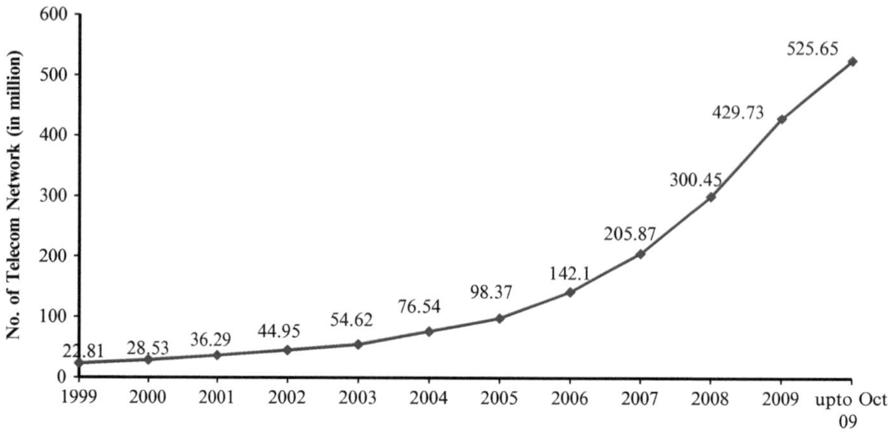

Fig. 14.2 Growth of telecom network in India

telecom sector during the last half decade. The entire Indian telecom sector has opened to unrestricted competition. The opening of the sector has not only led to rapid growth but also helped a great deal towards maximization of consumer benefits. The tariffs have been falling continuously across the board as a result of healthy and unrestricted competition and India today has one of the lowest tariffs in the world. To ensure fair play and benefit to the consumer, the government appointed an independent regulator called Telecommunication Regulatory Authority of India (TRAI) in 1997. A judicial body called TDSAT also exists to adjudicate disputes related to service providers. The role of TRAI is to regulate the telecom business in India [4] (Fig. 14.2).

In 1994, the first step to opening the telecom market to privatization was taken. The first private sector wireline and cellular licenses were issued in 1995. From then on, Indian telecom has seen several milestones crossed and many missteps that provided valuable lessons. The annual growth in the recent years has almost outpaced total combined growth from the beginning of telephony in India as we have crossed over 525.65 million subscriber base with an overall tele-density of about 44.87% by October 2009 [4]. The wireless subscribers are not only much more than the wireline subscribers in the country, but also increasing at a much faster pace.

The growth of wireless services has been phenomenal, with wireless subscribers growing at a compound annual growth rate (CAGR) of 75.7% per annum since 2003. Wireless subscriber base increased from 6.67 million in 2002 to 488.40 million as on October 31, 2009 at a monthly growth rate of 3.36%. Wireless Teledensity stands at 37.87. In April 2008, India overtook US as the second largest wireless market. Indian telecom network is the third largest telecom network and the second largest wireless network in the world after China. In India, about 10 million subscribers are added every month [6]. India contributed about 12% of world mobile subscriber base [7]. 3G services have also started by PSUs. Private operators are yet to start 3G services (Figs. 14.3 and 14.4).

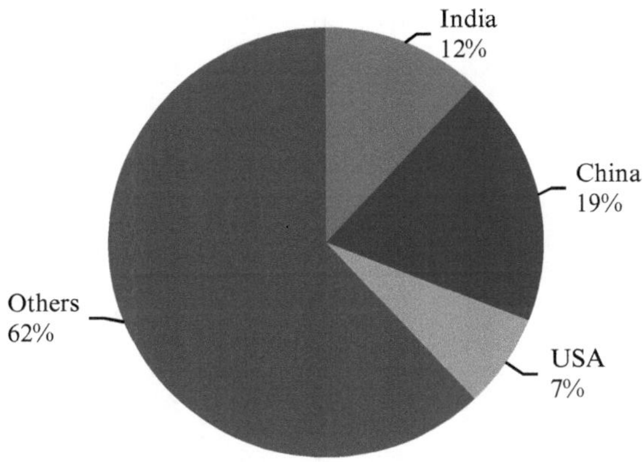

Fig. 14.3 World mobile subscriber base

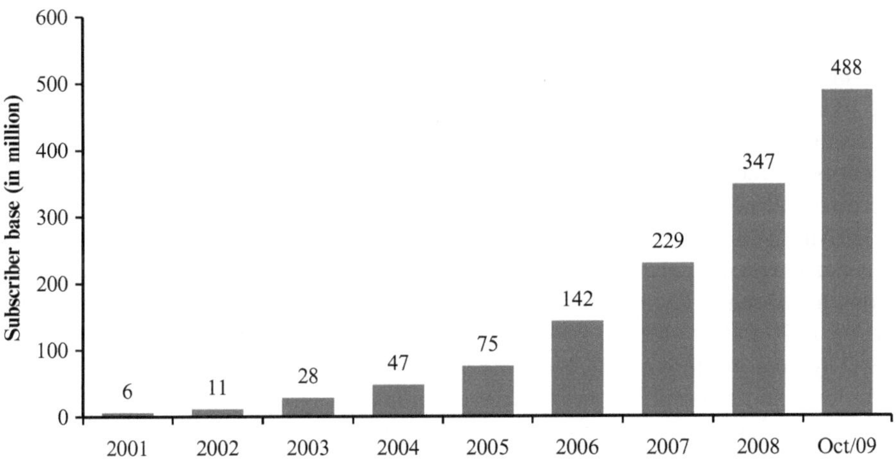

Fig. 14.4 GSM mobile subscriber base growth in India

The total wireline connection in india increased from 5.81 million in 1991–1992 to 37.25 million by October 2009 with overall Wireline teledensity is 3.18 [4]. Upto 2002, wireline subscriber base was increasing but thereafter it is continually declining due to significantly drop in mobile tariffs and increase in coverage of mobile services. The two PSU operators hold 85.46% of the Wireline market share and remaining hold by private operators [7]. Private operators are mainly active in major cities only. There is ample opportunity in wireline segment is available due to very low penetration in rural areas (Fig. 14.5).

Penetration internet/broadband in India is very low though it experiences significantly growth from the last 2–3 years. This growth is being driven by popularity of

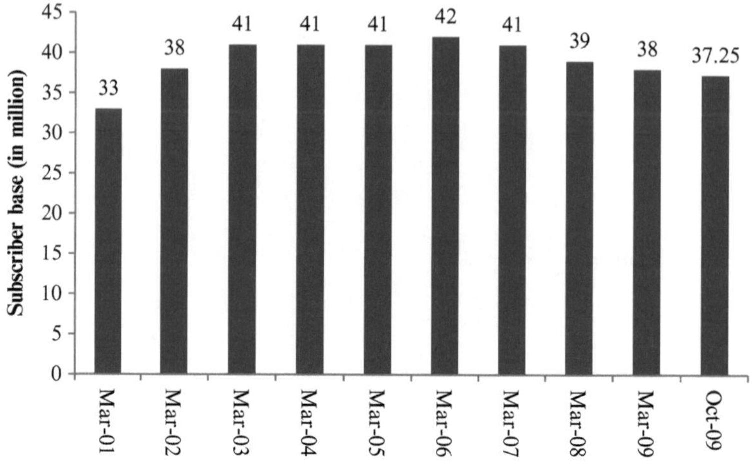

Fig. 14.5 Wireline subscriber base growth in India

broadband and continuous fall in PC prices and tariffs for access. The Government of India has set the target of 20 million, for broadband by 2010 from the total Broadband subscriber base is 7.40 million in October-09 [2]. DSL broadband is most preferred broadband services constitute about 87% of total broadband subscribers followed by cable modem technology contribute 7.4% [7].

Private and Public sector both are providing telecom services. Private sector has been growing very fast. The share of private sector in total telephone connections has steadily increased to 79.16% in March 2009 whereas share of public sector is 20.84% in March 2009. The private sector is mainly active in the wireless segment while wireline accounts for only about 1.17%. Seeing the tele-density level and the rapid growth of the economy, the growth in subscriber base is expected to continue at a rapid pace in the future. It is forecast that the number of subscribers could reach more than 868 million with teledensity more than 70% by 2013 [5].

India is one of the few countries where CDMA and GSM both the technologies are being used for mobile telecom services. GSM is more popular than CDMA as more than 75% subscribers are belongs to GSM service. The country has been divided into 22 service areas for the purpose of telecom services. The service providers have been given license for providing telecom services in these service areas. The service area has further been divided into various categories viz. 'Metro', 'A','B' and 'C' based on the population, traffic pattern and revenue generations capacity. Presently in each service area, more than ten GSM operators and three CDMA operators are providing their services. spectrum allotted to them is varies from 4.4 MHz to 10 MHz (max) while in other countries, no. of operators are not more than four to five and spectrum allotted to them is generally more than 10 MHz.. Due to stiff competition between operators, the tariff rates to hit an all time low. This has allowed mobile services to penetrate into the low income population attracting more potential users. Presently penetration rates in metros are more than 100%. However, there is still immense potential available in telecom circle A, B and C seeing the total tele-density is still less than 50% and population is more than 1.1 billion at present. With mobile number portability by March 2010, the existing tariff wars are expected to escalate further. Mobile operators are continuing to upgrade their technology and have started providing an increasing array of sophisticated services. The content service segment is growing at a rapid pace. It is believed to speed up further in the future and become the most lucrative part of the market.

Presently, India is preparing for auctioning of spectrum for 3G (2.1 GHz band) and BWA (2.3 and 2.5 GHz bands) services with objectives to maximize the revenue from auctions, ensure efficient use of spectrum and avoid hoarding, stimulate competition in the sector and help resolve congestion issues related to second generation ("2G") mobile services [7]. Auctions will be a two-stage simultaneous ascending e-auction, run over the Internet. Bidders will be able to access the Electronic Auction System ("EAS") using standard web browsing software. Anyone who already operating mobile services in India or having experience in 3G services in any other country can participate in the auction.

14.3 Auction Theory

An auction is a mechanism based on a pair of rules, namely the allocation rule that defines which good is allocated to whom and the payment rule that defines the charge of the auction winner(s) [8]. A well designed auction is the most likely method to allocate goods to those who can exploit it efficiently. There are many ways to classify the various auction mechanisms. Thus, auctions are referred to as simple or single-unit if only one good is auctioned and multi-unit if multiple units of a good (e.g. integral units of a link's bandwidth) are to be traded. Moreover, depending on whether bids are made in public or submitted as sealed envelopes, the auction is referred to as open or sealed respectively. Auctions that maximize seller's revenue are referred to as optimal and those that maximize social welfare are referred to as efficient. If the auction is conducted in rounds, then it is called progressive. There are following type of auctions are being used worldwide [9]:

Open auction, which has two parts; Ascending bid auction (also known as English bid auction), where bidder start with a minimum price and incremented it until only one bidder remain, and that bidder win the good by paying the final price. Descending bid auction (also known as Dutch bid auction), where the auctioneer starts with a very high price and decremented it until a bidder call out that he will accept the current price and win the object. Today, a large number of Internet auction sites have been set up to process both consumer–oriented and business–oriented transactions. Most auction sites e.g., eBay, Amazon support a basic bidding strategy through a proxy service for a single-unit auction where ascending bidding continues till a winner emerges.

First Bid sealed auction, in which each bidder submits a single bid without knowing the other bidders and the good, goes to highest bidder after paying the bid price. The main advantage of the first-price sealed-bid auction is that it is simple, and merely involves all interested parties submitting a single bid for the resource that is being auctioned. However, there are a number of problems also. There is no guarantee that the bidders will reveal their true value for the resource. This may give undervalue of resource to be auctioned. Another major problem of a sealed bid auction is that winner bid very high i.e. more than market value, which in result, winner unable to raise money from the resource. In some cases, the winner might actually go bankrupt. This is called the "Winner's Curse".

Uniform price Auction in which all winning bidders have to pay the same price per good as decided by the auctioneer irrespective of bid price. The uniform pricing is simple, and provides "fairness" to bidders and promotes aggressive bidding. However, uniform pricing is suspect to collusion among the bidders and for an unsettled market, it might be more dangerous with respect to the amount of revenue it generates.

Vickrey Auction, which considered as generalization of single unit second-price auction [8]. It has the same procedure as in uniform price auction but differs only in pricing strategy. Here the pricing strategy based upon the price of the next loosing bid (second bid) instead of highest bid price. In the Vickrey scheme what a winner pays for a won good is determined, not by the winning bid, but by the losing bids of its opponents. Thus, for the first won good, a bidder pays the highest losing bid submitted by the other bidders; for the next won good, the bidder pays the second-highest losing bid submitted by others, and so on. Thus, a bidder that has won 'x' good pays the sum of the 'x' highest losing bids submitted by the other bidders if these bids exceed reserve price, otherwise the reserved price [10]. Vickrey auctions is economically efficient and truth revealing as the winning bidder will pay the amount determined by competitors' bids alone and it does not depend upon any action the bidder undertakes. There is no incentive to misrepresent the true value of the resource. However, there are problems associated with Vickery auction also. There is nothing to prevent a general sense of risk aversion in which all bidders would "shade" their bids. In this case, certainly the auctioneer will receive lower revenue. This happened in spectrum auctions in New Zealand in 1990, the winning firm's bid was $100,000, but the second-price was $6 (in New Zealand dollars) and winner got the spectrum almost free of cost [11].

Due to negligible revenue realisation in New Zealand, Vickery method for auctioning of radio spectrum is seldom used in any other country. In practice, the simultaneous ascending auction (SAA) is the most common design in spectrum auctions. A SAA takes place over multiple rounds; in each round, bidders place increasing bids on licences. The auction closes when no new bids are submitted. The licence is awarded to the highest bidder. This auction format was originally developed by the FCC in the early 1990s and was used extensively in the recent 3G auctions in most of the countries. This is very similar to the English auction and is economically efficient in that the highest bidder gets the resource. The ascending nature of the simultaneous ascending auction is supposed to help bidders dynamically bid for various licenses as a bidder can re-evaluate future bids in up and coming rounds by utilizing information from the bids placed by other bidders during a current round. This prevents the winner curse. In addition, the ascending aspect does not allow bidders to submit new bids that are lower than previous bids. This feature allows for more revenue generation, which is obviously beneficial from the government's point of view. When there is a high degree of interest amongst bidders and a low risk of coordination, the SAA would be preferable. Conversely, when interest amongst bidders is lower, the First price sealed-bid auction can stimulate more competition than the SAA. This method is normally used for auction for long term period.

In the next section, we will discuss another method of Auctioning i.e. "Prasad Algorithm", which gives more accurate value as compared prevailing method.

14.4 Prasad Auction Method

This auction method is combination of Uniform price method and Vickrey method. It has the same procedure as uniform price auction but pricing strategy is different. The winner bidders will pay the uniform price, which is sum of highest loosing bid and reserve price (decided by the auctioneer). This methodology has been introduced by Prof. Ramjee Prasad, Director, CTIF, Alborg University, Denmark.

In this method, the auctioneer offers P goods along with reserve price 'r' and incremental price 'x'. All the bidders submit a P-valued bid for goods with reference to reserve price to the auctioneer. The first component means how much the bidder offers to pay for the first good assigned to the bidder. The Pth component means how much the bidder offers to pay for an additional good if $P-1$ goods have already been assigned to this bidder. The auctioneer chooses the P highest bids, which exceeding the reserve price and sequence wise goods are assigned to the winning bidders. The winner bidders will have to pay the amount equal of bid value of $P+1^{th}$ bidder and reserve price 'r'.

Prasad method is economically efficient as the winning bidder will pay the amount not only determined by competitors' bids but also reserve price. Disadvantage of Vickery method can be solved by adding reserve price in the amount to be paid by the winner. The critical point in this method is that the determination of reserve price. Both too much competition and too little competition affects the outcome of the auction. If auctioneer set too high, limited no. of firms would participate in the auction and this would lead to insignificant revenue from the auction. On the other hand, if auctioneer set too low, there is risk that a wrong firm would win the auction and also bidders would not bid aggressive which again not give the actual market price of the good. It is advisable that reserve price should be kept neither too high nor too low.

In communications systems an auction can occur periodically in a local region. Bidders, auctioneer and regulator are required for an auction.

(i) Bidders: Users are the bidders. Users submit the bid as per their preferences and purchasing power for goods offered by the operators. This information considered as bidding strategy (BIS) [12]. The bidders may also need to know some other information like knowledge of price development in the past, nos. of bidders, quality of goods offered in the cell etc. to make their BIS more effective. The best bidding strategy is the traffic aware bidding in which bidder distributes its budget among auction periods proportionally to its traffic demand.

(ii) Auctioneer: Auctioneer should be either an independent agency or a government agency, which would be responsible for conducting auction. Operators will offer their spectrum slot for auction to auctioneer through some transfer right while maintaining the ownership. The operators will also clarify the details of offered spectrum slot like bandwidth, data rate, etc. to auctioneer, which auctioneer would announce to bidders before start of auction. The auctioneer also decides the reserve price for their goods in consultation with regulator. The reserve price should normally be equal to minimum market value of spectrum slot.

(iii) Regulator: The role of regulator in the auction is very important. Regulator defines the auction mechanism i.e. rule of auction. The regulating agency will decide that how many no. of slots can be offered by the operators for an auction in a cell. The reserve price will be decided by the auctioneer's in consultation with regulator. A bidder should have to bid higher price than reserve price. To make the competition healthy, regulatory agency should put cap on the maximum bid price to avoid the winner curse tragedy. Once the bidder reaches at maximum price, he cannot enhance his bid price further. In case of many successful bidders bidding the same price, 'first come, first serve' basis will be applicable.

14.5 Cellular Spectrum Slot Auction

As already stated FPM model could not evaluate the correct price of spectrum when it is in demand. Since spectrum is nothing but air, it is very difficult to assess its value in money terms especially when all the blocks of radio spectrum are not have the same value, for example cellular band is valuable than TV band/Radar Band. Therefore, in such scenario, auction is best way to assess the best price for spectrum blocks as auction is not only to raise the money for spectrum but it also ensures the efficient use of spectrum. Auction of spectrum is not new phenomena; spectrum auction was first occurred in New Zealand and Australia in the late 1980s and early 1990s with very simple auction procedures. In 1994 the United States became the first country to use the more advanced simultaneous ascending auction [12]. After FCC's success, auctions have now been implemented by most of the countries from all over the world. Auction provides transparency in allotment of spectrum as in an auction the rules must be stated in advance to users and also give fair chance to new incumbent to participate in and also promotes competition in the market, which ultimately benefits the end users. There is misconception attached with spectrum auction is that auctions will raise the price of telecommunications services but it is not true as it has been experienced that the cost of winning bids at auctions will have no effect on the price customers pay for spectrum based telecommunications services.

Spectrum can be auctioned in two ways; static and dynamic. Static auction means that to auction the spectrum for long term basis and technology specific like in India 2.1 GHz bands has been proposed for auction for 3G services for providing services for 20 years. So far auction held all over the world are static in nature. In dynamic spectrum auction, spectrum rights are given to winner for short intervals matching the traffic dynamics and make the best use of spectrum. In the proposed case, we have taken dynamic auction as spectrum slot will be given to user for a short period and auction occurs regularly at certain interval. This interval is known as auction period [3]. Auction can be conducted in a periodic manner taking the service providers in that region into account in each telecom service. The bidder after winning spectrum slot, use it in form of services to the end users to make additional profits. The demands for spectrum and the revenue generated from the end users become the driving factors for the bidders to participate in the auction further.

Dynamic Spectrum Auction can be conducted in two ways [13]; sequential and concurrent. In sequential auction, spectrum slots are auctioned one after another. First, bidders submit their sealed bids for first spectrum slot and the winner is determined. Winner of first spectrum slot does not participate for the rest of the auction in that auction period. Remaining bidders then bid for second spectrum slot and so on till all the spectrum slots are auctioned. In concurrent auction, spectrum slots are auctioned concurrently where all the bidders submit their bids together at the beginning of an auction period and bids are arranged in descending order. If 'n' spectrum slot are placed for auction, 'n' nos. of bidders are winner. In the proposed case, concurrent auction has been taken.

There important issues behind any dynamic spectrum auction design are that [10] (i) attracting bidders (enticing bidders by increasing their probability of winning), (ii) preventing collusion thus preventing bidders to control the auction and (iii) maximizing auctioneer's revenue and Ensure efficient use of spectrum and avoid hoarding. The spectrum auction should be designed in such a way that it should take care of points mentioned above.

Consider a simple case in which N slots/channels have been offered within a cell for auction. The sealed bid auction occur every ΔT i.e. a periodically repeated spectrum allocation mechanism process. The bidders submit their bids containing their preferences (how much bandwidth they need for transmission as per their requirements like voice, data or video etc.) along with bid price for these 'N' slots. Auctioneer start the auction, the highest bidder won the first slot; second slot goes to next higher bidder and so on. The allotment of slots goes until all N slots get allotted. Thereafter, the process repeats. After a user win slot, he starts communication and communication lost when either user finishes their communication or certain time limit reached (decided by the regulator). During this communication, he can participate in further auction. Price to be paid by the winner can be decided using Prasad Algorithm as described in Section 14.3. Single sealed bid auction is most appropriate for such case as auction occurs at very short interval.

The Auction should be conducted on an electronic auction system (EAS) over the public Internet so that bidders will not need any specialist hardware or software and can access the system using a standard Internet browser. The auctioneer's computer will perform functions mainly, receiving bids from bidders, sorting the bids, allocates the appropriate spectrum slot and receiving payment from the winner. The bidder's computer will perform the following function, sense the status, generates the bid considering its paying capacity and requirement, submit bid to auctioneer and after winning authorise the slot for appropriate work.

14.5.1 Open Incremental Method

In this method bidders start with a minimum price and incremented it until only 'N' bidders remain, and that all the winner bidders will pay the price equal to highest lost bid price plus reserve price.

Consider an example, three slots in a cell has been offered for auction. Suppose six bids are participating in the auction. The reserve price and incremental price is '3' and '1' respectively and the maximum price any bidder can offer is 15. The bidder would start with higher than reserve price and incremented their bid value until three bidders will remain. These three bidders will be winner. They would pay the bid value of highest looser bidder plus incremental value. The price will be decided as given in Table 14.1.

This method cannot be considered as reliable as it consumes a lot of time and also there is chance that bidders need not to be bid aggressively when nos. of bidders are limited and they know the history of each other. In this situation, auctioneer would not get market price. Since in spectrum slot auction, auction period is very short, this method is not appropriate. It would be better to go for auction through 'single sealed bid auction method'.

Table 14.1 Open incremental method

→Bidder ↓Bid Price	A	B	C	D	E	F	Result
First round	3	5	7	9	6	4	C, D, E winner
Second round	5	9	7	13	10	8	(D, E, F) C quit
Third round	8	11	–	14	11	10	(D, B, E) A quit
Fourth round	–	12	–	15	13	11	(B, D, E) F quit

The winners are B, D, and E. These winners would pay 14 each i.e. 11 (highest lost bid value) +3 (reserve price). Therefore, the auctioneer would get 42 for three slots

14.5.2 Single Sealed Bid Auction Method

In this method bidders submit their bids for 'N' slots. The auctioneer chooses the highest 'N' bids and declares them winner. The all the winner bidders will pay the price equal to highest lost bid price plus incremental/reserve price (some fixed price, decided by the regulator).

Consider the case for 3 slots has been offered for auction. A bid (p, q, r) has been received. (p, q, r) means that bidder offer to pay p if he end up with a total of one slot; offer p+q for a total of 2 slots; and offer p+q+r for all three slots [8].

Let three bids have been received; A=(8, 7, 4) and B=(6, 4, 3), C=(10, 5, 2). The reserve price/fixed price is 3. The third bid of 'C' cannot be accepted as it is lower than the reserve price. The assignment proceeds as follows: one slot is assigned to bidder C (10 is the highest bid, over all components), the second slot is assigned to bidder A (second highest component bid is 8), the last slot is assigned to bidder A (the third highest bid) again. Therefore, A would get two slots and 'C' would get one slot. This method has been described in Table 14.2.

In case of any big event, where some of the bidders are desperately want slots, they may quote max price so that in any case they would get at least one slot. In above example, suppose 'A' want one at any cost, he will submit the bid (*, 7, 4). Here * signifies that whatever bid price come, bidder is ready to pay. In this case, first slot would be allotted to 'A' instead of 'C', second slot would be allotted to 'C' and third slot would be given to 'A' again. This has been represented in tabular form in Table 14.3.

The price calculation in above two methods of auction has been based on Prasad algorithm. The amount paid by the winner can also be calculated using Vickrey method. It can easily be observed that the averaged total gain is always smaller in Vickrey auction as compare to Prasad Algorithm. Therefore, the Prasad Algorithm is looking more accurate as it provides benefit to the auctioneer as compare to other two methods.

A situation may come when no. of slots is more than the no. of bidders, in this case auction will not take place and bidders would pay the floor price only.

Table 14.2 Single sealed bid method

Bid Value	Bidder	Result
10	C	Winner of first slot
8	A	Winner of second slot
7	A	Winner of third slot
6	B	Loser (highest)
5	C	Loser
4	A	Loser
4	A	Loser
3	B	Loser

Amount to be paid by the each winners = 6 + 3(reserve price) = 9
The auctioneer would get = 27 (9 × 3)

Table 14.3 Singled sealed with star

Bid Value	Bidder	Result
*	A	Winner of first slot
10	C	Winner of second slot
7	A	Winner of third slot
6	B	Loser (highest)
5	C	Loser
4	A	Loser
4	A	Loser
3	B	Loser

Amount to be paid by the each winners = 6 + 3(reserve price) = 9
The auctioneer would get = 27 (9 × 3)

Normally this situation will come when reserve price would be kept at higher side, which reduces the no. of bidders. Therefore, the reserve price should be kept at such level where sufficient no. of bidders can participate in the auction and at the same it should be more than fixed cost spend by the operators for maintaining slot so that operator can get their operating cost as well as match minimum market value of spectrum.

14.6 Implementation Issues

The concept of DSA is new one. So far only paper work has been done. We have discussed some of implementation issues. Some other implementation issues will come on surface when such auction would be practically realised.

A first major challenge is that there is chance that bidders may collude to avoid bidding up prices. This will not give the market value of spectrum slot. We can see the case of New Zealand, the winning firm's bid was $100,000, but the second-price was $6 (in New Zealand dollars) and winner got the spectrum almost free of cost in an auction of spectrum auctions in 1990 [11]. This can be avoided by proper designing of auction. In Prasad methodology, this has been taken care of by introducing reserve price, any bid below the reserve price cannot be considered and also winner will pay amount equal to second bid vale and reserve price, therefore, collusion may not affect very much on the revenue. Another way to avoid collusion is adopt for sealed bid auction instead of most famous ascending auction. In a sealed-bid auction, each bidder simultaneously makes a single "best and final" offer. As a result, firms are unable to retaliate against bidders who fail to cooperate with them, so collusion is much harder than in an ascending auction.

A situation may also arise in which a bidder, to punish other bidders, may submit bids for all spectrum slots with a very high value [9]. Under this situation, no other bidder will win spectrum slot, only one bidder will win all the spectrum slots by showing money power. This is another type of collusion. This collusion will have more impact as such things will ultimately destabilise the market and whole thing will get collapsed. There should be a cap on nos. of spectrum slots can be allotted to one bidder in an auction period.

Determination of reserve price is major issue for any auction design. As already stated high reserve price will attract less bidders and low reserve price will attract more bidders. In both cases, auctioneer would not get market value of spectrum. Therefore, reserve price should be kept neither not too high nor not too low. Reserve price should be in line with market value so that auctioneer would at least get minimum market price and it help to avoid collusion.

The size of auction cycle will also impact the performance of the auction [3]. Very small auction period will deter the bidders to participate in the auction and enhance the complication in computation. Large auction period will provide low revenue and fail to respond to demands thereby efficient use of spectrum cannot be guaranteed. Auction period will also depend upon event. Auction period will be different for different scenario for example, traffic volume during a cricket/football match cannot be similar to any festival/new year eve etc. and also traffic volume during busy hour in various cities are different. For determining auction period, it is suggested that first conduct few auction with different auction period thereafter taking into account experiences of these auctions, an optimum auction period can be setup for any occasion.

In such auction, process shall be computerised. As already stated above about the function of the auctioneer's computer and bidder's computer, an algorithm for auction should be developed in such a way that a uniformity and unison can be maintained in these activities. This is major challenge on the technical side otherwise auction would be considered as failure.

Another significant challenge is managing concurrent auction in adjacent cells. In the main cell, all the bidders may not be able to win a spectrum slot. For them, it would be better to conduct similar auction in adjacent cells concurrently also so that losers can participate. This would not only help to increase the revenue but also enhance the efficiency of utilisation of spectrum in adjacent cell also. Simultaneously conducting auction in main cell and adjacent cells is implementation issue. For this, Cognitive radio enabled system can be utilised. The cognability properties of CR technology can be used to manage the auction in adjacent cell simultaneously.

14.7 Extensions and Future Auctions

In the above two cases, we have considered only one service provider. There is case when more than one service providers may offer some of the slots in given area for auction. Now bidders have the choice, they may bid for all service providers or choose some of them according to quality of services offered by the service providers, reserve price and also their buying capacity. The Prasad Algorithm can be used to calculate the price to be paid by the winners. An algorithm is needed to be developed to consider more than one service providers. It would also be appropriate to give ratings to radio resource submitted for auction by the service providers according to their performance so that user can easily make their choice. We have focused on simple auction scenario. With repeated and simultaneous auction, the auction method can be engaged in more sophisticated bidding system adding additional element to make the auction process efficient.

As we are aware that cognitive radio technology is very fast emerging technology. Definition of CR [14] can be given as a radio system employing technology that allows the system to obtain knowledge of its operational and geographical environment, established policies and its internal state; to dynamically and autonomously adjust its operational parameters and protocols according to its obtained knowledge in order to achieve predefined objectives; and to learn from the results obtained.

Bidder can use experiences gained in the past auction in order to estimate the behavior of other bidders, quality of services offered by the various providers, his purchasing power and reserve price etc. to make its bid strategy more effective to win the spectrum slot. This requires the ability to learn about the environment. Therefore, gaining cognition about the environment and acting accordingly, are part of the CR technology. The cognability properties of CR technology can be used to make bidding more efficient. The ability of CR can be mainly categorized in three functions [2] i.e. gaining information from the environment, learn from the information and act accordingly. CR properties will also be utilized to optimize the revenue and efficiency of spectrum from the auctioneer side and from bidders side, his preferences, purchase power and his actions like risk neutral, risk averse, risk encouraged depending upon the situation can be categorized and also detail of demand in any cell, reserve price and outcome of previous auction and current auction can be easily maintained and this information can be utilized by the bidders to make their bid more effective and this ultimately enhance the success of auction system. In case, a user/bidder could not get succeed to acquire a spectrum slot in main cell. By using CR properties, he can gather information in the adjacent cell, where demand is certainly lower than the main cell, user can participate in auction in adjacent cell also.

Spectrum auction for short period is new trend in telecom sector. Nowhere, it has so far been exercised. Whether such approach will be beneficial for enhancing the spectrum utilization or make a new mesh in the existing spectrum management, which has already been criticized from every corner. Such method certainly increases the efficiency of spectrum utilization as in future wireless based service will govern the market and spectrum is the essential raw material. Therefore, spectrum in cannot be allocated abundantly for service. Presently there are several services, which depend upon the spectrum especially for data exchange between their various offices. During busy hours, they service would get disturb due non availability of spectrum, which in turn affect their business and also during some special occasions, there are so many spectrum based services like electronic media channel, newspapers, social networking services etc. which want to exploit the event to maximize their revenue. These are the potential buyer. In India, more than ten service providers are providing cellular services in a telecom service area and spectrum has been allotted to them varies from 4.4 MHz to 10 MHz max and seeing the teledensity, which is far low comparison to developed countries, war to consolidate market share among the service providers is going to be increased and certainly spectrum is going to play a major role in this war. Auction of spectrum slot would certainly give a new and efficient method to earn revenue in the leading telecom market such as India.

14.8 Conclusions

Network congestion during busy hours and on special occasion is very common. Auctioning of spectrum slots during busy hours is one of the solutions to overcome with call congestion. We have presented a simple auction scenario. Using DSA, efficiency of spectrum can be enhanced and also more revenue could be generated as compared to traditional method of spectrum auction. This is the demand of future as coming era is the wireless era and any wireless based service can survive only adequate spectrum is available. Prasad Algorithm for auction of slots in mobile service is more convincing as compared to other existing methods of calculation of revenue. This auction theory is based on the highest loser bidder bid value and reserve price, therefore, it provides more prices to the service providers and at the same time it also ensures that bidder not to bid erratically.

References

1. www.mtctelecom.com
2. Clemens Kloeck, Holger Jaekel, Friedrich K. Jondral, Multi-agent wireless system for dynamic and local combined pricing, allocation and billing. J. Commun. **1**(1), 48–59 (2006)
3. Xia Zhou, Shravan Mettu, Heather Zheng, Elizabeth M. Belding, Traffic-driven dynamic spectrum auctions. Sensor, Mesh and Ad Hoc Communications and Networks Workshops, 2008. SECON Workshops '08. 5th IEEE Annual Communications Society Conference, San Francisco, CA
4. www.trai.org.in. Press Release No 78/2009 of Telecom regulatory Authority of India (TRAI)
5. Indian Mobile Market Dynamics and Forecast, ROA Group Research Report No. 090801, www.researchonasia.com
6. www.dot.gov.in. Annual Report 2008–09, Auction of 3G & BWA Spectrum-Information Memorandum and www.wpc.gov.in
7. The Indian Telecom Success Story, www.dot.gov.in
8. Manos Dramitinos, George D. Stamoulis, Costas Courcoubetis, Auction-based resource allocation in UMTS high speed downlink packet access (HSDPA). 0-7803-8901-8/05/(c) 2005 IEEE
9. V. Krishna, Auction theory. (Academic, April 2002), ISBN 0-12426297-X
10. Virgilio Rodriguez, Klaus Moessner, Rahim Tafazolli, Market driven dynamic spectrum allocation over space and time among radio-access networks: DVB-T and B3G CDMA with heterogeneous terminals. Mobile Netw. Appl. **11**, 847–860 (2006). doi 10.1007/s11036-006-0053-2
11. G. Anandalingam, Robert H. Smith, School of Business, University of Maryland College Park, MD 20742. Vickrey Auctions for Spectrum Allocation in Developing Countries. Computers & Sociaty. arXiv:cs/0109065v1 [cs.CY]
12. Ian Munro, Charles River Associates, Auctions as a spectrum management tool. Presentation to the ITU Radiocommunication Bureau Seminar Geneva, Switzerland, 8 Nov 2000

13. Shamik Sengupta, Mainak Chatterjee, Designing auction mechanisms for dynamic spectrum access. Mobile Netw. Appl. **13**, 498–515 (2008). doi 10.1007/s11036-008-0081-1. Springer Science+Business Media, LLC 2008
14. Chairman's report of Working Part 1B of Radiocommunication Sector (ITU-R) of International telecom Union (www.itu.int)

K. Sridhara an Indian Telecom Service officer who graduated in 1971, did his B.E. (Hons.) in Electronics and Communication Engineering in 1970 and Post-graduation in Engineering in Computer Technology in 1972 from College of Engineering, Guindy, Chennai with distinction. Later, he did his M.B.A. (Finance) from FMS, Delhi, L.L.B. from Delhi University and CFA. During his career in Indian telecom spanning about 35 years, Mr. Sridhara has held various positions in DoT covering research and development, standardization, network planning, operations and maintenance, office automation and computerization of telecom network, costing, setting up of production facility, finance advice, policy planning, licensing, spectrum management, induction of new technology and regulatory affairs and his entire career is marked by many milestones. In his responsibility as Member (Technology), Department of Telecom, the apex body of Government of India responsible for framing of telecom policy, he was associated with licensing of telecom services, spectrum management, research and development, standardization, interface with international bodies, policy regarding induction of new technologies, formulation of network security policy and co-ordination with service providers on security matters, regulatory affairs, etc. Mr. Sridhara is widely travelled. He has represented Government of India at various international forums and has made very effective contributions. He is at present Director General, Global ICT Standardization Forum of India (GIFSI). Currently, he is keenly involved in introduction of New Generation technologies in Indian telecom network and in framing Wi-MAX policy.

P.S.M. Tripathi an Indian Engineering Services officer of 1998 batch, did his B.E. in Electronics from M.M.M. Engineering College, Gorakhpur, India. He is having technical experience of 8 years in the field of Radio Communications/Radio Spectrum Management. He is Engineer in Wireless Planning and Coordination Wing of Department of Telecommunications (DOT), Ministry of Communications and IT, Government of India. In DOT, he is associated with spectrum management activities, including in spectrum planning and engineering and policy regarding regulatory affairs for new technologies and related research and development activities and ITU-R related matter. He has also been associated with implementation of a very prestigious World Bank Project on National Radio Spectrum Management and Monitoring System (NRSMMS). This project includes automation of Spectrum Management processes and design, supply, installation/commissioning of HF/VHF/UHF fixed monitoring stations; V/UHF mobile monitoring stations; and LAN/WAN communications network, etc. His area of interest includes Radio Regulatory affairs for new technologies and cognitive radio. He was also worked as research fellow in department of electronics, University of Tor Vergata, Rome, Italy.

Ambuj Kumar did his B.E. from Birla Institute of Technology, Ranchi (India). Currently he is working as Research Associate at CTIF, Aalborg, Denmark on the topic of Cognitive Radio. The research studies include comparison of Routing in CR with that of IP networks and MANET. He has worked for a period of 2 months in the year 1999 at the Institute of High Frequency Technology, Technical University (RWTH), Aachen (Germany). Joined the Institute's research study Group on 'Indoor Mobile Communications' and contributed in carrying out a series of radio wave propagation measurements. He had also worked at senior position at Alcatel- Lucent, M/S Vodafone and Lucent Technologies Hindustan Pvt. Ltd. He has more than six research publications in the areas of indoor mobile communications, Cognitive radio

Dr. Ashok Chandra an Indian Engineering Services officer of 1976 batch, Wireless Adviser to the Government of India, did his Post graduate in Electronics and Ph.D. in Electronics and Doctorate of Science (D.Sc.) in Radio Mobile Communications. He has also worked as a Guest Scientist at the Institute of High Frequency Technology, Technical University (RWTH), Aachen, Germany during 1995 and 1999. During 2002, worked as a Guest Scientist at Bremen University, Bremen (Germany), where he undertook a series of research studies in the area of radio mobile communications. Dr. Ashok Chandra is having Technical Experience of 30 years in the field of Radio Communications/Radio Spectrum Management including about 7 years of experience dealing with Technical Education matters of Indian Institutes of Technology, Indian Institute of Science etc. particularly their various research projects also in the areas of telematics/radio communications and delivered over 25 research presentations at International Conferences in the areas of EMI, Radio Propagation etc. He has visited various technical Institutions and Universities namely Technical University of Aachen, Germany, Aalborg University, Denmark, Bremen University, Germany, and University of Lisbon, Portugal etc. and took several lectures in the area of radio mobile communications at Bremen University and Aalborg University. He has chaired various Technical Sessions at the International Conferences on Wireless Communications and he is on the panel of Reviewers of an International Journal-Wireless Personal Communications. In his current responsibility as Wireless Adviser to the Government of India, he is associated with spectrum management activities, including in spectrum planning and engineering, frequency assignment, frequency coordination, spectrum monitoring, policy regarding regulatory affairs for new technologies and related research and development activities, etc. He was also associated with implementation of a very prestigious World Bank Project on National Radio Spectrum Management and Monitoring System (NRSMMS). This project includes automation of Spectrum Management processes and design, supply, installation/commissioning of HF/VHF/UHF fixed monitoring stations; V/UHF mobile monitoring stations; and LAN/WAN communications network, etc. Recently, he has been elected Vice-Chairman, Study Group 5 of ITU-Radio Sector. His area of interest includes Radio Regulatory affairs for new technologies, Radio mobile communications and cognitive radio.

Chapter 15
Wavelet Radio: A Future Vision for Wireless Communications

Homayoun Nikookar

15.1 Introduction

The convergence of information, multimedia, entertainment and wireless communications has raised hopes of realizing the vision of ubiquitous communication. To actualize this there is a challenge of developing technologies and architectures capable of handling large volumes of data under severe constraints of interference and resources such as power and bandwidth. The design of an intelligent wireless communication system that estimates the channel and adaptively reconfigures to maximize resource utilization and interference mitigation is therefore highly desirable.

Wavelet or Wavelet transform is a method to decompose a signal of interest into a set of basis waveforms which provides a way to analyze the signal by examining the coefficients (or weights) of wavelets. This method is used in various applications and is finding tremendous popularity among technologists, engineers and mathematicians alike. In most of the applications, the power of the transform comes from the fact that the basis functions of the transform are localized in time (or space) and frequency, and have different resolutions in these domains. Different resolutions often correspond to the natural behaviour of the process one wants to analyze, hence the power of the transform. Wavelets provide promising potential applications in wireless communications, ranging from source and channel coding to transceiver design and from wireless physical channel to network and higher layers. The major property of wavelets for these applications is their ability and flexibility to characterize signals with adaptive time-frequency resolution. Wavelets are thus uniquely qualified to address the objectives of intelligent communication design. With their strong advantage of being generic schemes their characteristics can be widely customized to fulfil the various requirements and constraints of intelligent wireless communication systems.

H. Nikookar (✉)
Delft University of Technology, The Netherlands
e-mail: H.Nikookar@irctr.tudelft.nl

R. Prasad et al. (eds.), *Globalization of Mobile and Wireless Communications: Today and in 2020*, Signals and Communication Technology,
DOI 10.1007/978-94-007-0107-6_15,

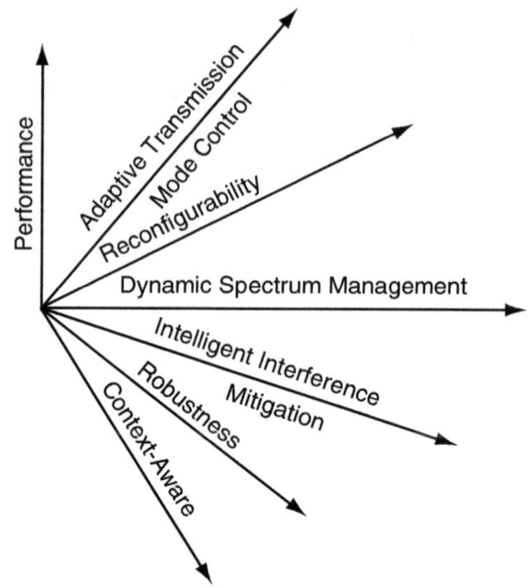

Fig. 15.1 Goals of the wavelet radio

In this chapter we propose to utilize the wavelet technology as the choice for smart and resource aware radio system. Titled "Wavelet Radio," the major thrust of the this chapter is to explain the importance of wavelets in the future wireless radio systems which opportunistically and efficiently use available resources to guarantee highly reliable communication whenever and wherever needed. This is achieved by implementing adaptive transmission bandwidths, data rates, power adaptation and modulation schemes taking into cognizance transceiver parameters such as radio frequency spectrum, transmission power, user behaviour, as well as channel state and network state information. The keystones of the proposed system are – adaptation, intelligence, context awareness, reconfigurability and robustness (Fig. 15.1).

15.2 Wavelets and Their Applications

The word wavelet derives from the French researcher, Jean Morlet, who used the French word ondelette - meaning a "small wave" [1]. A little later it was transformed into English by translating "onde" into "wave" to thus arrive at the name wavelets. As the name suggests, wavelets are small waveforms with a set oscillatory structure that is non-zero for a limited period of time (or space) with additional mathematical properties. The wavelet transform is a multi-resolution analysis mechanism where an input signal is decomposed into different frequency components, and then each component is studied with resolutions matched to its scales. The Fourier transform also decomposes signals into elementary waveforms, but these basis functions are sines and cosines. Thus, when one wants to analyze the local properties of the input signal, such as edges or transients, the Fourier transform is not an efficient analysis tool. By contrast the wavelet transforms which use irregularly shaped wavelets offer better tools to represent sharp changes and local features. A comprehensive study of the subject can be found in references [1–6].

The wavelet transform is used in various applications and is finding tremendous popularity among technologists, engineers and mathematicians alike. In most of the applications, the power of the transform comes from the fact that the basis functions of the transform are localized in time (or space) and frequency, and have different resolutions in these domains. Different resolutions often correspond to the natural behaviour of the process one wants to analyze, hence the power of the transform. These properties make wavelets and wavelet transforms natural choices in fields as diverse as image synthesis, data compression, computer graphics and animation, human vision, radar, optics, astronomy, acoustics, seismology, nuclear engineering, biomedical engineering, magnetic resonance imaging, music, fractals, turbulence, and pure mathematics. An exhaustive survey of the application of wavelets in these fields can be found in [7] and [8].

Recently the wavelet transform has also been proposed as a possible analysis scheme to design sophisticated digital wireless communication systems, with advantages such as flexibility of the transform, lower sensitivity to channel distortion and interference and better utilization of spectrum [9–11]. Wavelets have found beneficial applicability in various aspects of wireless communication systems design including channel modelling, design of transceivers, data representation, data compression, source and channel coding, interference mitigation, signal de-noising, energy efficient networking. Fig. 15.2 gives a graphical representation

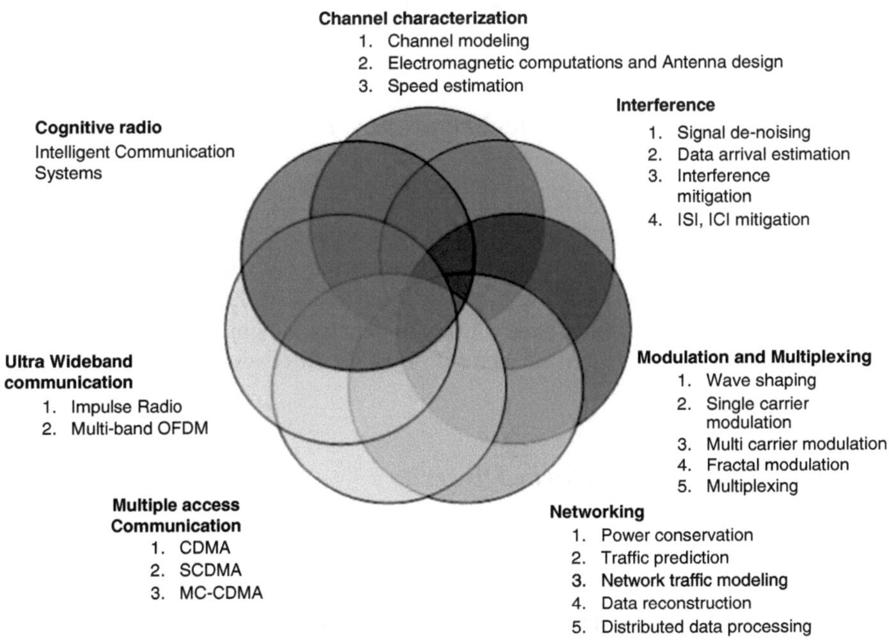

Fig. 15.2 The spectrum of wavelet applications for wireless communication

of some of the facets of wireless communication where wavelets have found place. A thorough review of the applications of wavelets in the field of wireless communication can be found in [12].

15.3 Motivation for Using Wavelets for Design of Intelligent Wireless Devices

There are several advantages of using wavelets for wireless communication systems. Here we mention major motivations of using wavelets in wireless radio transmission:

15.3.1 Semi-arbitrary Division of the Signal Space and Multi-rate Systems

Wavelet transform can create subcarriers of different bandwidth and symbol length. Since each subcarrier has the same time-frequency plane area, an increase (or decrease) of bandwidth is bound to a decrease (or increase) of subcarrier symbol length. Such characteristics of the wavelets can be exploited to create a multirate system. From a communication perspective, such a feature is favourable for systems that must support multiple data streams with different transport delay requirements.

15.3.2 Flexibility with Time-Frequency Tiling

Another advantage of wavelets lies in their ability to arrange the time-frequency tiling in a manner that minimizes the channel disturbances. By flexibly aligning the time-frequency tiling, the effect of noise and interference on the signal can be minimized. Wavelet based systems are capable of overcoming known channel disturbances at the transmitter, rather than waiting to deal with them at the receiver. Thus, they can enhance the quality of service (QoS) of wireless systems.

15.3.3 Signal or Waveform Diversity

Wavelets give a new dimension – “Waveform diversity,” to the physical diversities currently exploited, namely, space, frequency and time-diversity. Signal diversity which is similar to spread spectrum systems, could be exploited in a cellular communication system, where adjacent cells can be designated different wavelets in order to minimize inter-cell interference. Another example is the Ultra Wideband

(UWB) communication system where a very large band with reduced interference can be shared by users by clever use of transmitting pulse wave shapes.

15.3.4 Sensitivity to Channel Effects

The performance of a modulation scheme depends on the set of waveforms that the carriers use. The wavelet scheme therefore holds the promise of reducing the sensitivity of the system to harmful channel effects like Inter-symbol interference (ISI) and Inter-carrier interference (ICI) as well as other interferences on the system.

15.3.5 Flexibility with Sub-carriers

The derivation of wavelets is directly related to the iterative nature of the wavelet transform. The wavelet transform allows for a configurable transform size and hence a configurable number of carriers. This facility can be used, for instance, to reconfigure a transceiver according to a given communication protocol; the transform size could be selected according to the channel impulse response characteristics, computational complexity or link quality.

15.3.6 Power Conservation

Wavelet-based algorithms have long been used for data compression. In the context of mobile wireless devices, which are mostly energy starved, this has an added significance. By compressing the data, a reduced volume of data is transmitted so that the communication power needed for transmission is reduced.

15.4 Importance of Wavelet Radio

Wireless interoperable communication networks are now a reality. They have spawned many new and exciting applications like mobile entertainment, mobile internet access, healthcare and medical monitoring services, data sensing in sensor networks, smart homes, combat radios, disaster management, automated highways and factories. With each passing day newer and newer such services are being launched even while existing services continue to flourish. Demand for wireless services is therefore likely to continue so for the foreseeable future. However, with increasing popularity of the wireless services the demands on prime resources like battery power and radio spectrum are put to great test. For example, currently most spectrum has been allocated, and it is becoming increasingly difficult to find frequency bands that can be made available either for new services or to expand existing ones. Even while available frequency bands appear to be fully occupied, a FCC study conducted in 2002 revealed that much of the available spectrum lies fallow most of the time (20% or less of the spectrum is used) and that spectrum congestions are more due to the sub-optimal use of spectrum than to the lack of free spectrum. Furthermore, wireless systems frequently have to contend with disruption of services due to interference. Interferences can be due to many reasons – unintentional, intentional (Jamming), overlap of symbols due to temporal spreading (Inter-symbol interference or ISI), adjacent channel interference (Inter-channel interference or ICI), multiple access interference posing many challenges in system design and construction.

Thus in a wireless environment the system requirements, network capabilities, and device capabilities have enormous variations giving rise to significant network design challenges. The design of an intelligent communication system that estimates the channel and adaptively reconfigures to maximize resource utilization and interference mitigation is therefore highly desirable. Current developments in wireless communications are therefore towards developing novel signal transmission techniques that allow for significant increases in wireless capacity with good interference mitigation capabilities. With such strict demands one is therefore entitled to wonder about the possible improvements that advanced schemes such as wavelet based systems could bring compared to the conventional configurations.

15.5 Objectives of the Wavelet Radio

The future wavelet wireless systems aim intelligent radio transmission system based on wavelet technology to handle the multi-faceted challenge of maximizing resource utilization and interference mitigation. For successful implementation of wavelet radio for wireless communications the following steps have to be taken:

- Developing wavelet based wireless communication systems with capabilities of adaptive transmission bandwidths, data rates, modulation, power according to channel situation.
- Replacing Fourier bases with wavelets for achieving higher performances.
- Observing and gathering the intelligence required for wavelet radio.
- Designing suitable wavelets.
    - For maximizing spectrum utilization.
    - For minimization of interference to primary users of the spectrum.
    - For opportunistic operation in spectrum holes.
- Mitigating interference with wavelet radio.
- Building multiple-access with wavelet radio.
- Establishing multirate configurations for high speed data transmission.

- Establishing power control protocols for adaptive power loading based on channel state information.
- Designing wavelet radio for specific applications such as radio positioning or wireless ad-hoc networking.
- Addressing compatibility of operation with conventional and existing radios.
- Evaluating the performance of the system.
- Conducting feasibility studies on the applicability of wavelet radio for 3G and beyond.
- Developing a proof of concept to demonstrate the feasibility of the wavelet based system.

These challenges are pictorially represented in Fig. 15.3.

15.6 Description of Wavelet Radio System

The system model of the proposed system is illustrated in Fig. 15.4. The major blocks in the transmitted include a spectrum and channel estimator, a transmission waveform shaper, a modulator. And at the receiver the main blocks include decorrelator, integrator and detector.

The three fundamental tasks of the system are

15.6.1 Spectrum Estimation – Gauge the Radio Spectrum Scenario and Perform Radio Scene Analysis

This block gauges the channel and performs a radio scene analysis. The tasks include estimation of the interference and detection of spectrum holes. To estimate the electromagnetic spectrum methods including periodogram, autoregression and wavelet based approach are commonly used. In order to quantify and manage interference a new metric called interference temperature is defined. The interference temperature for a desired band of transmission is defined as the acceptable level of interference that can be tolerated by the system [13]. From this estimate, through the use of adaptive thresholds and notches, sub-bands containing the interference are effectively cancelled.

15.6.2 Channel State Estimation and Predictive Modelling

Accurate and timely channel state information (CSI) at the transmitter is an important component of the system. Channel information is important for accurate power control, prediction of channel capacity, scheduling and detection of data. Existing wireless communication channel models are based on statistical impulse response

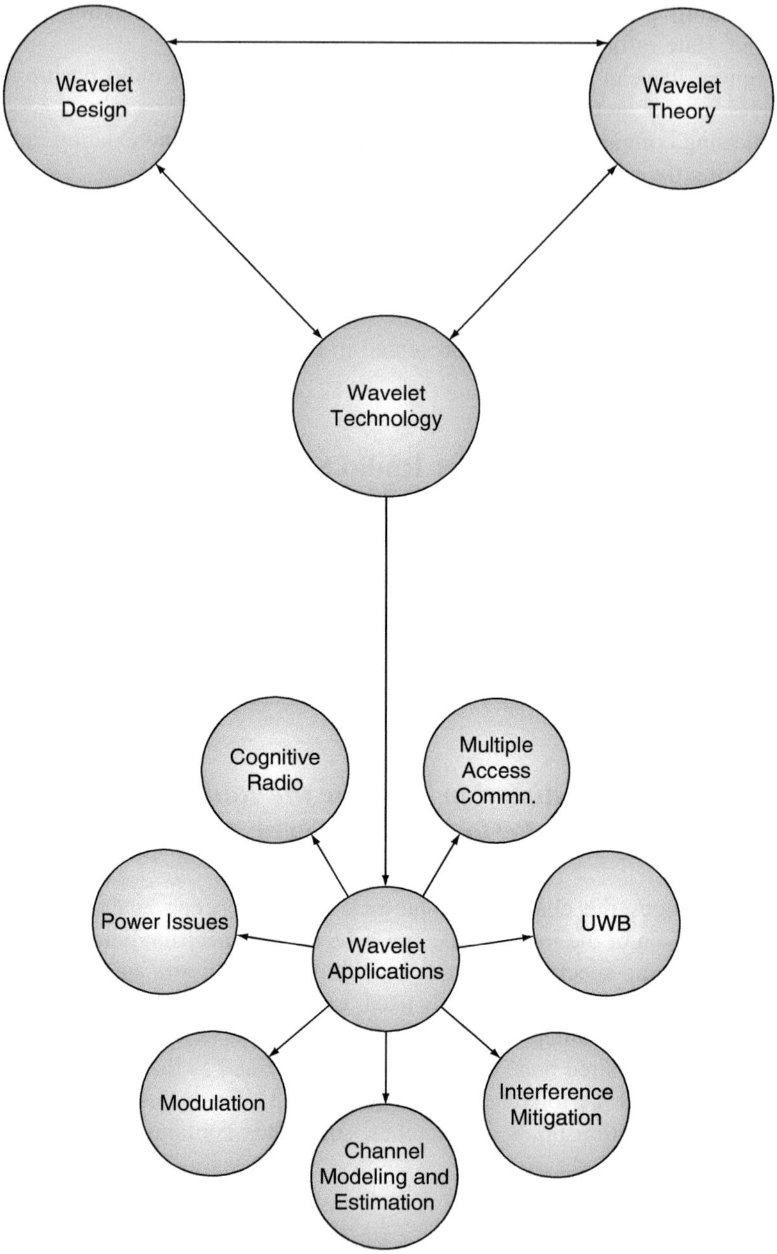

Fig. 15.3 Objectives of wavelet radio

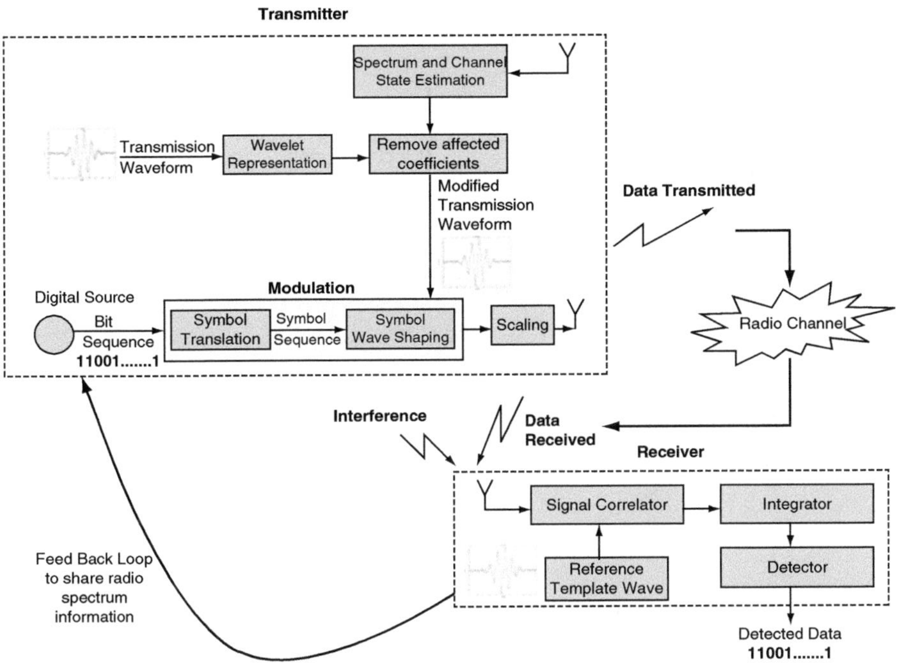

Fig. 15.4 System model of the wavelet radio. Transmitter: The major blocks include the spectrum and channel state estimator, adaptive pulse shaper (Wavelet transform and transmission waveform shaping) and modulator. Receiver: The major blocks include the Signal decorrelator, integrator and detector

models derived from empirical results. While these models perform adequately for time invariant channels, they fail to accurately map time varying channels. Due to their inherent joint time-frequency localization property and their ability to accurately characterize the time-varying nature of the estimation problem, the wavelets offer various advantages for channel modelling. Some of them are: accurate characterization of time-varying as well as frequency selective multipath fading channels, fast convergence of estimate, representation of channel with fewer number of coefficients, small output error, and clear interpretation of modelling error.

15.6.3 System Reconfiguration

Based on the radio spectrum scenario and the channel state information, dynamically adapt the transmission parameters such as power, spectrum, modulation scheme and data rate [14–24].

At the transmitting end, first a data source generates an arbitrary stream of data derived from the source alphabet. The spectrum estimator senses the spectrum and detects the presence of interference regions and spectrum holes. Based on the

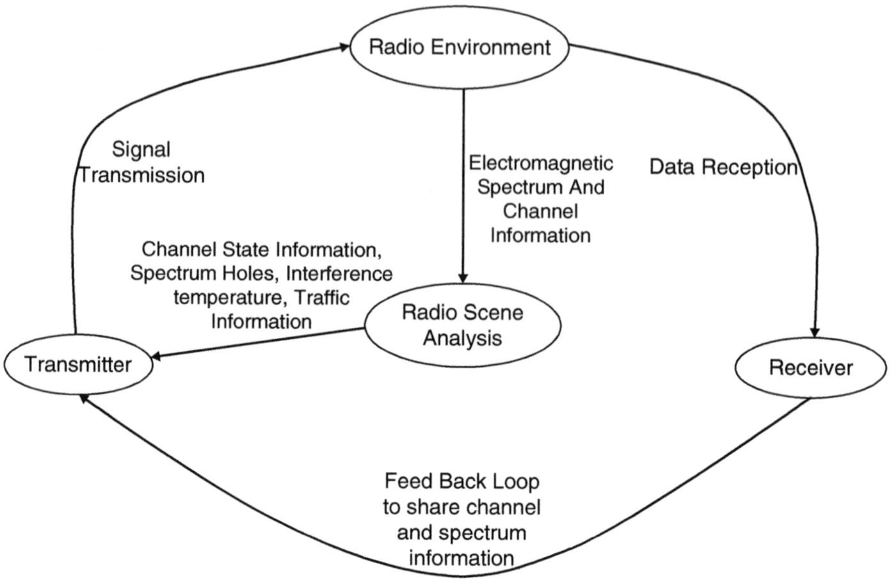

Fig. 15.5 Wavelet radio signal transmission cycle

spectrum estimates, a pulse waveform that has little or no energy in the interference domains is constructed by a proper choice of wavelet based codes. The channel estimator gauges the channel to derive the channel state information (CSI). Based on this estimate the transmission mode is adapted in accordance with the channel. Depending on the condition of the channel, the transmitter could be adapting one or more of the following: constellation size, code rate, and power.

The stream of data is then linearly modulated by the pulse waveform to obtain a transmission signal. The signal is then transmitted through the channel.

At the receiver, the signal, corrupted by interference, is decorrelated with a similar wavelet packets based template waveform. An integrator follows the correlator. And the data is detected using a suitable detector. A feed back loop mechanism exists between the transmitter and receiver to co-ordinate the cognitive modules in the two ends to act harmoniously.

Fig. 15.5 illustrates the signal transmission process by which the system interacts with the environment and communicates.

15.7 Challenges in the Development of Wavelet Radio

A few challenges that dictate and guide the system design are as stated below –

- Translating the engineering requirements into a mathematical condition.
- Defining the metric that gauges performance improvements.
- Observing and gathering intelligence required for analyzing radio environment.
- Utilizing available information.

- Designing wavelets and waveforms that maximize the spectrum utilization.
- Operating opportunistically in a spectrum hole without causing interference to primary users.
- Altering the weights of the transmission waveforms such that they map channel requirements and handle power constraints adequately.
- Designing wavelets which facilitate transmission with minimum interference from and to co-existing users.
- Handling problems like inter channel interference (ICI) and inter symbol interference (ISI), channel fading, multi-path effects.
- Changing transmission parameters seamlessly without affecting system performance.
- Tackling problems related to multiple-access.
- Handling spectrum coordination for broadcast and multicast with multiple cognitive systems competing for available resources.
- Ensuring backward compatibility with existing technologies and systems
- Making the system generic and flexible so that it can be easily scalable.

15.8 Conclusion

The convergence of information, multimedia, entertainment and wireless communications has raised hopes of realizing the vision of ubiquitous communication. To actualize this there is a challenge of developing technologies and architectures capable of handling large volumes of data under severe constraints of resources such as power and bandwidth. Wavelets are uniquely qualified to address this challenge. They have strong advantage of being generic schemes whose actual characteristics can be widely customized to fulfil the various requirements and constraints of advanced mobile communications systems.

In this chapter the perspective of wavelet technology for future smart resource aware radio systems was suggested. The wavelet radio aims intelligent radio transmission system based on wavelet technology to handle the multi-faceted challenge of maximization of resource utilization and interference mitigation. The benefits of wavelet radio cannot be fully exploited with the current radio systems and technologies. It is therefore foreseen that wavelet technology will have an eminent position in the future of wireless communications.

Acknowledgement The author acknowledges the contribution and input of M.K. Lakshmanan (a PhD candidate at IRCTR-TUDelft) in making this chapter.

References

1. M. Vetterli, I. Kovacevic, *Wavelets and Subband Coding* (Prentice Hall PTR, Englewood Cliffs, New Jersey, 1995)
2. I. Daubechies, *Ten Lectures on Wavelets* (SIAM, Philadelphia, 1992)
3. G. Strang, T. Nguyen, Wavelets and filter banks, Cambridge University Press (1996)

4. P.P. Vaidyanathan, *Multirate Systems and Filter Banks* (Prentice-Hall, Upper Saddle River, NJ, 1993)
5. S. Mallat, A Theory for Multiresolution Signal Decomposition: the wavelet representation. IEEE Trans. Pattern Anal. **11**, 674–693 (July 1989)
6. A. Cohen, J. Kovacević, Wavelets: The Mathematical Background. Proc. IEEE **84**(4), 514–522 (1996)
7. Proceedings of the IEEE, Special Issue on Wavelets, vol. 84, no. 4, April 1996
8. IEEE Trans. on Circuits and Systems, Special Issue on Multirate Systems, Filterbanks, Wavelets and Applications, vol. 45, no. 8, Aug. 1998
9. A. Jamin, P. Mahonen, Wavelet Packet Modulation for Wireless Communications, Wireless Communications & Mobile Computing Journal, Vol. 5, Issue 2, pp. 123–137, John Wiley and Sons Ltd., March 2005
10. B.G. Negash, H. Nikookar, Wavelet-based Multicarrier Transmission over Multipath Wireless Channels, IEE Electronics Letters, Vol. 36 No. 21, pp. 1787–1788, October 2000
11. G. Wornell, Emerging Applications of Multirate Signal Processing and Wavelets in Digital Communications, Proc. IEEE, vol. 84, pp. 586–603, April 1996
12. M.K. Lakshmanan, H. Nikookar, A Review of Wavelets for Digital Wireless Communication. Wireless Pers. Commun. **37**(3–4), 387–420(34) (May 2006). Springer
13. S. Haykin, Cognitive radio: brain-empowered Wireless Communications. IEEE JSAC **23**(2), 201–20 (February 2005)
14. M.K. Lakshmanan, H. Nikookar, UWB Interference mitigation using multi-carrier templates, European Conference Wireless Technology, September 2006, Manchester, UK, pp.115–118
15. H. Nikookar, R. Prasad, Waveshaping of Multicarrier Signal for Data Transmission over Wireless Channels, IEEE Proc. 6th International Conference on Universal Personal Communications, ICUPC'97, San Diego, CA, USA, Oct. 1997, pp. 173–177
16. H. Nikookar and R. Prasad, Optimal Waveform Design for Multicarrier Transmission over a Multipath Channel, IEEE Proc. 47th IEEE Vehicular Technology Conference, Phoenix, Arizona, USA, May 1997, pp. 1812–1816
17. H. Nikookar, B. Negash, Frequency Offset Sensitivity Reduction of Multicarrier Transmission by Waveshaping, IEEE International Conference on Personal Wireless Communications (ICPWC'2000), pp.444–448. Hydarabad, India, Dec. 2000
18. H. Nikookar, Improved Weighted OFDM for Wireless Multipath Channels, International Conference on Telecommunications, Acapulco, Mexico, May 2000
19. H. Nikookar, R. Prasad, Weighted OFDM for wireless multipath channels. IEICE Trans. Commun. **E83-B**(8), 1864–1872 (August 2000)
20. H. Nikookar, R. Prasad, On the Sensitivity of Multicarrier Transmission over Multipath Channels to Phase Noise and Frequency Offset, IEEE Proc. 7th International Symposium on Personal Indoor Mobile Radio Communications, pp.68–72. Taipei, Taiwan, Oct. 1996
21. H. Nikookar, R. Prasad, Multicarrier Transmission with Nonuniform Carriers in a Multipath Channel, IEEE Proc. 5th International Conference on Universal Personal Communications, ICUPC'96, pp.628–632. Boston, MA, USA, Sept. 1996
22. H. Nikookar, R. Prasad, OFDM Performance Evaluation over Measured Indoor Radio Propagation Channels, IEEE Proc. 4th International Conference on Universal Personal Communications, ICUPC'95, pp.968–972. Tokyo, Japan, Nov. 1995
23. H. Nikookar, R. Prasad, Performance Evaluation of Multicarrier Transmission over Measured Indoor Radio Propagation Channels, IEEE Proc. 6th International Symposium on Personal Indoor Mobile Radio Communications, pp.61–65. Toronto, Canada, Sept. 1995
24. H. Zhang, H. Fan, Wavelet Packet Waveforms for Multicarrier CDMA Communication, Proc. International Conference on Acoustics Speech and Signal Processing. Orlando, FL, May 2002

Homayoun Nikookar received his B.Sc. and M.Sc. (with distinction) in Electrical Engineering from Sharif University of Technology, Teheran, Iran in 1986 and 1987, respectively and Ph.D. in Electrical Engineering from Delft University of Technology (TUDelft), The Netherlands, in 1995. From 1996 to 1998 he was a post doctoral research fellow at the International Research Centre for Telecommunications and Radar (IRCTR) of the Department of Electrical Engineering, Mathematics and Computer Science of TUDelft, and since 1999 serves as a faculty member of the IRCTR of TUDelft where he is currently an Associate Professor. Dr. Nikookar is also the leader of the Radio Advanced Technologies and Systems (RATS) Program of IRCTR, leading a team of researchers carrying out cutting-edge research in the field of radio transmission. He has conducted active research in many areas of wireless communications, including wireless channel modeling, UWB, MIMO, multicarrier transmission, Wavelet-based OFDM and Cognitive Radio. He is the co-recipient of the 2000 IEEE Neal Shepherd Best Propagation Paper Award for publication in the March issue of Transactions on Vehicular Technology. A number of his papers have been listed in the IEEE Xplore Top 100 documents. He is also recipient of several paper awards of International Conferences and Symposiums. In 2007 Dr Nikookar served as the Chair of the 14th IEEE Symposium on Communications and Vehicular Technology (SCVT) in Benelux and in 2008 was the Chairman of the European Wireless Technology Conference (EuWiT). He has been a technical program committee member of several international conferences. He was the technical program chairman of European Conference on Wireless Technology in 2004. Dr Nikookar served as a Guest Editor of the IEEE Microwave and Wireless Components Letters in 2005 and a Guest Editor of the Journal of Communications, Special issue on Advanced Wireless Technology, in Aug. 2009. He is a senior member of the IEEE and the coauthor of the Book, Introduction to Ultra Wideband for Wireless Communications, Springer, 2008.

Chapter 16
Cooperative Localization in a Hybrid WiMAX/WiFi System: A Future-Proof Framework

Francescantonio Della Rosa, Simone Frattasi, and João Figueiras

16.1 Introduction

Mobile and wireless communications are global phenomena of large social relevance. During the last decades we have observed a constant increase of mobile subscribers worldwide, where, thanks to globalization and huge demands, the costs have decreased so much that the mobile phone has become a device affordable by almost everybody. In particular, mobile and wireless communication systems have opened up new scenarios for developing countries, which have finally the possibility to have global access from remote areas at low costs, much lower than when deploying wired systems. At the same time, we have experienced a metamorphosis of the mobile phone with respect to the services it provides to its users and its social relevance; indeed, from being a device merely used for voice calls and SMSs, it is now used for exchanging e-mails, taking pictures, retrieving location-based information, etc. Practically, it has turned into an advanced social-networking tool, which is nowadays exploited for enlarging one's social sphere. Along with this change of skin there have been increasing research and development efforts at a global scale for mobile phones and mobile and wireless communications, in general. At the present time, we are actually heading to the fourth generation (4G), where the main drivers are higher data rates and the convergence among the several technologies available today. In this chapter, we exploit the second track in order to develop a localization solution that will provide 4G and beyond users more advanced services, like indoor navigation and cooperative location-based services.

F.D. Rosa (✉)
Tampere University of Technology, Finland
e-mail: francescantonio.dellarosa@tut.fi

S. Frattasi and J. Figueiras
Aalborg University, Denmark

R. Prasad et al. (eds.), *Globalization of Mobile and Wireless Communications: Today and in 2020*, Signals and Communication Technology,
DOI 10.1007/978-94-007-0107-6_16,

Location information and location-based services have been receiving a growing attention in the last years. The reason is the business potential behind these kinds of applications, which encompass emergency, security, monitoring, tracking, logistics, mobile yellow pages, radio planning, cellular system management, etc. The Global Positioning System (GPS) is still nowadays the most popular commercial localization solution on the market. It provides location information by exploiting time of arrival (TOA) measurements from downstream satellite links. As a technological consequence of the market demands, third generation (3G) handsets have started to embed GPS receivers. Therefore, they have got more flexible, sharing higher number of systems and services; however, their battery drains faster, and their cost and complexity have increased [1, 2]. Furthermore, it has to be pointed out that the GPS is not always the most suitable localization solution for all scenarios and locations [3]. Several drawbacks emerge especially in outdoor urban canyons, indoor environments or underground, where it is not possible to obtain a clear view of at least four satellites. These conditions, caused essentially by signal blocking and multipath, make it difficult, if not impossible, to obtain adequate location information [2].

The straightforward alternative to the GPS in wide area scenarios is the exploitation of time difference of arrival (TDOA) measurements from the serving cellular system, where synchronization in time between transmitter and receiver is not needed, and no hardware modifications are required. Different types of measurements – e.g., received signal strength (RSS) measurements - are instead considered in local area scenarios, as the short distances involved imply that small errors in time cause significant errors in space to be considered useful for the localization process. Valid and interesting categories of localization systems using RSS measurements are those featuring WiFi Hot Spots deployed in big cities. In this case, the user terminal can predict its location by means of the known positions of the various detected access points (APs). Location information is provided by a database commonly placed on a remote server (e.g., managed by universities, WiFi clubs, wardriving activities, etc.), which needs to be constantly updated in order to provide valid and up-to-date position information. However, in less populated areas, the lack of WiFi APs may cause a lack of estimated positions [4]. Therefore, the main focus (at least regarding outdoor environments) is still on cellular-based localization technologies, as few base stations (BSs) are enough to perform positioning and tracking, and cellular systems can still provide high data rates with wide area coverage.

As the 4G is approaching, there exists a pressing need for providing a robust localization solution, which will be continuously available, regardless of the specific environment (i.e., outdoor and indoor), and which would overcome or complement the shortcomings of GPS-based and GPS-free systems, without adding more hardware into the devices [4]. Among the several emerging wireless systems, researchers have already selected Mobile Worldwide Interoperability for Microwave Access (Mobile WiMAX) as one of the best candidates for satisfying all of the requirements placed on 4G (e.g. inter-operability, ubiquity, groundbreaking services, high data rates, wide coverage). However, there are still too few contributions regarding its positioning and tracking performance.

In this chapter, we carry out such an investigation, where results show that the positioning accuracy under non-line-of-sight (NLOS) conditions is poor. Consequently, we propose an augmentation solution exploiting the "cooperation" among closeby users [3]. In particular, we study here the following case: TDOA measurements performed by the Mobile WiMAX network are combined with the RSS measurements performed within the Wi-Fi ad-hoc network. This requires the handsets to be equipped both with WiMAX and Wi-Fi modules. Thus, the main purpose of this work is to enhance the location estimation accuracy of Mobile WiMAX users by exploiting the possibility of performing relative localization among mobiles in close proximity. The foundation of this enhancement is that peer-to-peer measurements have usually lower absolute errors than those on long-range cellular links. Note that although we consider in this paper WiMAX and WiFi, since at the moment these are most popular technologies that complement each other well, the proposed framework is future-proofed, in the sense that it will always be possible to use it by replacing one or both these technologies with others or even newer ones.

The chapter is organized as follows: Section 13.2 presents the data fusion method that we use to combine TDOA and RSS measurements, which is based on the Extended Kalman Filter (EKF); and Section 13.3 shows the simulation results both for positioning and tracking for standalone Mobile WiMAX localization and for the proposed augmentation solution. Finally, our concluding remarks are given in Section 13.4.

16.2 Data Fusion and Filtering

Estimating the position of a mobile from noisy measurements requires the use of robust algorithms. In this chapter, we propose using an Extended Kalman Filter (EKF), which is an algorithm from the Bayesian filtering family of algorithms, to derive an accurate position of a mobile. It is robust and simple to implement. By definition, the EKF is able to predict the state vector X, which is assumed hidden (i.e. non-measurable), given the relation that X has with a second observable state vector Z (i.e. measurable). In physical terms, the variable X is commonly defined as a vector of the geographical coordinates of the mobile and its corresponding derivatives with respect to time. The variable Z is a vector of all the quantities that can be measured and have a relation with the position and its derivatives, such as received power, time of arrival, accelerometer measurements. Based on this idea, the most important models to define are the motion model and the observation model. While the motion model tells us how X evolves over time, the observation model tells us the relationship between X and Z. Since the EKF is a minimum mean-square error estimator, the noise components are filtered out, resulting in an estimation of the expected value of X. Given the EKF formulae [5], both state vectors, X and Z, can be corrupted by a noise component, which by definition is gaussian distributed with zero mean and known standard deviation. An important aspect is that these noise components shall be independent and identically distributed.

16.2.1 Non-Cooperative Approach

In order to define the state space of the position to be estimated, we assume that devices are moving freely in x and y coordinates with velocities $\hat{\dot{x}}^{(i)}$ and $\hat{\dot{y}}$, respectively. From now on, x and y are taken as the coordinates of the mobile in a generic referential. In this way, the state space X for each mobile i is defined as follows:

$$\hat{X}^{(i)} = \left[\hat{x}^{(i)}\hat{\dot{x}}^{(i)}\hat{y}^{(i)}\hat{\dot{y}}^{(i)}\right]^{\mathrm{T}},$$

where the notation ^ symbolizes that we are dealing with estimators, i.e., the mobile coordinates that we intend to estimate by using the Kalman filter. For simplicity and further comparison with the cooperative case, all the estimators of the users are integrated into a single extended state space:

$$\hat{X} = \left[\left\{\hat{X}^{(1)}\right\}^T \ldots \left\{\hat{X}^{(n)}\right\}^T\right]^T.$$

For the motion model and the observation model, we use respectively the constant velocity movement model [8] and the model defined by the Mobile WiMAX standard [7]. The localization technique is grounded by the evaluation of the TDOA measurements obtained from the Mobile WiMAX BSs:

$$T^{(i)} = \left[\tau^{(i)[2]} \ldots \tau^{(i)[N]}\right]^T,$$

where $\tau^{(i)[j]}$ is the measurement obtained by cross-correlating the signals coming from BS_j and BS_1 at mobile i. The full set of measurements is then represented by grouping all the measurements from all the mobiles in the same vector:

$$Z \equiv T = \left[\left\{T^{(1)}\right\}^T \ldots \left\{T^{(N)}\right\}^T\right]^{\mathrm{T}}.$$

Another important step in the design of the EKF is the determination of the process and measurement noise covariance matrices. These matrices are considered diagonal with an adequate variance on each entry of the matrix.

16.2.2 Cooperative Approach

The core of the enhanced localization technique is the integration of the RSS information retrieved on mobile-to-mobile links into the estimation algorithm previously defined for TDOA measurements. With respect to the non-cooperative case, differences exist only on the matrices related to the observations. Thus, the first step is to define the observations for the short-range links:

$$P^{(i)} = \left[P^{(i)(1)} \ldots P^{(i)(i-1)} P^{(i)(i+1)} \ldots P^{(i)(n)} \right]^T,$$

where $P^{(i)(j)}$ represents the power measurements between mobile i and j at mobile i $(i, j \leq n)$. Grouping all the power measurements for all the available mobiles we obtain:

$$P = \left[\{P^{(1)}\}^T \ldots \{P^{(n)}\}^T \right]^T$$

Finally, we can integrate all the available types of measurements (i.e., TDOA and RSS) in a single vector Z, which is defined as:

$$Z = \left[\{T\}^T \ldots \{P\}^T \right]^T.$$

The measurements noise is defined as a diagonal matrix like in the non-cooperative case. However, it considers also in the present case the noise existing in the measurements of RSS. Thus it is augmented with the noise component of the RSS measurements.

16.3 Results and Analysis

16.3.1 Mobile WiMAX Localization

Results for standalone Mobile WiMAX localization (i.e., non-cooperative approach) are carried out by taking into account the scenario depicted in Fig. 16.1a. Each mobile station (MS) measures TDOAs by performing the cross-correlation on the preamble signals received by the BSs (where BS_1 is the reference BS) [6]. Fig. 16.2 shows a plot of the output from the cross-correlator. The peak is in correspondence with the sample index value, which represents the estimated delay introduced by the channel. The sensitivity in the delay detection is given by the sampling time of the system, which is 175ns, and thus represents a lower bound in the estimation. The upper bound is given instead by the delay spread characteristic of the specific tap delay model used in the simulations [7].

Table 16.1 shows the system parameters used in our investigation [9, 10]. While shadowing is not considered, a Rayleigh channel, causing the signal to be delayed before arriving at the receiver side, has been adopted for simulating the multipath. Three TDOA data sets (i.e., BS_1-BS_2, BS_1-BS_3 and BS_1-BS_4) for each MS are the input values of the EKF, where the initial guesses are given by a least square algorithm [1], whose results are shown in Table 16.2. Moreover, the performance evaluation is carried out considering cells of radius 3 km, where BS_1 is placed at (0,0) m, and MS_1 and MS_2 are respectively placed at (500,10) m and (500,–10) m, respectively.

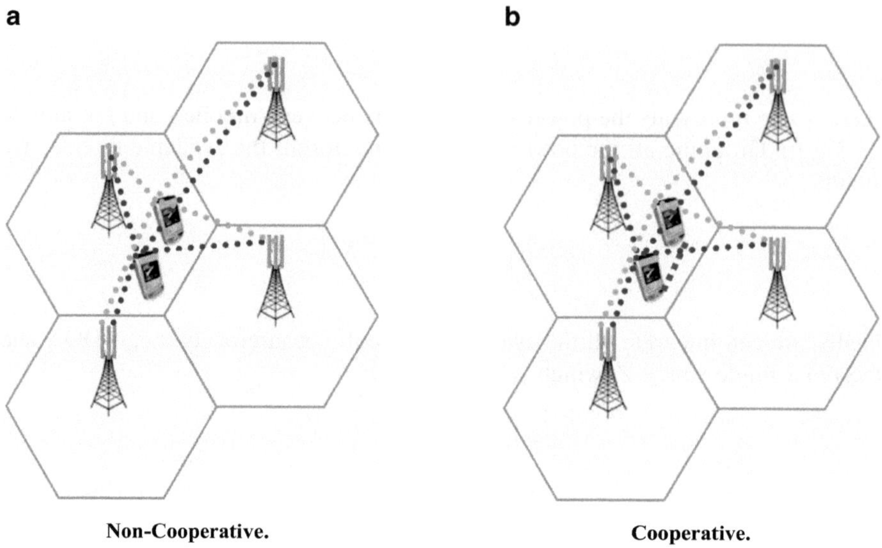

Fig. 16.1 Scenarios

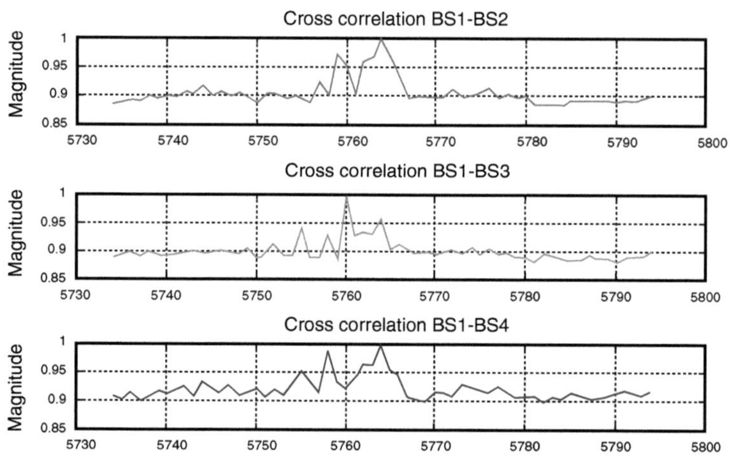

Fig. 16.2 TDOA peak detection

Table 16.1 802.16e system parameters

System bandwidth (MHz)	5
Carrier frequency (GHz)	3.5
Sampling frequency (MHz)	5.6
FFT size	512
Sub-carrier frequency spacing (KHz)	10.94
Useful symbol time (μs)	91.4
Guard time (μs)	11.4
Symbol Duration (μs)	102.9

Table 16.2 Initial guesses

MS	X Estimation (m)	Y Estimation (m)
MS1	0.42	60.67
MS2	1278.88	53.7

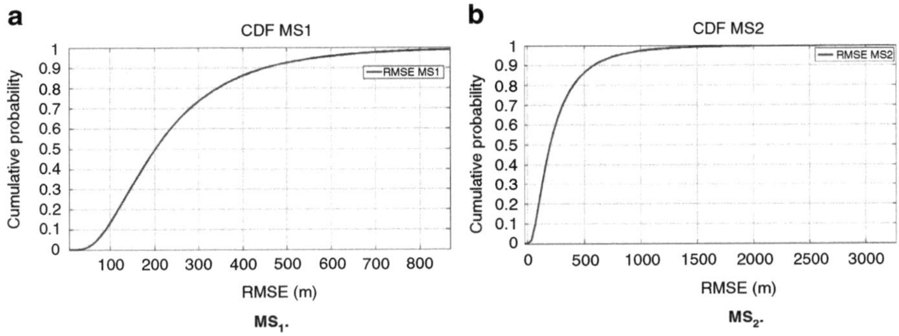

Fig. 16.3 Cumulative distribution function (CDF) of the average root mean square error (RMSE)

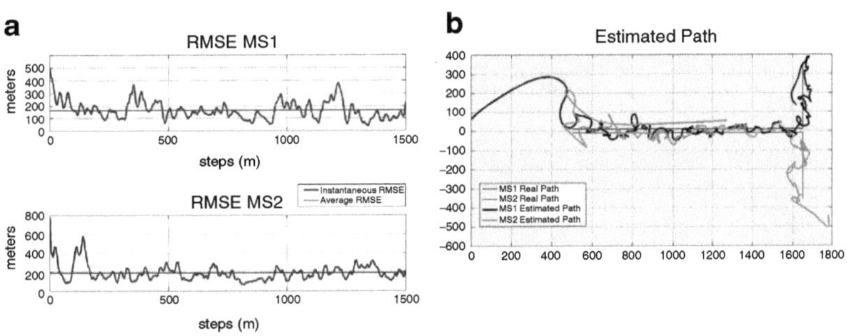

Fig. 16.4 RMSE and estimated path

1. *Static Case*: As the curves show in Fig. 16.3a and b, there are huge errors in estimation, up to the value of the radius of the cell, thus resulting in a non useful solution for real cases.
2. *Dynamic Case*: Similar errors are obtained also in case of mobility, where the two MSs start from (500, 10) m and (500,–10) m, respectively, move towards right parallel to the x-axis with a constant velocity of 3 km/h until they make a turn of 90° and go in opposite directions. While Fig. 16.4a shows the instantaneous and the average value of the RMSE, Fig. 16.4b shows the estimated path.

16.3.2 Hybrid Mobile WiMAX/WiFi Localization

Results for the hybrid Mobile WiMAX/WiFi localization [10] (i.e., cooperative approach) are carried out by taking into account the scenario depicted in Fig. 16.1b. Besides TDOAs, each MS now measures the RSS of the signals received by the other MS. In particular, Table 16.3 shows the system parameters used in our investigation for the WiFi communications.

1. *Static Case*: To ascertain the benefits observed in our scheme, the CDFs of the RMSE for MS_1 and MS_2 are plotted in Fig. 16.5a and b, where the non-cooperative

Table 16.3 802.11a system parameters

Channel Bandwidth (MHz)	20
Subcarriers	54
Subcarrier spacing (KHz)	312.5
Guard time (μs)	0.8
FFT period (μs)	3.2
Symbol rate (μs)	4
Modulations	BPSK,4,16,64 QAM
Data subcarriers (μs)	52
Sampling rate (ns)	50

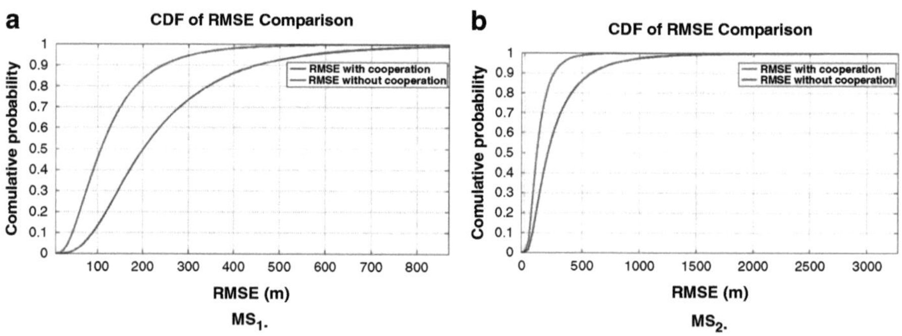

Fig. 16.5 CDF of the RMSE

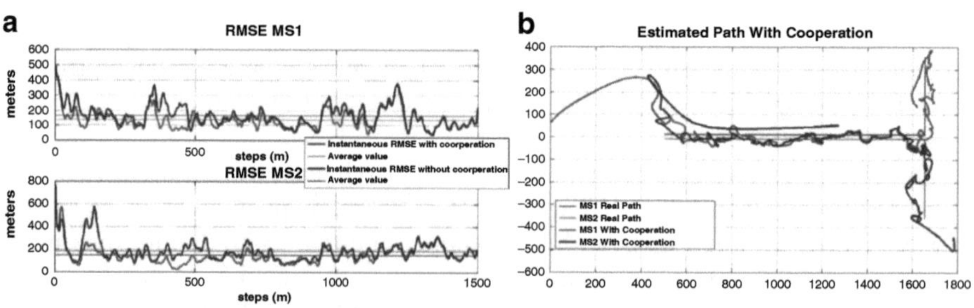

Fig. 16.6 RMSE and estimated path

and the cooperative cases are compared. It is evident that our proposal outperforms the standalone solution by greatly enhancing the localization accuracy of the system.
2. *Dynamic Case*: Similar errors are obtained also in case of mobility, where the two MSs move along the same paths as in Section 16.3.1. While Fig. 16.6a shows the instantaneous and the average value of the RMSE, Fig. 16.6b shows the estimated path. It is interesting to note the effect of cooperation with respect to the distance between the two mobiles: when the latter go in opposite directions, the distance between them increases, up to the point where they are not in range anymore, and the effect of cooperation suddenly vanishes (the RMSEs completely overlap).

16.4 Conclusions

This chapter has analyzed cooperative schemes under the realm of wireless location. The purpose of these schemes when applied to positioning applications in cellular systems is to increase the accuracy of user position estimation with respect to individualist schemes. This concept has been proven in this chapter with the implementation of a hybrid scenario and cooperative handsets. In particular, this chapter proposes an innovative solution for positioning and tracking estimation in hybrid 4G wireless networks by introducing the Cooperative Mobile Positioning System (COMET[1]). A simulation framework was implemented by modeling Mobile WiMAX and WiFi networks combined in the handsets. Basically, both graphical and numerical results show that cooperation enhances the location estimation accuracy when compared to conventional non-cooperative positioning techniques in stand-alone cellular networks. Thus, this work has demonstrated that the wireless cooperation has a beneficial impact on wireless location computation and that, although we consider in this chapter WiMAX and WiFi, as at the moment they are cutting-edge technologies and they complement each other, the proposed framework is future-proofed, in the sense that it will always be possible to use it by replacing one or both these technologies with others or even newer ones.

References

1. F. Della Rosa, G. Simone, P. Laurent, R. Charaffedine, N.H. Mahmood, B. Pietrarca, P. Kyritsi, N. Marchetti, G.P. Perrucci, J. Figueiras, S. Frattasi, Emerging Directions in Wireless Location: Vista from the COMET Project, in *Invited Paper in Proceedings of the 16th IST Mobile and Wireless Communications Summit* (Budapest, Hungary, July, 2007)
2. A.H. Sayed, A. Tarighat, N. Khajehnouri, Network-based wireless location: challenges faced in developing techniques for accurate wireless location information. IEEE Signal Proc. Mag. **22**(4), 24–40 (2005)

[1] COMET has been funded by Det Obelske Familiefond and Center for TeleInFrastruktur (CTIF).

3. S. Frattasi, J. Figueiras, Ad-Coop Positioning System: Using an Embedded Kalman Filter Data Fusion. Invited Paper in Encyclopedia of Wireless and Mobile Communications, pp. 120–131. CRC Press, April 2008
4. G. Borriello, M. Chalmers, A. La Marca, Delivering real-world ubiquitous location systems. Commun. ACM. Special Issue on The disappearing Computer **48**(3), 36–41 (2005)
5. M.S. Grewal, A.P. Andrews, *Kalman Filter: Theory and Practice Using MATLAB*, 2nd edn. (Wiley, New York, 2001)
6. C.H. Knapp, G.C. Carter, The Generalized Correlation Method for Estimation of Time Delay. IEEE Trans. Acoust. Speech **24**(4), 320U–327 (1976)
7. G. Senarath, et al., Multi-hop relay system evaluation methodology (Channel model and performance metric), IEEE 802.16j Relay Task Group, October, 2006
8. Y. Bar-Shalom, Tracking and data association, Mathematics in Science and Engineering. Academic, vol. 179, 1988. IEEE 802.16j Relay Task Group, Channel Models and Performance Metrics for IEEE 802.16j
9. WiMAX Forum, Mobile WiMAX U Part I: A Technical Overview and Performance Evaluation, June, 2006
10. F. Della Rosa, Cooperative Mobile Positioning and Tracking in Hybrid WiMAX/WLAN Networks. M.Sc. Thesis. Aalborg University (AAU), Aalborg, Denmark, June 2007

Francescantonio Della Rosa received the M.Sc. degree in Electrical and Electronic Engineering, with specialization in Mobile Communications, from Aalborg University (AAU), Aalborg, Denmark, in June 2007. Currently, he is a research scientist at Tampere University of Technology (TUT), Department of Computer Systems, working on hybrid data fusion and software development of GNSS receiver prototypes and mobile devices within the EU FP7 project GRAMMAR (Galileo Ready Advanced Mass Market Receiver). In 2006 he has been involved within the EU FP6 project MAGNET (My personal Adaptive Global NET) Beyond and in the COMET (Cooperative Mobile Positioning) project in 2007 at AAU. Since 2007 he has been working as software engineer and consultant in R&D departments for companies in the area of wireless communications and embedded systems. He is author and co-author of technical and scientific papers in the area of wireless communications and localization techniques for GNSS receiver prototypes, and reviewer for conferences and international journals. Currently he is the chairman of the IEEE GOLD Finland Tampere Activities Section and the IEEE Student Branch at Tampere University of Technology. His research interests include, but are not restricted to, Global Navigation Satellite Systems and receiver prototypes, wireless systems, hybrid and cooperative localization techniques.

Simone Frattasi received his Ph.D. in Wireless and Mobile Communications from Aalborg University, Aalborg, Denmark, in 2007, and his M.Sc. degree cum laude and his B.Sc. degree in Telecommunications Engineering from "Tor Vergata" University, Rome, Italy, in 2002 and 2001, respectively. In 2009, he worked as a patent attorney in the Intellectual Property Consultancy company Plougmann & Vingtoft (P&V), Århus, Denmark. Before joining P&V, he was employed from 2002 to 2005 as a Research Assistant at Aalborg University, where he was working on two European projects (STRIKE and VeRT) and one industrial project (JADE) in collaboration with the Global Standards & Research Team, Samsung Electronics Co. Ltd., Korea. From 2005 to 2007 he was appointed as an Assistant Professor and, besides still contributing to the JADE project, he was leading a Danish-funded project named Cooperative Mobile Positioning (COMET). From 2007 to 2008 he was project leader of an industrial project (LA-TDD) in collaboration with Nokia Siemens Networks, Aalborg, Denmark. Dr. Frattasi is author of numerous scientific and technical publications. His book "Mobile Positioning and Tracking: from Conventional to Cooperative Techniques" has been published by Wiley-Blackwell in June 2010. He has served as a TPC chairman for several conferences and as a Guest-Editor for several magazines and journals. Moreover, he was the main instructor for a half-day tutorial on wireless location at IEEE PIMRC'07. Simone is the co-founder of the International Workshop on Cognitive Radio and Advanced Spectrum Management (CogART) and the International Symposium on Applied Sciences in Biomedical and Communication Technologies (ISABEL). Finally, he had the honor to be appointed as a Chairman for the IEEE GOLD AG Denmark in 2008.

João Figueiras received his Ph.D. in Wireless Communications from Aalborg University, Aalborg, Denmark, in 2008 and his Licenciatura in Electrical Engineering and Computer Science from the Instituto Superior Técnico, Technical University of Lisbon (IST-UTL), Portugal, in 2004. From 2004 to 2008, he was employed by Aalborg University to work on the Danish-funded projects WANDA and COMET, and the European-funded projects MAGNET Beyond and WHERE. His work was in the area of positioning and tracking solutions for wireless networks, and his results have been used by several of the project partners, such as Blip Systems and Deutschen Zentrums für Luft- und Raumfahrt (DLR). His research has been published in several papers and presented at conference workshops (at IEEE PIMRC'07). During 2007, he was a research visitor at the Electrical Engineering Department of UCLA, Los Angeles, USA and an intern software engineer in the Mobile Team at Google Inc., Mountain View, USA. Since 2009 he has been employed by ZTE Portugal as the after-sales manager for the mobile market in Portugal. From 2003 and in parallel with his professional activities, he has been privately and continuously involved in several innovation projects in the area of cooperative communications and Internet services. Since 2006 he has been an active volunteer of IEEE, where he currently holds the position of IEEE Region 8 (EMEA) GOLD (Graduates of Last Decade) committee chair; his main duties are to increase the involvement of members in IEEE activities, coordinate communication among IEEE young members and leverage the benefits of IEEE in order to meet the needs of young members' careers. He has been the main organizer of several conferences (AISPC'08) and several worldwide IEEE meetings (1st World GOLD Summit 2008).

Chapter 17
Perspectives on Energy-Harvesting Wireless Sensor Networks

Mary Ann Ingram, Lakshmi Thanayankizil, Jin Woo Jung, and Aravind Kailas

17.1 Introduction

An energy-harvesting wireless sensor network (EHWSN) is an application-specific collection of wirelessly connected, highly resource-constrained radios, which are independent and capable of sensing, storing, processing and forwarding data, and capable of extracting energy from their environment. The network typically will have a connection to at least one less-constrained "Gateway" or "Sink" radio that is connected to the internet either wirelessly or by wire. Wireless sensor networks (WSNs) are used to remotely monitor physical quantities, such as temperature, sound intensity, acceleration, light intensity, or pressure. They may also monitor chemical quantities, such as glucose and oxygen concentrations in blood, or the presence of toxic compounds in air or water. However, chemical sensors typically have very short lifetimes compared to physical sensors, because of problems like bio-fouling [52], so EHWSNs are usually associated with physical sensing. The ability of the radios to scavenge or harvest energy implies that the network can operate much longer than a network of nodes that have conventional non-rechargeable batteries, before humans are required to replace the batteries. This has a significant financial consequence, especially for applications in remote or inaccessible areas. EHWSNs are a "green" technology, mainly because they enable better management of resources through better monitoring and control. They are also green because they make WSNs last much longer- potentially for decades.

Examples of WSN applications that can benefit from long network life and therefore from energy harvesting are those that need to last more than a few months and have so many nodes or are in such a location, as to make battery replacement very expensive. Examples include habitat and environmental monitoring in wilderness areas, and monitoring of seismic (e.g. volcano) activity [79]. Home automation [67] is an example, especially when the network is a retrofit for an existing structure,

M.A. Ingram, L. Thanayankizil, and J.W. Jung
Georgia Institute of Technology, Charlotte, USA
e-mail: mai@ece.gatech.edu

A. Kailas
Department of ECE, University of North Carolina, Charlotte, USA

R. Prasad et al. (eds.), *Globalization of Mobile and Wireless Communications: Today and in 2020*, Signals and Communication Technology,
DOI 10.1007/978-94-007-0107-6_17, © Springer Science+Business Media B.V. 2011

because of the low cost and ease of deployment and maintenance. Industrial monitoring and control is an example because a lot of extra wires cannot be accommodated, and energy, such as from motor vibration, is readily available. Structural health monitoring (e.g. fatigue detection) in large structures, such as for long-span bridges [9], requires hundreds of sensors [18]. Another application is security monitoring, such as in airports, sports arenas, or remote areas in a war zone [1,5].

There are many types of energy that can be used for EHWSNs, including light [49], electromagnetic [72], thermal [22], wind, and vibration [38]. As examples, solar energy can be collected by a photovoltaic cell array; vibration energy from a motor can be collected by a piezo-electric transducer; and energy from the thermal gradient on the exhaust pipe of an airplane can be collected using a thermoelectric transducer. The forms of energy available will depend on the application. For example, for structural health monitoring of bridges, some sensors must be under the bridge deck; this means that they will not have access to solar energy, but they will have access to vibration energy, and also perhaps to wind energy.

The study of EHWSNs is an emerging field of research. Although WSN processing and communication has received the attention of researchers for many years, much of the work has focused on power management on the radio platform and communication protocols that minimize energy consumption or maximize network lifetime, given that all nodes start with all the stored energy they will ever have. When energy harvesting is added to the design, the goal shifts toward "energy neutrality" [26], and network lifetime becomes less of a function of the "residual energy" on the nodes, and more of a function of the "cycle life remaining" on the nodes [23]. Attention must be given to energy source and energy storage characteristics. The storage device called the supercapacitor or ultracapacitor may revolutionize the EHWSN field, but models for it have only recently appeared. The communication protocols are affected, for example, because using up the residual energy on the node only temporarily disables the node, therefore the penalty for using this energy should perhaps not be as severe as it is when the node has a non-rechargeable battery. Another way protocols are affected is that, because nodes may not always have the energy to operate, duty cycling schemes and sleep protocols become more problematic. Some researchers have reported on energy-harvesting-aware protocols, however, since protocol assessment and design depends on the models for the energy sources and storage, there is much more work to be done.

This chapter will review the state of the art in EHWSN communication, with emphasis on energy sources, storage and consumption in EHWSNs, and medium access control (MAC) and routing protocols that are energy-harvesting aware.

17.2 Harvester and Storage Models

In modeling the harvested energy on a WSN node, one must consider the type of energy being harvested, the efficiency of the harvesting circuit, and the type of storage technology.

17.2.1 Storage Technologies

The storage technologies that are used with energy harvesting systems are the rechargeable battery (RB), the supercapacitor (SC), and the hybrid storage system, which comprises an RB and an SC. While the fuel cell is being considered for portable electronics, such as MP3 players, it is not appropriate for energy harvesting systems, because the fuel cell must be "refueled" using a hydrogen cartridge [51].

The cycle life of the storage device is defined to be the number of discharge-recharge cycles before the device fails to hold 80% of its original capacity (energy) [19]. Since a WSN node dies when its storage device no longer holds enough energy to run the circuits, the cycle life determines the life of the WSN, and is therefore a very important quantity in the study of EHWSNs.

17.2.1.1 Rechargeable Battery

The predominant rechargeable storage technology is the rechargeable battery (RB). While there are a number of battery chemistries, features that RBs have in common, compared to SCs, are:

- High energy density (Joules/kilogram)
- Low leakage
- Short cycle life
- Low power density
- Constant, high voltage
- Sensitivity to depth of discharge (DoD)
- Temperature sensitivity
- Sensitivity to value of discharge current

The cycle life of the RB varies with battery chemistry, but is typically between 300 and 500, corresponding to a maximum of a couple of years of operation if the node is discharged daily.

All batteries suffer from self-discharge or leakage. For example, Nickel-based batteries tend to suffer from relatively high levels of self discharge, losing around 10% of their capacity in the 24 h after charging, dropping to around 10% per month after this. Self-discharge is particularly sensitive to temperature (increasing with increased temperature), as shown for Ni-MH batteries in [19]. In other words, if an RB is not used, that is, it sustains only self-discharge (i.e. leaking) and is kept fully charged, then it can maintain its capacity up to 40 years or more depending on the chemical composition of the RB [12,19,36].

How an RB is discharged affects its cycle life. During discharge, positively charged ions (e.g. Li+) are created at the anode and diffuse over to the cathode, where they combine with negatively charged ions (e.g. Cl-). The combinations occur at fixed "reaction sites" on the cathode [30]. When no more reaction sites are available, the RB is discharged. Large discharge currents can cause the "Rate

Capacity" and "Recovery" effects [30]. Both effects reduce the effective capacity of the RB, but the mechanisms are different. In the Rate Capacity Effect, a large discharge current causes internal reaction sites on the Cathode to be permanently obscured [30], which causes an increase in the internal resistance of the RB and a progressively smaller capacity with each cycle [58]. In the Recovery Effect, a large discharge current causes a dearth of positively charged ions at the cathode, because diffusion of ions through the electrolyte cannot keep up with the consumption of ions at the cathode. Pausing the discharge gives the diffusion a chance to catch up with the consumption, enabling the RB to "recover" [30]. Low temperature reduces chemical activity in the cell, which increases the RB's internal resistance and exacerbates the recovery effect [58]. Both effects are avoided by keeping the discharge current sufficiently low and steady.

For small electronics, the leading battery chemistry choices are Nickel-Cadmium (NiCd), Nickel Metal-Hydride (NiMH), and Lithium Ion/Polymer. NiCd has a relatively low energy density and a long cycle life (50 Wh/kg and 1,200+ cycles [31]), but it has two disadvantages that have diminished its use recently: the "memory effect," which reduces capacity without periodic deep cycling (full discharge) [31, 29], and toxicity [29, 30]. NiMH also has the memory problem, but it has a higher energy density (70 Wh/kg) than NiCd, and is not toxic. However, NiMH is more sensitive to overcharging and has a shorter cycle life (300 cycles [31]) than NiCd [29, 30]. Lithium Ion RBs, the most popular choice for energy harvesting WSNs, have no memory problem and have the highest energy density and lowest leakage (100+ Wh/kg and 8% per month [31]) of the three choices [29, 30], however, Lithium RBs are more sensitive to variations in discharge current and are more expensive than NiMH RBs [30]. Also, Lithium RBs can be a fire risk without careful management, however, newer technologies have mitigated this problem [29, 43].

Predicting the remaining cycle life of an RB is challenging because the cycle life depends on many variables, including discharge rate, temperature, and DoD history [59]. RB cycle life can be extended by limiting its "depth of discharge" (DoD), which is the fraction of energy removed of the total capacity of the RB, at the time of recharging. For example, the number of cycles yielded by the Nickel Metal Hydride (NiMH) battery goes up exponentially by the reduction of the depth of discharge [12, 43].

17.2.1.2 Supercapacitor

Most of the SC features are the opposite of the RB features. The SC is characterized by

- Low energy density
- High leakage
- Long cycle life
- High power density
- Variable voltage
- Low maximum voltage

In the context of WSNs, the most important feature of the SC is its long cycle life, which is on the order of a million cycles [64, 33, 31]. If SC DoD is kept below 50%, SC operational lifetime can be as high as 20 years [64]. This has led some researchers to consider the SC as the only storage technology on the node [64, 74, 40].

An SC is a capacitor whose plates are "molecularly" close together, yet the porous carbon plates have very large area [14]. SCs can be folded up into very small packages [14], so they can be used in small wireless devices. SCs can have capacitances measured in Farads, as opposed to the microfarads of common capacitors. However, the disadvantage of the size and proximity of the plates is that SCs leak more than RBs and will breakdown if the voltage across the SC is greater than a few volts [64,40]. Unlike an RB, SC voltage drops as the energy is taken from it, according to the standard capacitor equation $E_{SC} = 0.5CV^2$. Since a WSN node fails to run if the voltage provided to it drops below a certain level, e.g. 2 V, not all the energy on the SC can be used. To provide the voltages of 3–5 V that are required to run the node, typically two SCs are arranged in series [40]. When this is done, a balancing circuit [39] is necessary to ensure that the SCs receive equal charge during harvesting.

For example, two 5 F SCs in series, each with a 2.5 V maximum voltage, can supply a maximum of 5 V to the load. The series combination holds a maximum energy of about 31 J. If the combination is directly connected to the load (i.e. to the node electronics), then the minimum voltage of 2 V implies that a maximum of 30 J is available to the node.

The leakage of an SC depends on its stored energy and the period of time that the SC has been kept charged to its maximum voltage [78]. The leakage generally increases with stored energy [31,78] and decreases with time kept fully charged [78], with typical leakage powers between 0.1 and 0.6 mW [78], allowing SCs to hold charge (without harvesting) for a maximum of a few weeks [64].

17.2.1.3 Hybrid Energy Storage Systems

SCs are sometimes used in hybrid combinations with RBs to protect the RBs from delivering or absorbing short, high pulses of power [31, 64], which shorten the life of the RB because of the Rate Capacity Effect discussed above. We refer to the hybrid combination of RB and SC as a hybrid energy storage system (HESS). HESS applications range from electric vehicles to portable electronics [37].

The HESS combination can take several forms. In the simplest case, the SC and RB are simply connected in parallel; [15] show that in pulsed loads, the SC meets the initial current demand by providing almost a step increase in current at the start of the pulse, and then has a first-order decay, whereas the RB provides the complementary component of the current pulse, rising slowly from zero. Then, in the period between pulses, the RB recharges the SC [15].

Fig. 17.1 shows a popular alternative HESS architecture, which has more control circuitry [31]. The aim in [31] is to make the SC large enough for routine

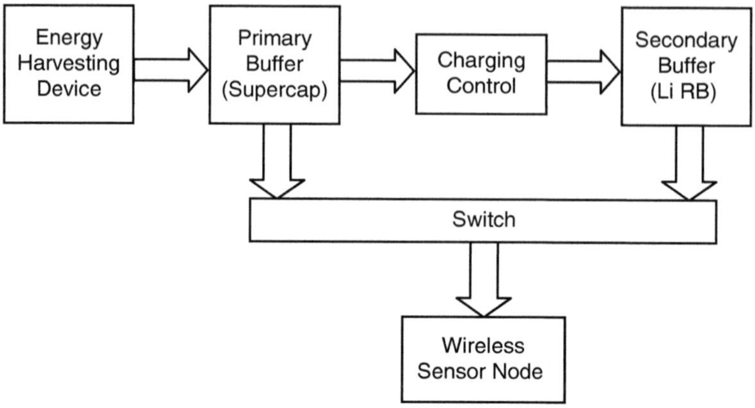

Fig. 17.1 System architecture of a hybrid energy storage system with a charge controller between the SC and RB, and a switch that selects just one storage device to drive the load

operation, and keep the RB as a "reliable emergency backup," thereby avoiding daily cycling of the RB. In the context of solar energy, "large enough" means that the SC can supply adequate voltage to drive the load at a given low duty cycle during the dark periods of the day.

17.2.2 Energy Transducers

A wide variety of ambient energy sources are possible, including light, vibration, strain, wind, temperature difference, and radio waves (RF). Each form of energy requires a different sort of transducer. For example, light energy is harvested by photovoltaic cells and vibration can be harvested by piezo-electric or magnetic coil devices. The energy harvester devices produce either alternating current (AC) or direct current (DC). The DC sources include photovoltaic and thermal generators, while the AC sources include the vibration and wind energy harvesters. Digital circuits use DC power, so rectifiers are used with AC sources [56].

The transducer that converts the ambient energy into electrical energy can be modeled by a Thevenin's equivalent circuit, as shown in Fig. 17.2a, where V_I is the Thevenin's equivalent ideal voltage source, and R_s is the Thevenin's equivalent source resistance. These two Thevenin's parameters are both functions of the size and type of harvester and the amount of ambient energy. For example, under a certain set of vibration conditions, a harvester consisting of a magnetic circuit coupled to a mechanical resonator can be represented by a 9.3 V battery in series with a 36 Kohm resistor [39].

It is well known that for the source to transfer the maximum power to a load resistance, R_L, we must have that the load resistance is matched to the source resistance, or $R_L = R_S$. It follows that the optimal voltage across R_L is $V_S = V_I/2 = V_{MPP}$,

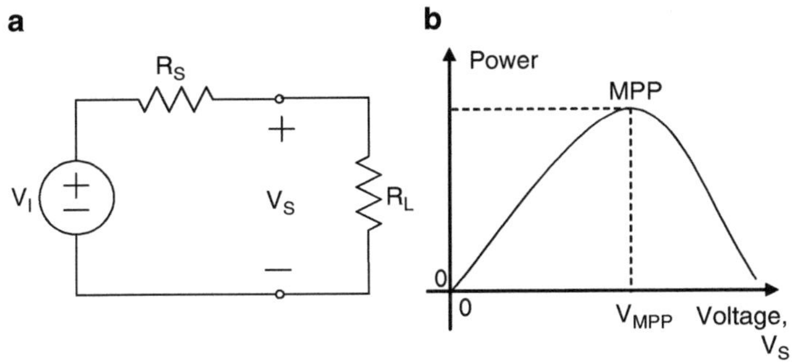

Fig. 17.2 (**a**) Thevenin's equivalent of the transducer, and (**b**), the typical relationship between the transducer's voltage and power

where MPP stands for "maximum power point," and the maximum power is $V_{MPP}^2 / R_L = V_I^2 / 4R_S$, corresponding to about 0.6 mW for the vibration energy harvester mentioned above.

If the voltage, V_S, is not equal to V_{MPP}, then the output power will be lower than the MPP, by as much as much as two orders of magnitude [50]. A sketch of a typical power vs. voltage curve is shown in Fig. 17.2b; examples of curves like this can be found in [4, 7, 39, 48, 50]. If the ambient energy conditions change, for example if the intensity of sunlight on a solar cell changes, the curve generally shifts and the MPP and V_{MPP} change [56].

Table 17.1 shows the typical optimum powers of various ambient energy sources that have been reported in the literature. Full-sun solar energy is the strongest source, with a 3.3 square inch panel generating 250 mW [32]. However, even solar generators will produce submilliwatt powers from very small solar cells, e.g. sizes of 0.8 cm^2 [7]. Indoor light energy is weaker than full sun by a couple of orders of magnitude. RF energy depends on the aperture size of the harvester antenna, and is submilliwatt unless the harvester is very close to a powerful transmitter.

17.2.3 MPP Tracking

In harvester design, the goal is efficient power transfer between the transducer, such as the solar panel, and the storage device, such as the supercapacitor. In this context, the storage device is the "load." Unfortunately, simply connecting the transducer directly to the storage device usually results in inefficient and slow charging because the harvesting device is forced to operate at a suboptimum voltage. Ideally, the problem is addressed through maximum power point tracking (MPPT), provided by an adaptively controlled voltage regulator that attempts to keep V_S approximately equal to V_{MPP}. In very low energy scenarios, careful attention must be paid to regulator design, because the regulator itself can consume a significant

Table 17.1 Example sources of ambient energy

Type of energy	Authors or source	Typical power	Description
Solar	Ambient Micro LLC [32]	~250 mW	3.3 square inches of photovoltaic collector in bright sunlight
	[4]	0.94 mW	Two 2 × 3 cm solar panels, 1 m from a 60 W Incandescent bulb
	[4]	76 □W	Two 2 × 3 cm solar panels, 2 m from a 34 W Fluorescent bulb
	[55]	15 mW/cm^2	Outdoors, noon
	Solar cells [7]	200□W -275□W	0.8 cm^2 solar cells in low and high intensity scenarios, respectively
RF	Ambient Micro LLC [32]	0.5 mW	
	[72]	0.1–3.0 mW/m^2	25–100 m from a GSM base station
Vibration	Shenck and Paradiso [65]	8.4 mW	Piezoelectric – from a PVDF stave in the insole of a shoe at a 0.9 Hz walking pace
	Lumedyne Technologies, Inc., [38]	0.43□W	On a car that is parked on the side of the road
		19.2□W	On a car that is idling
		0.64–1.45 mW	On a car being driven on the freeway
		18–47□W	Walking (harvester in the shoe)
	[76]	1 pW	2 mm^2 from nanowires and piezo
	[55]	116 µW/cm^3	Small microwave oven
		330 µW/cm^3	Shoe inserts, piezoelectric
Thermal	Hi-Z Technology, Inc. [22]	4.8 mW	40°C difference; open circuit voltage of 2.93 V, 26-couple device, 1 inch diameter. Quantum Well (QW) design
	[28]	~200 µV/°C	"The accepted bulk value for Bi2Te3 alloys"
	[55]	40 µW/cm^3	Generated from a 10°C gradient
	[49]	30□W	At 3 V, 0.5 cm^2 in area by 1.6 mm thickness, 5°K difference
Acoustic	[37]	960 nW/cm^3	100 dB of acoustic energy

portion of node energy [83]. The regulator design determines how long the node takes to recharge, which impacts the performance of protocols at the link and network layers. As switched regulators are more efficient (i.e. consume less energy) than linear regulators [83], linear regulators are not discussed in this section.

Fig. 17.3 shows three harvesting circuits, one diode regulator and two switching regulators, in order of increasing complexity and overhead. Fig. 17.3a has the simplest, diode-only design [55]; in this design, which does not do MPP tracking, harvesting occurs only when the harvester's voltage exceeds the storage device voltage (an RB was used in [55]) plus the diode voltage (i.e. 0.7 V). The result is that harvester energy is wasted unless its voltage is high enough [50]. The other problem with this design if it were applied to a discharged SC, is that the SC would

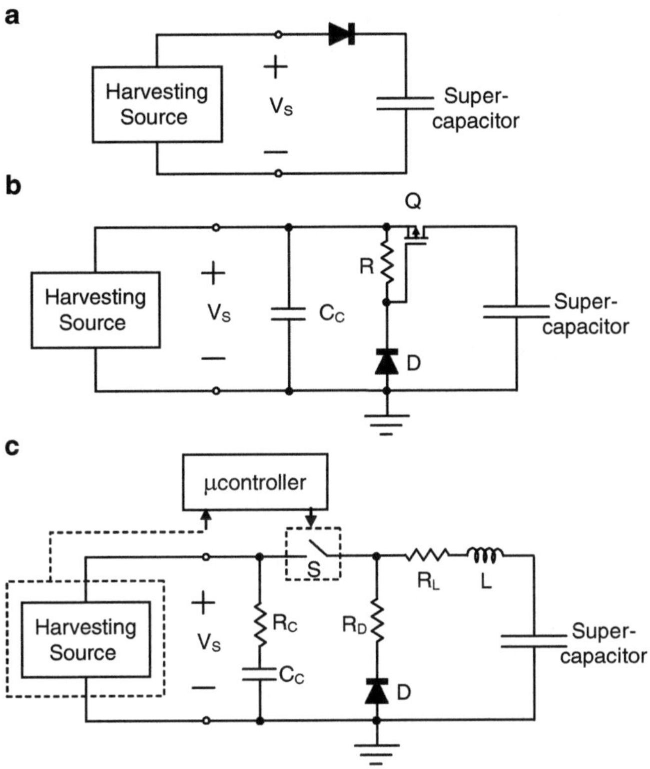

Fig. 17.3 Charging circuits with increasing complexity

appear as a very low resistance, pulling the voltage of the transducer far below its optimal value. Fig. 17.3b, which was described for a machine vibration harvester in [39], assumes a fixed MPP and uses a conventional charging capacitor (C_C) that charges very quickly and a switch (Q) that transfers the charge onto the SC when V_S reaches its optimal value. Fig. 17.3c performs MPP tracking using a microcontroller and regulates the voltage using a Pulse Frequency Modulator (PFM) circuit [64]. The dashed rectangle enclosing the Harvesting Source, represents that the microcontroller periodically measures the Thevenin's equivalent parameters, either by measuring the open circuit voltage and the short circuit current, or using the "hill climbing" method [64]. The inductor smooths the current that charges the SC, and prevents shorting when the SC is discharged. A fourth solution, not shown in Fig. 17.3, achieves MPP tracking without the overhead of the microcontroller; instead this solution infers the MPP from simple sensor measurements, for example, using a photo-resistor in the case of solar energy and a rotor's rotating frequency in the case of wind energy [50]. The sensor's signal controls a Pulse Width Modulator (PWM) voltage regulator, in an all-analog implementation [50].

One interesting feature of ambient energy harvesting systems, even for MPP tracking, is that charging SCs can take from minutes to hours, depending on the size of the SC and the strength of the harvester. Examples including 10 min for a 10 F SC from a nearby halogen light source [64], 1.5 h for a 140 mF SC from a vibrating motor [39], and "less than two hours" for a 37 × 82 mm solar panel under direct sunlight with a 11 F SC. This implies that if the energy demand exceeds the harvested energy, then "temporary partitions" lasting for a significant amount of time are possible in multi-hop networks that use only SC storage.

17.3 Load Models

The energy that is stored in the SC and RB drives the wireless sensor node. Therefore, an understanding of the energy consumption of a node is essential in the design and analysis of EHWSNs. Communication protocol designers like to use "energy per bit" models [27, 35]. However, such models depend on many factors, such as microprocessor type, power management on the radio platform, type of modulation and demodulation, how much energy the node radiates in transmission, and higher layer protocols such as medium access control (MAC) and routing protocols.

This section will focus on three parts of the load: the processor, the transceiver and the sensor device. Given a reasonably low duty cycle on the order of 1%, radio power consumption must be limited to approximately 1 mW in order for nodes (powered by a single battery or scavenged ambient energy) to last for years [44]. Therefore, power management techniques and communication protocols that can ensure the low duty cycle and reduce the energy consumption during the node is idle are essential part of the energy-efficiency.

17.3.1 Processor

The processor for the sensor node controls the device and executes algorithms/protocols. WSN Processors are designed and manufactured by several companies such as Atmel, Intel, Texas Instrument, etc. The power consumption of the processors is represented in Watts, but as [66] pointed out, the low-power should not be confused with energy-efficiency. That is, ATMega128L @ 4 MHz consumes 16.5 mW, but the number of instructions per Watt it can perform is 242 MIPS/W (4 nJ/Instruction), whereas ARM Thumb @ 40 MHz requires 75 mW, but can perform 480 MIPS/W (2.1 nJ/Instruction) [66]. Therefore, to evaluate the energy-efficiency of the processor, the energy per instruction rather than just power has to be taken into account. Table 17.2 lists some of the off-the-shelf processors used in popular sensor nodes along with their energy-efficiencies.

Many authors have pointed out that the energy-efficiencies of the off-the-shelf processors are still too high and inappropriate for WSN nodes, and a number of

Table 17.2 Energy-efficiency of the off-the-shelf processors

Processor	Voltage	Efficiency (pJ/inst)	Description
Atmel Mega128L [16]	3V	1,500	Used in MICA2 [41] and MICAz [42]
Intel XScale [16]	1.3–1.65 V	890–1028	Used in Rockwell sensors, Intel Mote
MSP430 [45]	1.8–3.6 V	600	Used in Telos [53] and ANT [3]

works have been done to improve the energy-efficiencies of the processors [16, 46, 77]. In [16], the authors devised a new asynchronous microprocessor that uses no clock in any component. They have shown that their design will only take 24 pJ/instruction. Similar efforts (reducing the energy per instruction) have been made in [46] and [77], which reduced the energy usage to 12pJ/instruction and 1.6pJ/instruction respectively.

The processor usually supports different operating modes (sleep, active, etc.) and the number of operating modes may differ depending on the processor model. Since the power consumption of the active mode (on the order of mW) is much higher than that of the sleep mode (on the order of μW) [53], the argument that the node should be switched to the sleep mode whenever is necessary seems reasonable. However, one must also note that changing between operating modes involves a power and latency overhead [54] and must be taken into account. How fast the nodes can wake up and become active is referred to as wakeup time, and it is desirable to have short wakeup time, because it reduces latency overhead. To summarize, both wakeup time and the power consumption during sleep mode are important factors for energy-efficiency and they are considered when choosing and designing the low-power processor.

17.3.2 Transceiver

The transceiver (transmitter + receiver) sends and receives the data and known as the most energy-consuming portion of the node. As is the case of the processor, the transceiver chip also has different modes (generally transmit, receive, idle and sleep) and turn-on time. Similar to the processor, transitioning between modes needs careful consideration, because it can consume a significant amount of power [54]. Low energy consumption and appropriate behavior of sleep/powerdown mode are essential in achieving long network lifetime. It is important to note that even though the radio chip satisfies these needs, these models are useless without appropriate protocols that trigger and utilize them, which will be introduced in Sections 17.4 and 17.5.

Many people believe that most of the power consumption for communication is caused by transmitting the data. However, as introduced in [66] (see Table 17.2 in [66]), this may not be always true. In short range application, the power required for receiving data can be higher than transmitting the data. Therefore, RX power consumption must not be ignored when considering energy-efficiency of the radio.

Some of the popular radio transceivers that are being used in sensor nodes are listed in Table 17.3. In the "Tx Power" column, the first number is the current drawn in mA from the storage device during transmission, and the second number is the corresponding radiated power in dBm ("Max" meaning the maximum radiated power and "Min" the minimum radiated power).

The choice of modulation is important in the design of the wireless sensor transceiver. Modulation schemes that can be non-coherently detected, such as Pulse Position Modulation (PPM) and Frequency Shift Keying (FSK), can reduce energy consumption, because they do not require a Phase Locked Loop (PLL). In the transmitter, the power amplifier plays important role in the energy consumption. If a modulation scheme with high Peak-to-Average Ratio (PAR) is used, such as orthogonal frequency division multiplexing (OFDM), a linear power amplifier is needed, which leads to high power consumption. Therefore, constant envelope modulation schemes such as FSK can reduce the energy consumption of the transmitter and have been considered as good candidates for WSNs [68, 75]. In [75], the authors compared the transmit power of M-FSK, M-PSK and M-QAM and found that M-FSK with non-coherent detection can use less transmit power than M-PSK/M-QAM for large M (M>8), sacrificing the bandwidth efficiency. Additionally, [75] addressed the importance of minimizing the start-up time of the transmitter which could be longer than actual transmission time. [68] compared the power efficiencies of FSK and PPM considering nonlinear characteristics of the batteries (stored energy in the battery can be wasted during its discharge process). The authors showed that when the circuit power consumption is negligible, FSK has power efficiency advantage over PPM regardless of constellation size and transmit distance, whereas the opposite trend can be observed when the circuit power consumption dominates the total power consumption of sensors. OOK (on-off keying) which is a very simple modulation scheme (1: signal, 0: no signal) has been considered, because the transmitter can be idle when 0 is transmitted and the demodulation is very simple [2, 17]. In [2], the authors addressed the issue of the synchronization having significant power consumption and OOK was considered for designing low power synchronization because it does not require frequency and phase synchronizations. Note that Zigbee standard [82] specifies O-QPSK (2.4 GHz) and BPSK (sub-1 GHz) as its modulation schemes, which are supported by CC2420 (Table 17.3).

The burst data rate (i.e. the rate within the period of the packet) also plays role in energy consumption. With a higher burst data rate, the node can transmit the data fast, which means that it can spend more time in idle mode [10]. However, higher burst data rate does not guarantee the energy-efficiency because as the data rate goes up, required transmit power for successful communication also increases. Therefore, choosing the radio with highest maximum burst data rate may not always be an energy-efficient strategy.

There have been a lot of works to design and develop low-power transceivers. [17] introduces results from the WiseNet project, the goal of which is to optimize both the MAC and physical layers to obtain a low-power solution for sensor networks. This project designed an ultra low-power transceiver and optimized the transceiver for the WiseMAC protocol [13] specially developed for WSN.

Table 17.3 Radio transceivers

RF chip	RX power (mA)	TX power (mA/dBm)	Voltage (V)	Data rate (kbps)	Modulation	Description
Chipcon CC2420	19.7	17.4/0 (Max) 8.5/–25 (Min)	2.1–3.6	250	DSSS OQPSK	Used in Telos [53] and MICAz [42]
Chipcon CC1000	9.6	26.7/10 (Max) 5.8/–20 (Min)	2.1–3.6	76.8	FSK	When 433MHz is assumed Used in MICA2 [41]
RFM TR1000	3.8	12/1.5 (Typical)	2.2–3.7	30 (OOK) 115.2 (ASK)	OOK/ASK	Used in Mica [41]
Semtech XE1205	14	62/15 (Max) 26/0 (Min)	2.4–3.6	152.3	FSK	Used in TinyNode [69]
RFM TR1001	3.8	12/1.5 (Typical)	2.2–3.7	30 (OOK) 115.2 (ASK)	OOK/ASK	Used in ScatterWeb Embedded Sensor Board [61]
Nordic nRF24AP1	22	16/0 (Max) 13/–20 (Min)	1.9–3.6	1,000	GFSK	Used in ANT [3]

Table 17.4 Energy-per-bit for off-the-shelf nodes, when the operating voltage is 3V

RF chip	nJ/bit for TX (maximum Tx power)	nJ/bit for TX (minimum Tx power)	nJ/bit for RX
Chipcon CC2420	208.8	102	236.4
Chipcon CC1000	1,043	226.6	375
RFM TR1000	1,200 (OOK) 312.5 (ASK)		380 (OOK) 99 (ASK)
Semtech XE1205	1,221.3	512.1	275.8
RFM TR1001	1,200 (OOK) 312.5 (ASK)		380 (OOK) 99 (ASK)
Nordic nRF24AP1	48	39	66

In [47], the authors modeled a system architecture for an integrated, CMOS, impulse ultra-wideband transceiver suitable for sensor network. [11] proposed energy-efficient OOK transceiver that achieves an energy-per-bit as low as 0.5 nJ/bit for the receiver and 3.8 nJ/bit for the transmitter.

We will end this subsection with Table 17.4, which shows the "energy-per-bit" values for the transceivers listed in Table 17.3 obtained by dividing the power consumption by the data rate. As mentioned in the beginning of the section, "energy-per-bit" is an important parameter when energy-efficiency of the wireless sensor node is evaluated.

17.3.3 *Sensor Device*

While the energy consumption of the sensor device is relatively smaller than the consumption of the processor and the transceiver for low-rate and low-quality sensors, this may not be true for some commonly used sensor devices [57]. Reducing the energy consumption of the sensor device is done by adjusting the number of the samples measured by the sensor device. Although the strategies and approaches to reduce the energy consumption of the sensor are different, the principle behind the energy-efficient sensing is simple: reduce the number of the unnecessary samples and focus on the number of important/required samples while maintaining the sensing fidelity. Some of the energy-aware sensing strategies are introduced in [57], which can be summarized and incorporated into the above principle as follows. In the case of the adaptive sampling, an initial coarse sampling is followed by a finer-grained sampling in the region where the signal variations are higher, focusing on the required samples while minimizing the unnecessary samples. Likewise, in the case of triggered sensing, important samples are obtained using high-quality (thus high power) sensor only if an interesting event is detected by low-power sensor.

17.4 MAC Protocols

In wireless sensor networks, the medium access control (MAC) protocol controls access to the channel among radios that whose transmissions would interfere or collide at their intended receivers. MAC protocols play a key role in determining channel capacity utilization, fairness, network delays and more importantly, power consumption.

There are several challenges in designing MAC protocols for EHWSNs. Traditionally wireless devices (like 802.11) use contention-based MAC schemes like CSMA/CA. However this is not suitable for WSNs as they are severely energy constrained and hence duty-cycle-based schemes like S-MAC [81] have been proposed for them. Duty cycling is the energy saving scheme where the radio is off by default but wakes up periodically to participate in potential network communication. Even though these schemes have been shown to save large amounts of energy in traditional WSNs they cannot be easily extended to EHWSNs. This is because in EHWSNs, unlike WSNs, it is a non-trivial task to design duty cycling algorithms, as the dynamics of the energy sources being harvested may not be easily predictable.

17.4.1 *Overview of EHWSN MAC Protocols*

[63] analyzed the performance of the conventional MAC schemes like CSMA and ID Polling on WSNs with ambient energy harvesting capabilities. They showed that, any scheme that involves some form of backoff and retransmission is very

likely to be non-optimal because timing schedules cannot be strictly enforced when a node has no energy to operate, and the amount of harvested energy may not be sufficient for retransmissions. Furthermore, unnecessary waiting (to synchronize with time slots) or retransmissions can be counter-productive; it has been shown in [62] that a slotted CSMA MAC performs worse than an unslotted scheme because energy is consumed during the slot synchronization process, resulting in longer harvesting periods thereby reducing throughput. A polling-based MAC protocol has been proposed for use in sensors powered by ambient vibrations but the scheme has not been shown to be optimal [62].

Some works have been proposed lately [24–26, 71], which attempt to model the energy source and adjust the node's duty cycle in anticipation of expected incoming energy or lack thereof. [24] presents algorithms for dynamically adapting the duty cycle of a node so as to maximize both lifetime and performance, however this work only considers periodic energy sources. [26] analyzed the requirements for "energy neutral" WSN operation. By characterizing the nominal energy harvesting rate and its maximum deviation and by characterizing the nominal energy consumption rate and its maximum deviation, [26] determined bounds on the nominal consumption, the required battery capacity, and the required starting battery stored energy. Consumption rate is adapted by changing the duty cycle or the transmit power of the nodes [26]. A more general model was considered in [71], which considers both periodic and aperiodic sources of energy. They use results from adaptive control theory to solve the problem of dynamic duty cycling and achieve better performance on a broader class of energy source datasets.

17.5 Routing Protocols for EHWSNs

A routing protocol is a procedure for deciding the path, or sequence of links, that packets will follow in a multi-hop network. Even though a lot of work has been put in to the design and development of EHWSNs, only a few routing protocols have been implemented for these networks. The problem of routing in EHWSNs is a nontrivial one and is very different from the problem of routing in traditional WSNs, which has been extensively studied. Routing protocols proposed so far for EHWSNs can be classified as multi-hop-based and two-tier cluster-based. In a multi-hop scheme, which is the most widely researched of the two, the energy-constrained and energy harvesting sensing modules transmit the message to the Sink over a multi-hop route (see Fig. 17.4a). However finding a reliable multi-hop route is very challenging in these networks. Since the wakeup time of any sensor cannot be estimated accurately as the exact rate of energy harvested fluctuates with time and other environmental factors, it is very difficult to ensure that the next-hop node is awake to receive a packet. The uncertainty in how long it takes a node to harvest enough energy before it can function again makes existing sleep-wake scheduling schemes for WSNs unusable since a node may not have harvested sufficient energy at the scheduled wakeup time. Furthermore, if it has depleted all its energy in its previous cycle, it may loose its

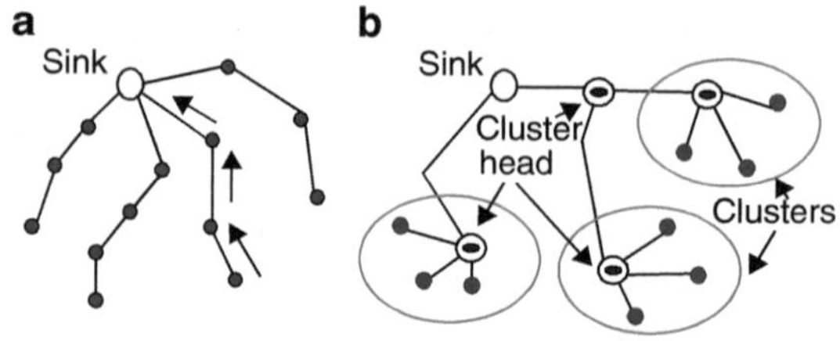

Fig. 17.4 (**a**) multi-hop-based and (**b**) cooperation-based [84]

timing reference when it wakes up again. Additionally, broadcasting -based schemes, which have been proposed for traditional WSNs to avoid the overhead involved in finding routes, are not suitable for EHWSNs. This is because broadcasting results in many duplicates of the same packet if many nodes are awake; therefore, some form of duplication-suppression would be needed so that the harvested energy is not wasted on delivering duplicates.

17.5.1 Overview of EHWSN Routing Protocols

The majority of the routing proposed for EHWSNs are multi-hop-based. First we describe the most prominent multi-hop-based protocols proposed for EHWSNs and then the cluster-based architecture.

17.5.1.1 Multi-Hop-Based Schemes

When energy harvesting techniques were less attractive, there was some research on integrating a few harvesting sensors into an otherwise battery-powered network. In [73], the well known WSN multi-hop routing protocol Direct Diffusion [21] was modified to prolong network life by putting heavier work-loads on the energy-harvesting nodes. That is, the routing protocol would choose harvesting nodes as relays, however if no harvesting node was available, a battery-powered node would be chosen. But such a network cannot operate indefinitely as most of the nodes in the network are battery powered.

The first full-scale approach to the utilization of environmental energy was developed in [24]. They proposed that the spatio-temporal properties of the energy-harvesting rate across the network should be considered for task allotment within the network. They made the first step towards solving the harvesting problem by providing a distributed framework, referred to as the Environmental Energy Harvesting Framework (EEHF) to adaptively learn the energy properties of the environment and harvesting/renewal opportunities in a distributed fashion. This knowledge of the harvesting rate obtained through EEHF along with the remaining battery energy is incorporated in to the cost metric while calculating routes. The proposed cost metric, C, is expressed as follows:

$C = f(T, Em, Ecm, \eta, \varphi, Eb, B)$, where T is the time epoch over which energy prediction is made, Em is the mean energy expected in T time, Ecm is the mean expected energy consumed over this period, η is the prediction confidence, which is a number between 0 and 1, φ is the information regarding when the next recharging opportunity is expected within next time, Eb is the current energy remaining and B is the maximum battery capacity (in one charging cycle). This function f should be such that scheduling decisions based on C automatically selects nodes with higher environmental energy to the maximum extent required for optimizing the lifetime. For instance, when Em-Ecm and η are high at a node the scheduler may use this node even to its last Joule as this node will become alive within next T, to save energy at nodes with low Em .

Following this work by [25], a large number of works were proposed where the harvesting rate, residual energy and other parameters were incorporated into the cost metric [24, 34, 35, 80]. [35] proposed the Energy-opportunistic Weighted Minimum (E-WME) algorithm where the cost metric takes into account the nodes' residual energy, harvesting rate and energy requirement for routing the packet. They provided a mathematical framework using which they showed that the algorithm asymptotically achieves the best achievable performance by any "online algorithm" (An "online" algorithm does not know future packet routing requests at decision time. In contrast, an offline algorithm knows the arrival times and packet sizes of all the packet routing requests, including those in the future) They also showed that the E-WME algorithm can be seamlessly integrated with distance vector-based ad-hoc routing schemes and can be integrated with on-demand-based schemes with minor modifications.

Duty-Cycle-Aware Routing Protocols for EHWSNs

The routing schemes described above either assume fixed, known duty cycles or leave the duty cycle management to the MAC layer. [14] proposed a 'duty-cycle aware multi-hop-based' geographical routing scheme where a node is assumed to be aware of its neighbors duty cycles and uses this information to find the optimum route. However this scheme incurs a lot of delay, as a transmitting node will wait till its intended receiver is awake to transmit its next message. To avoid the excessive delay in these low duty-cycle applications a broadcast-based routing scheme was proposed in [62]. In this scheme, any sensor node that is nearer to the sink than the sender has to rebroadcast the packet. By doing this, this scheme alleviates the delay incurred in waiting for the next hop to be awake in very low duty-cycled applications, since there is a greater probability that at least one of the overhearing relay nodes is in receive mode. However, broadcasting results in many duplicates of the same packet if many nodes are awake; therefore the proposed scheme employs a type of duplication-suppression so that the harvested energy is not wasted on delivering duplicates. A relay node discards all other packets with the same ID that it has already received. This ensures that a relay doesn't transit the same packet more than once.

Communication Using Hybrid Energy Storage System (CHESS)

The protocols described above assume that the RBs have infinite cycle life, which as discussed in Section 17.2.1.1, is not the case. In response to this problem, the authors have proposed a new Communication Using Hybrid Energy Storage System (CHESS) routing metric that distinguishes between the energy states of the SC and the RB in a switched HESS architecture like that shown in Fig. 17.1. CHESS, which was proposed and analyzed in [23], is also sensitive to the depth of discharge of the RB. The general aim of CHESS is similar to that of, as discussed in Section 17.2.1.3: to use SC energy for routine and low data-rate transmissions that can often tolerate high latency, and save the RBs for "alarm" situations. There are several examples of the sort of "fat tailed" traffic distribution to which this strategy might apply: in structural health monitoring (SHM) of a bridge, while each routine measurement has a lot of data, the measurements can be done infrequently, on the order of once per week, and the application can tolerate high latency, so the rate of transmission can be low. However, if fatigue is detected, then the civil engineer may wish to query the monitoring system for much more data that couldn't be supported by SC energy alone. Another example is video camera surveillance: it is often not desired or necessary to transmit video from all cameras all the time. Instead, low data rate sensors, such as sound sensors, operate routinely; when they detect an alarm situation (e.g. a fast rise in sound volume), all the cameras would be turned on, and the RBs would be necessary to transmit the video.

The CHESS metric is an extension of the Lin et al. metric [35] that was discussed in the previous section. The CHESS metric is calculated for each node. Like the Lin et al. approach and other "shortest path" algorithms, the selected CHESS route is the one with the smallest sum of CHESS metrics for each node along the route. The CHESS metric is zero for a node that has sufficient energy on its SC to route a packet. If the SC has insufficient energy, then a non-zero value will be calculated that is based on the residual energy of the RB as well as the remaining cycle life of the RB. In particular, the CHESS metric is a product of two "energy depletion" functions for the RB – one like Lin et al's replenishable RB metric, which discourages selection of a node (within a single discharge cycle) that is near its specified depth of discharge, and another that is like Lin et al's non-replenishable RB metric, where the capacity of the battery is replaced by the RB's cycle life [23].

The authors did a preliminary analysis in [23] that compared a non-HESS (i.e. RB-only) scenario using the Lin et al. protocol, to a HESS scenario that uses the CHESS protocol; this study is summarized next. The analysis was performed for a simple four-node, two-hop network: source-cluster-destination, where the cluster contains two nodes,, as shown in Fig. 17.5. The destination is assumed to be out of decoding range of the source.

In order to exploit the cycle-life vs. depth of discharge (DoD) data that is available in the literature, which assumes a constant (DoD) [8, 12]. We assume that the RB is not connected to the harvesting source (i.e. not re-charged) until the RB has discharged down to the specified DoD. A simplified block diagram of the HESS architecture assumed for analysis is shown in Fig. 17.6. The switches, S1 and S2,

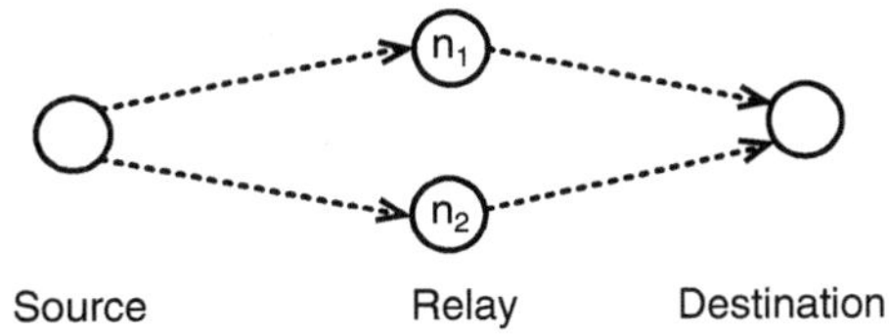

Fig. 17.5 Two-hop relay network

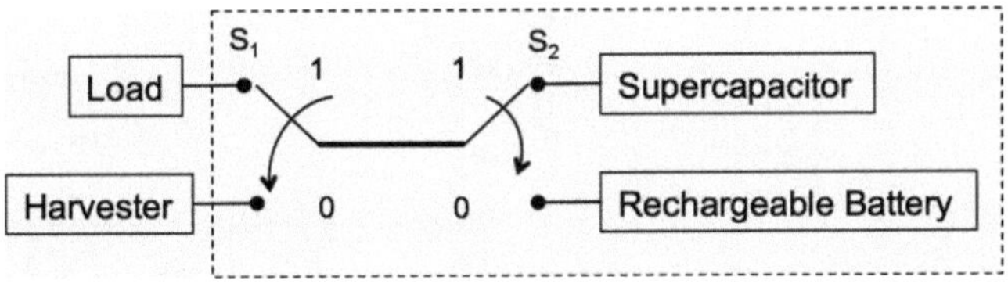

Fig. 17.6 Simplified block diagram of the switched hybrid energy storage system

respectively represent the assumptions that a node cannot harvest energy and transmit a packet at the same time, and that the SC is never connected to the RB. A node that is selected to relay sets S1 to the "1" or "Load" state. Unselected nodes set Switch S1 to the "0" or "harvesting" state. The RB is recharged when its residual energy falls below a pre-set threshold, which is dependent on the DoD and the maximum capacity of the RB. In other words, For example, a DoD of 0.3 means that only 30% of the maximum available energy on the RB is used before recharging. If the RB energy is above the threshold, then the SC is charged using the harvested energy, which corresponds to Switch S2, being set to "1."

The algorithm first attempts to route the packet using the node with the highest SC energy. If that is not possible, then the CHESS metric is evaluated for both nodes in the cluster. The node with the lowest metric then routes the packet.

Fig. 17.7 is a plot of the residual energies in Joules on the SCs and RBs for both nodes versus time for the HESS and non-HESS node architectures from [23]. The simulation setup assumed a very simple daily periodic solar energy model of 12 h of un-interrupted sun, with constant intensity, and 12 h of total darkness. The data traffic was modeled as a sequence of periodic packets for as long as there is energy to route them.

For the network with CHESS, when there is sunlight, there is enough energy to route all the packets using only the SCs, and the RBs on both the nodes simply leak. Once the sunlight stops, the nodes use the SCs to route the packets until their residual energies are no longer enough to route a packet. Subsequent routing of packets involves computation of the CHESS metrics for the relay nodes, and the node with lowest cost function is selected as the relay node. Since there are only two relay nodes, the CHESS routing metric, in this case, merely alternates between the nodes. Route requests are accepted until the residual energies on the nodes go below the specified DoD. Since there is no sunlight, the nodes cannot recharge the RB, and no further packets can be routed. The batteries on both the nodes leak during the rest of the non-harvesting period.

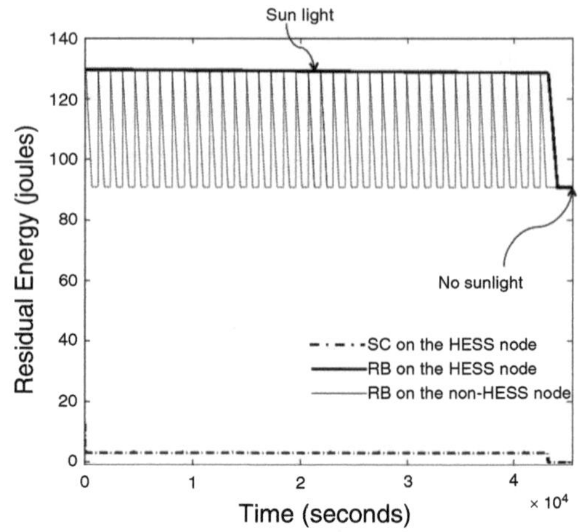

Fig. 17.7 Comparison of the residual energies for the CHESS and non-CHESS cases over a single harvesting period

For the non-HESS scenario, when there is sunlight, the RBs on both nodes route packets and have several charge-discharge cycles. It was found that every charge-discharge cycle in the CHESS case corresponded to 40 charge-discharge cycles for the non-HESS case, which implies a network life extension of approximately 40 times by CHESS relative to non-CHESS. These results suggest the extreme life extension that is possible with HESS technology.

Extensions of this study should include a more realistic charging model that would allow the SC and/or RB to be continuously recharged during packet relaying. Having such continuous charging would reduce the number of charge-discharge cycles in the RB-only case, and diminish the CHESS improvement relative to the non-HESS case. Another interesting extension would be to compare CHESS for switched and non-switched HESS [15] architectures.

17.5.1.2 Cluster-Based Two-Tier Architecture

In addition to the multi-hop-based strategies described above there have been a few cluster-based two-tier architecture schemes (see Fig. 17.4b) proposed for EHWSNs [60, 70]. The most prominent two-tier architecture-based scheme was proposed in [60]. In this work, the distributed energy-harvesting sensor units are divided into clusters, similar to the cells in a cellular network. A cluster head is assigned to each cluster to coordinate the sensors in its cluster and to collect the data from them during monitoring. The clustering of sensors is done based on the proximity (i.e., every node is within one-hop of the cluster head.) The sensor units operate on battery power (rechargeable batteries or super capacitors) and low power operation is the primary criterion in their design. The cluster heads operate on power mains supply, and so do not have any constraints on power. They are provided with a battery backup in case of power failure. The clusters of sensors communicating with their corresponding cluster heads form the lower tier and the network of cluster heads forms the upper tier.

The cluster heads receive the data from the sensors and route the data to the Sink by multi-hop routing over the network formed by the cluster heads.

17.6 Conclusion

EHWSNs are an emerging technology that has the potential to dramatically expand the scope of application of WSNs. Users may be more likely to make the capital investment for applications that have extremely large numbers of nodes or nodes in difficult-to-reach places, if the networks will be reliable and maintenance-free. While the cycle-life of the rechargeable battery (RB) continues to be the barrier to extremely long network life, supercapacitor-only or hybrid energy storage (supercapacitor (SC) and RB) may be able to extend network life to decades. However, the design of communication protocols that can optimally exploit SC energy remain an open problem.

References

1. R. Aguilar-Ponce, A. Kumar, J.L. Tecpanecati-Xihuitl, M. Bayoumi, A network of sensor-based framework for automated visual surveillance. J. Netw. Comput. Appl. **30**(3), 1244–1271 (2007)
2. M.J. Ammer, J. Rabaey, Low power synchronization for wireless sensor network modems, in *Proceedings of the IEEE Wireless Communications and Networking Conference*, 2005
3. ANT. http://www.thisisant.com
4. W.W. Arms, C.P. Townsend, D.L. Churchill, J.H. Galbreath, S.W. Mundell, Power management for energy harvesting wireless sensors, in *Proceedings of SPIE – 15th Annual International Symposium on Smart Structures and Materials*, Mar 2005
5. T. Bokareva, W. Hu *Wireless Sensor Networks for Battlefield Surveillance* (The University of New South Wales, Sydney, Australia, Oct 2006), pp. 1–8
6. D. Brunelli, Energy scavenging techniques (2007), http://www- micrel.deis.unibo.it/MPHS/slidecorso0607/en_scav.pdf. 25 Jan 2007
7. D. Brunelli, L. Benini, Designing and managing sub-milliwatt energy harvesting nodes: opportunities and challenges, in *Proceedings of the Wireless VITAE*, May 2009
8. I. Buchmann, *Batteries in a Portable World*, 2nd edn. (Cadex Electronics Inc., 2001). http://www.buchmann.ca/default.asp
9. K. Chintalapudi, T. Fu, J. Paek, N. Kothari, S. Rangwala, J. Caffrey, R. Govindan, E. Johnson, S. Masri (2006). Monitoring civil structures with a wireless sensor network. IEEE Internet Comput. **March/April**, 26–34 (2006)
10. S. Cui, A.J. Goldsmith, A. Bahai, Energy-efficiency of MIMO and cooperative MIMO techniques in sensor networks. IEEE Trans. Wireless Commun. **22**(6) 1089–1098 (Aug 2004)
11. D.C. Daly, A.P. Chandrakasan, An efficient OOK transceiver for wireless sensor networks. IEEE J. Solid-State Circuits, 42(5), 1003–1011 (May 2007)
12. Ed. Decker, C. Millsaps, Rechargeable battery cycle life issues (2001), Motorola Energy Systems Group Testing Laboratories, Battery Power Products & Technology, http://www.motorola.com/testservices/article1.htm. Mar 2001

13. A. El-Hoiydi, J.-D. Decotignie, WiseMAC: an ultra low power MAC protocol for the downlink of infrastructure wireless sensor networks, in *Proceedings of the Ninth International Symposium on Computers and Communications*, 2004, pp. 244–251
14. T. DeLurio, Storing power with super capacitors, Power Electronics Technology Conference, 2007
15. R.A. Dougal, S. Li, R.E. White Power and life extension of battery-ultracapacitor hybrids. Proc. IEEE Trans. Compon. Packaging Technol. **25**(1), 120–131 (Mar 2002)
16. V. Ekanayake, I.V. Clinton Kelly, R. Manohar, An ultra low-power processor for sensor networks, in *ASPLOS-XI: Proceedings of the 11th International Conference on Architectural Support for Programming Languages and Operating Systems*, New York, NY, USA, pp. 27–36, 2004
17. C. Enz, A. El-Hoiydi, J.-D. Decotignie, V. Peiris, WiseNet: an ultralow-power wireless sensor network solution. Computer **37** (Aug 2004)
18. M.V. Gangone, M.J. Whelan, K.D. Janoyan, R. Jha, Field deployment of a dense wireless sensor network for condition assessment of a bridge superstructure, SPIE 08
19. G.P. Batteries, *Nickel Metal Hydride Technical Handbook*, 2006
21. C. Intanagonwiwat, R. Govindan, D. Estrin, Directed diffusion: A scalable and robust communication paradigm for sensor networks, in *Proceedings of the Sixth Annual International Conference on Mobile Computing and Networking (MobiCOM '00)*, Aug. 2000
22. V. Jovanovic, S. Ghamaty, J.C. Bass, Design, fabrication and testing of a novel energy-harvesting thermoelectric power supply for wireless sensors, in *Proceedings of the PWR2006 2006 ASME Power*, 2006
23. A. Kailas, M.A. Ingram, Y. Zhang, A novel routing metric for environmentally-powered sensors with hybrid energy storage systems, in *Proceedings of the Wireless VITAE*, May 2009
24. A. Kansal, M.B. Srivastava, An environmental energy harvesting framework for sensor networks, in *International Symposium on Low Power Electronics and Design*, pp. 481–486
25. A. Kansal, J. Hsu, M. Srivastava, V. Raghunathan, Harvesting aware power management for sensor networks, in *Proc. Design Automation Conference*, 2006 43rd ACM/IEEE, pp. 651–656, July 2006
26. Aman Kansal et al., Power management in energy harvesting sensor networks. ACM Trans. Embed. Comput. Syst. **6**(4), Article 32 (Sept 2007)
27. K. Kar, M. Kodialam, T.V. Lakshman, L. Tassiulas, Routing for network capacity maximization in energy-constrained ad-hoc networks, in Proceedings of IEEE INFOCOM, 2003
28. M. Koplow et al., Thick film thermoelectric energy harvesting systems for biomedical applications, in *Proceedings of the 5th International Workshop on Wearable and Implantable Body Sensor Networks*, 2008
29. B.T. Kuhn, R.S. Balog, Design considerations for long-term remote photovoltaic-based power supply, in *Proceedings of the Twenty-Third Annual IEEE Applied Power Electronics Conference and Exposition (APEC 2008)*, Feb. 2008
30. K. Lahiri, A. Raghunathan, S. Dey, D. Panigrahi, Battery-driven system design: a new frontier in low power design, in *Proceedings of the 7th Asia and South Pacific Design Automation Conference (ASP-DAC 2002)*, Jan 2002
31. X. Jiang, J. Polastre, D. Culler, Perpetual environmentally powered sensor networks, in *Proceedings of the 4th International Conference on Information Processing in Sensor Networks*, 2005, pp. 463–468
32. J. Langley et al., Multi-source energy harvesting, nanoPower Forum (2007), http://www.maxwell.com/ultracapacitors/products/small-cell/pc-5.asp
34. L. Lin, N.B. Shroff, R. Srikant, A distributed poweraware routing algorithm with logarithmic competitive ratio for sensor networks. Technical report, Purdue University, 2002. Available on http://yara.ecn.purdue.edu/~shroff/PRF/Lin02.pdf
35. L. Lin, N. Shroff, R. Srikant, Asymptotically optimal power-aware routing for multihop wireless networks with renewable energy sources. IEEE/ACM Trans. Netw. **15**(5), 1021–1034 (Oct 2007)
36. D. Linden, T.B. Reddy, *Handbook of Batteries*, 3rd edn. (McGraw-Hill, New York, 2002)

37. F. Liu, A. Phipps, S. Horowitz, K. Ngo, L. Cattafesta, T. Nishida, M. Sheplak, Acoustic energy harvesting using an electromechanical Helmholtz resonator. J. Acoust. Soc. Am. **123**(4) 1983–90 (Apr. 2008)
38. http://www.lumedynetechnologies.com/
39. P. Mars, Energy harvester perpetually powers wireless sensors, Electronic Design, November 17, 2008
40. G.V. Merrett et al., An empirical energy model for supercapacitor powered wireless sensor nodes, in *Proceedings of the 17th International Conference on Computer Communications and Networks (ICCCN '08)*, Aug. 2008
41. Crossbow. Mica2 Data Sheet, http://www.xbow.com/Products/Product_pdf_files/Wireless_pdf/MICA2_Datasheet.pdf
42. Crossbow. Micaz Data Sheet, http://www.xbow.com/Products/Product_pdf_files/Wireless_pdf/MICAz_Datasheet.pdf
43. Application Notes: Battery Technologies used in Alternative Power Systems, http://www.modernoutpost.com/gear/app_power_batteries.html. Downloaded Apr 2009
44. A. Molnar, B. Lu, S. Lanzisera, B. Cook, K.S.J. Pister, An ultra-low power 900 MHz RF transceiver for wireless sensor networks, in *Proceedings of IEEE 2004 Custom Integrated Circuits Conference (CICC)*, pp. 401–404, Oct 2004
45. Texas Instruments Inc., http://focus.ti.com/mcu/docs/mcugettingstarteddetail.tsp?sectionId=97&tabId=1511&familyId=342
46. L. Nazhandali, B. Zhai, J. Olson, A. Reeves, M. Minuth, R. Helfand, S. Pant, T.M. Austin, D. Blaauw. Energy optimization of subthreshold-voltage sensor network processors, in *ISCA* (IEEE Computer Society, 2005), pp. 197–207
47. I. O'Donnell, R. Brodersen, An ultra-wideband transceiver architecture for low power, low rate, wireless systems. IEEE Trans. Veh. Technol. **54**(5), 1623–1631 (Sept 2005)
48. G.K. Ottman et al., Adaptive piezoelectric energy harvesting circuit for wireless remote power supply. IEEE Trans. Power Electron. **17**(5), 669–676 (Sept 2002)
49. J.A. Paradiso, T. Starner, Energy scavenging for mobile and wireless electronics. Pervasive Comput. **4**(1), 18–27 (2005)
50. C. Park, P.H. Chou, AmbiMax: autonomous energy harvesting platform for multi-supply wireless sensor nodes, in *3rd Annual IEEE Communications Society Conference on Sensor and Ad Hoc Communications and Networks (SECON '06)*, Sept 2006
51. A.-F. Pele, ST, CEA concoct tiny fuel cell for wireless phones, EE Times Europe, http://www.eetasia.com/ART_8800527704_765245_NT_52a65943.HTM
52. A. Pereira, J. Mendes, L.F. Melo, Using nanovibrations to monitor biofouling. Biotechnol. Bioeng. **99**(6) (Apr 2008)
53. J. Polastre, R. Szewczyk, D. Culler, Telos: Enabling ultra-low power wireless research, in *Proceedings of The Fourth International Conference on Information Processing in Sensor Networks: Special track on Platform Tools and Design Methods for Network Embedded Sensors (IPSN/SPOTS)*, Apr 2005
54. V. Raghunathan, C. Schurgers, S. Park, M.B. Srivastava, Energy-aware wireless microsensor networks. IEEE Signal Process. Mag. **19**(2), 40–50 (Mar 2002)
55. V. Raghunathan, A. Kansal, J. Hsu, J. Friedman, M. Srivastava, Design considerations for solar energy harvesting wireless embedded systems, in *Proceedings of the Fourth International Symposium on Information Processing in Sensor Networks (IPSN 2005)*, April 2005
56. V. Raghunathan, P.H. Chou, Design and power management of energy harvesting embedded systems, in *Proceedings of the ISLPED*, 2006
57. V. Raghunathan, S. Ganeriwal, M. Srivastava, Emerging techniques for long lived wireless sensor networks. IEEE Commun. Mag. **44**(4), 108–114 (Apr 2006)
58. R. Rao, S. Vrudhula, D.N. Rakhmatov, Battery modeling for energy aware system design. Computer **36**(12), 77–87 (Dec 2003)
59. P. Rong, M. Pedram, An analytical model for predicting the remaining battery capacity of lithium-ion batteries. IEEE Trans. VLSI Syst. **14**(5) (May 2006)

60. E. Sazonova, K. Janoyan, R. Jhac, Sensor network application framework for autonomous structural health monitoring of bridges, in *Proceedings of Structural Materials Technology (SMT): NDE/NDT for Highways and Bridges*, Buffalo, NY, 2005
61. J. Schiller, A. Liers, H. Ritter, ScatterWeb: A wireless sensornet platform for research and teaching. Elsevier Comput. Commun. **28**(13), 1545–1551 (2005)
62. Z.A. Eu, H.P. Tan, W.K.G. Seah, Routing and relay node placement in wireless sensor networks powered by ambient energy harvesting, accepted by IEEE WCNC 2009, Budapest, Hungary, 5–8 Apr 2009
63. Z.A. Eu, W.K.G. Seah, H.P. Tan, A study of MAC schemes for wireless sensor networks powered by ambient energy harvesting, in *Proceedings of the Fourth International Wireless Internet Conference (WICON 2008)*, Maui, Hawaii, USA, 17–19 Nov 2008
64. F. Simjee, P.H. Chou, Everlast: Long-life, supercapacitor-operated wireless sensor node, in *International Symposium on Low Power Electronics and Design (ISLPED)*, 2006
65. N.S. Shenck, J.A. Paradiso, Energy scavenging with shoe-mounted piezoelectrics. IEEE Micro. **21**(3), 30–42 (2001)
66. M. Srivastava, Sensor node platforms & energy issues, Tutorial, Mobicom 2002
67. A.M. Tabar, A. Keshavarz, H. Aghajan, Smart home care network using sensor fusion and distributed vision-based reasoning, in *Proceedings of ACM Multimedia Workshop on VSSN* (2006)
68. Q. Tang, L. Yang, G.B. Giannakis, T. Qin, Battery power efficiency of PPM and FSK in wireless sensor networks. IEEE Trans. Wireless Commun. **6**(4), 1308–1319 (Apr 2007)
69. H. Dubois-Ferrière, L. Fabre, R. Meier, P. Metrailler, TinyNode: a comprehensive platform for wireless sensor network applications, in *Proceedings of the Fifth International Conference on Information Processing in Sensor Networks*, Apr 2006
70. V.A. Kottapalli, A.S. Kiremidjian, J.P. Lynch, E. Carryer, T.W. Kenny, K.H. Law, Y. Lei, Two-tiered wireless sensor network architecture for structural health monitoring, in *SPIE 10th Annual International Symposium on Smart Structures and Materials*, San Diego, CA, USA, 2–6 Mar 2003
71. C. Vigorito, D. Ganesan, A. Barto, Adaptive control for duty-cycling in energy harvesting-based wireless sensor networks, in *Proceedings of the Fourth Annual IEEE Communications Society Conference on Sensor, Mesh, and Ad Hoc Communications and Networks (SECON 2007)*, San Diego, CA, June 2007
72. H.J. Visser, A.C.F. Reniers, J.A.C. Theeuwes, Ambient RF energy scavenging: GSM and WLAN power density measurements, *European Microwave Conference*, Amsterdam, The Netherlands, 2008
73. T. Voigt, H. Ritter, J. Schiller, Utilizing solar power in wireless sensor networks, in *Proceedings of LCN*, Bonn/Konigswinter, Germany, pp.416–422, Oct 2003
74. Z.A. Eu, W.K.G. Seah, H.-P. Tan, A study of MAC schemes for wireless sensor networks powered by ambient energy harvesting, submitted to WiCon 2008
75. A. Wang, S-H. Cho, C.G. Sodini, A.P. Chandrakasan, Energy-efficient modulation and MAC for asymmetric RF microsensor systems, in *Proceedings of ISLPED*, 2001
76. X. Wang et al., Direct current nanogenerator driven by ultrasonic waves. Science **316**(6) (Apr 2007)
77. B.A. Warneke, K.S.J. Pister, An ultra-low energy microcontroller for smart dust wireless sensor networks, in *International Solid-State Circuits Conference*, Feb 2004
78. A.S. Weddell, G.V. Merrett, N.R. Harris, N.M. White, Energy devices for sensor networks: properties for simulation and deployment, in *Proceedings of the wireless VITAE*, May 2009
79. G. Werner-Allen, K. Lorincz, M.C. Ruiz, O. Marcillo, J.B. Johnson, J.M. Lees, M. Welsh, Deploying a wireless sensor network on an active volcano. IEEE Internet Comput. Special Issue on Data Driven Appl. Sens. Netw. **10**, 18–25 (Mar 2006)
80. J. Wieselthier, G. Nguyen, A. Ephremides, Energy limited wireless networking with directional antennas: the case of session-based multicasting, in Proceedings of the IEEE Infocom, 2002
81. W. Ye, J. Heidemann, D. Estrin, Medium access control with coordinated adaptive sleeping for wireless sensor network. IEEE Commun. Lett. **12**(3), 493–506 (June 2004)

82. Zigbee alliance, ieee 802.15.4, zigbee standard. ZigBee Alliance. [Online]. Available at http://www.zigbee.org
83. V. Raghunathan, P.H. Chou, Design and power management of energy harvesting embedded systems, in *Proceedings of the 2006 International Symposium on Low Power Electronics and Design*, pp. 369–374
84. M. Tacca, P. Monti, A. Fumagalli, Cooperative and reliable ARQ protocols for energy harvesting wireless sensor nodes. IEEE Trans. Wireless Commun. **6**(7), 2519–2529 (July 2007)

Mary Ann Ingram received the B.E.E. and Ph.D. degrees from the Georgia Institute of Technology (Georgia Tech), Atlanta, in 1983 and 1989, respectively. After graduating in 1989, she joined the faculty of the School of Electrical and Computer Engineering at Georgia Tech, where she is currently Professor. Her early research areas were optical communications and radar systems. In 1997, she established the Smart Antenna Research Laboratory (SARL) at Georgia Tech, which emphasizes the application of multiple antennas to wireless communication systems. In 2006, she received the Georgia Tech ADVANCE Professorship for the College of Engineering. She was a Visiting Professor at Aalborg University, Aalborg, Denmark in the summers of 2006–2008. The SARL performs system analysis and design, channel measurement, and prototyping relating primarily to wireless local area, ad hoc and sensor networks, with focus on the lower layers of communication networks. Professor Ingram is a Senior Member of the IEEE and has authored or co-authored over 120 refereed journal and conference papers, including four conference papers that have won “Best Paper” awards.

Lakshmi Thanayankizil received her M.S. (2008) and B.S. (2004) degrees in Electrical and Computer Engineering from Georgia Institute of Technology and University of Calicut respectively. Currently, she is working towards the Ph.D. Degree in Electrical and Computer Engineering Department at Georgia Tech. Since 2006, she has been a Graduate Research Assistant at Georgia Tech, where she is working on designing reliable and robust routing schemes for sensor and ad hoc wireless networks. She has also held Industry positions at IBM Research, where her work was on vehicular ad hoc networks and smart-grid-based home sensor networks and at the Indian Space Research Organization (ISRO). She was awarded the IEEE SENSORCOMM Best Paper Award (2008) and NSF ADVANCE Fellowship for Professional Training (2009), and was elected the Workshop Co-Chair for IEEE/ACM N2W (2010).

Jin Woo Jung received the B.S. and M.S. degrees from the Yonsei University, South Korea, in 1999 and 2002, respectively. After graduating in 2002, he joined Datagate International, where he developed security software. In 2004, he joined Samsung Electronics, where he developed digital cable set-top boxes in the Telecommunication Group. He joined the Georgia Institute of Technology (Georgia Tech), Atlanta, in 2007, and he is pursuing Ph.D. degree since then. He is a graduate research assistant of Professor Mary Ann Ingram in the Smart Antenna Research Laboratory (SARL), and his research is focused on the energy-efficient communication protocols for wireless sensor networks. Beginning May 2010, he is working for Apple, Inc. as an intern, which will end in August 2010.

Aravind Kailas received his Ph.D. in Electrical and Computer Engineering (ECE) from Georgia Institute of Technology in 2010, and his doctoral research work addressed the key issue of network sustainability with the design of novel, simple decentralized cooperative diversity-based protocols and intelligent exploitation of energy scavenging using hybrid energy storage systems. He received his M.Sc. in Applied Mathematics from Georgia Institute of Technology in 2010, M.S. in ECE from the University of Wisconsin-Madison, Madison in 2005, and B.E. in Electronics and Telecommunications from Mumbai University, India in 2002 with the highest honors.

Aravind is a recipient of the ECE Graduate Research Assistant Excellence Award (in 2010) and the Colonel Oscar P. Cleaver Award (in 2007) at the Georgia Institute of Technology. He has been a recipient of many other academic scholarships and awards for his Teaching throughout his undergraduate and graduate years. He was a Visiting Research Scientist at the DOCOMO USA Labs, Palo Alto, CA from January to July, 2010, and a Visiting Researcher at the Center for TeleInfraStruktur (CTIF), Aalborg University, Aalborg, Denmark during the summers of 2007–2008. He has also held industry positions at QUALCOMM Inc. and General Electric (GE).

Chapter 18
Coherent Transmission: A Technique for Stopping Global Warming and Climate Change

Shinsuke Hara

18.1 Introduction

"Green ICT" means Information and Communications Technology friendly for the globe. To stop global warming and climate change, ICT, with low emission of carbon dioxide thus low consumption of energy, is required. When there was no constraint of low energy consumption in communications, one solution for making us happier and more comfortable from the information-transmission perspective was to have a "stupid network" in which communication entities can handle just higher data transmission rates with no intelligence. However, once the constraint is imposed, an "intelligent network" is required as another solution in which the entities can intelligently reduce their energy consumption as much as possible while satisfying users' needs.

In wireless communication, a technique to reduce energy consumption is also required, which is called "Green Wireless." Emission of wireless signal from antenna is dominant in energy consumption for wireless devices, so a shortcut for reducing energy consumption is to reduce the power consumption in wireless signal transmission. In principle, the power required for wireless signal emission is proportional to the distance between a pair of communicating transceivers, so this can be achieved by shortening the distance between them. Therefore, in a cellular system, one solution for reducing energy consumption is to put a lot of relay stations to forward information between a base station and mobile stations in each cell.

Assume the distance between a base station and a mobile station is d. For this single hop case, the power consumption is denoted by

$$P_s \propto d^{\beta} \tag{18.1}$$

where β is the power decay factor >2. Now assume that a relay station to forward information is located between the base station and the mobile station with distance

S. Hara (✉)
Osaka City University, Japan
e-mail: hara@info.eng.osaka-cu.ac.jp

R. Prasad et al. (eds.), *Globalization of Mobile and Wireless Communications: Today and in 2020*, Signals and Communication Technology,
DOI 10.1007/978-94-007-0107-6_18,

to the base station of d_r. For this double hop case, the power consumption is written as

$$P_d \propto d_r^{\beta} + (d - d_r)^{\beta} \leq d^{\beta} = P_s. \tag{18.2}$$

Equation (18.2) is satisfied as long as $\beta > 1$.

Recently, "coherent transmission" concept, which can achieve much less energy consumption, has been proposed [1, 2]. Figure 18.1 shows the concept of a coherent transmission system which is composed of a mobile station, K relay stations and a base station. Let us pay attention to the link from the relay stations to the base station. When the relay stations forward the information from the mobile station to the base stations, in a conventional "incoherent" transmission, they transmit their signals with no care about their amplitudes and phases, so the power of the signals combined at the base station is just in proportion to the number of relay stations, namely, K. On the other hand, in a coherent transmission, they transmit their signals so that the signals combined at the base station can be coherently added in "amplitude," so the combined signal power is in proportion to the squared number of relay stations, namely, K^2. This fact clearly shows that the coherent transmission has the power gain of K; in other words, to have the same received power at the base station, as compared with the incoherent transmission, the coherent transmission can reduce the transmission power at each station by a factor of K; therefore, it can significantly reduce the energy consumption.

Furthermore, coherent transmission is categorized into two classes, such as regenerative and non-regenerative. The pre-processing at relay stations which once demodulates the received signal before forwarding is called "regenerative transmission," whereas the one which does not demodulate (just amplifies and forwards) the received signal is called "non-regenerative."

This chapter introduces the coherent transmission as a Green Wireless ICT. Its principle is shown with some computer simulation results and some technical challenges to realize it are discussed. Finally, a realization scenario of the coherent transmission is suggested.

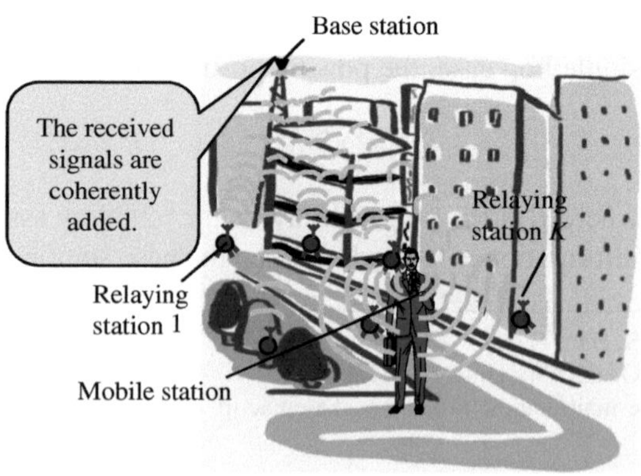

Fig. 18.1 Concept of a coherent transmission

18.2 Coherent Transmission System

A non-regenerative coherent transmission contains a regenerative coherent transmission as a subclass; therefore, in this chapter, we derive the optimal weights at relay stations for a non-regenerative coherent transmission system. Let us look at Fig. 18.2, where the communication link between a transmitter (*TX*) and a receiver (*RX*) uses K non-regenerative relay stations (*RSs*). First, the transmitter sends a signal over frequency non-selective fading channels for the relay stations (Link_1). The received signal at the kth relay station is written as ($k = 1, \cdots, K$)

$$y_k = h_{1k} x + n_{RSk} \tag{18.3}$$

where x is the transmitted signal, and h_{1k} and n_{RSk} are the complex-valued channel gain between the transmitter and the kth relay station and the complex-valued Gaussian noise added at the kth relay station, respectively. They have the following statistical properties:

$$E[x] = 0 \tag{18.4}$$

$$E[n_{RSk}] = 0 \tag{18.5}$$

$$E[x^H x] = P_{TX} \tag{18.6}$$

$$E[n_{RSk}^H n_{RSk}] = \sigma_{RSk}^2 \tag{18.7}$$

where $E[(\cdot)]$ and $(\cdot)^H$ denote the statistical average and Hermitian transpose of $(\cdot)$ respectively. Therefore, the average power of y_k is given by

$$E[y_k^H y_k] = |h_{1k}|^2 P_{TX} + \sigma_{RSk}^2. \tag{18.8}$$

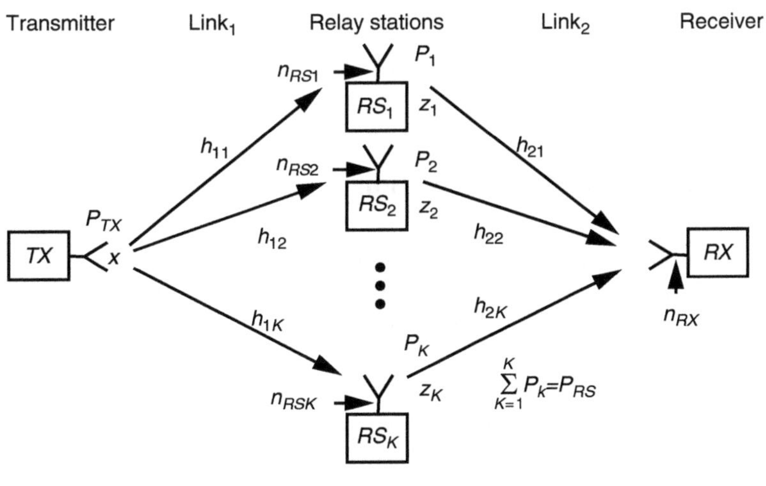

Fig. 18.2 A non-regenerative transmission system model

Then, after phase adjustment and power amplification, the kth relay station forwards the stored signal to the receiver. The transmitted signal from the kth relay station is written as

$$z_k = \left(\frac{\sqrt{P_k}}{\sqrt{|h_{1k}|^2 P_{TXk} + {}^{2}_{RSk}}} \right) y_k. \tag{18.9}$$

Taking into consideration that the signal power from the total K relay station is constrained, P_k is written as

$$\sqrt{P_k} = w_{RSk} \sqrt{P_{RS}} \tag{18.10}$$

$$\sum_{k=1}^{K} w_{RSk}^{H} w_{RSk} = 1 \tag{18.11}$$

where P_k is the total signal power from the K relay stations and w_{RSk} is the weight at the kth relay station to adjust the phase and amplitude of the transmitted signal.

Finally, the signals transmitted from the relay stations arrive at the receiver over non-frequency selective fading channels (Link$_2$). The received signal at the receiver is written as

$$v = \sum_{k=1}^{K} h_{2k} z_k + n_{RX} \tag{18.12}$$

where h_{2k} and n_{RX} are the complex-valued channel gain between the kth relay station and the receiver and the complex-valued Gaussian noise added at the receiver, respectively. They have the following statistical properties:

$$E[n_{RX}] = 0 \tag{18.13}$$

$$E[n_{RX}^{H} n_{RX}] = \sigma_{RX}^{2}. \tag{18.14}$$

18.3 Optimal Weights Derivation

Larsson beautifully derived the optimal weights at relay stations for maximizing the *SNR* at a receiver by means of Cauchy-Schwarz inequality. Here in this section, we try to derive it with matrix/vector notations [3].

Let us define the following vectors and matrices as

$$\mathbf{w}_{RS} = [w_{RS1}, \ldots, w_{RSK}]^T \tag{18.15}$$

$$\mathbf{H}_1 = diag\{h_{11}, \ldots, h_{1K}\} \tag{18.16}$$

$$\mathbf{h}_2 = [h_{21}, \ldots, h_{2K}]^T \tag{18.17}$$

$$\mathbf{N}_{RS} = diag\{n_{RS1}, \ldots n_{RSK}\} \tag{18.18}$$

$$\mathbf{Q} = diag\left(\frac{\sqrt{P_{RS}}}{\sqrt{|h_{11}|^2 P_{TX} + \sigma_{RS1}^2}}, \cdots, \frac{\sqrt{P_{RS}}}{\sqrt{|h_{1K}|^2 P_{TX} + \sigma_{RSK}^2}}\right) \tag{18.19}$$

$$\mathbf{z} = [z_1, \cdots z_K]^T \tag{18.20}$$

where $(\cdot)^T$ denotes the transpose of $(\cdot)$ and $diag\{\cdots\}$ denotes the diagonal matrix. With (18.22)–(18.26), (18.27) is written as

$$\mathbf{z} = \mathbf{Q}\left(x\mathbf{H}_1 + \mathbf{N}_{RS}\right)\mathbf{w}_{RS} \tag{18.21}$$

so the received signal at the receiver is written as

$$v = x\mathbf{h}_2^T\mathbf{Q}\mathbf{H}_1\mathbf{w}_{RS} + \mathbf{h}_2^T\mathbf{Q}\mathbf{N}_{RS}\mathbf{w}_{RS} + n_{RX}. \tag{18.22}$$

From (18.22), the signal power is calculated as

$$\begin{aligned} S &= E\left[\left(x\mathbf{h}_2^T\mathbf{Q}\mathbf{H}_1\mathbf{w}_{RS}\right)^H \left(x\mathbf{h}_2^T\mathbf{Q}\mathbf{H}_1\mathbf{w}_{RS}\right)\right] \\ &= \mathbf{w}_{RS}^H\left(\mathbf{H}_1^H\mathbf{Q}^H\mathbf{h}_2^*\sqrt{P_{TX}}\sqrt{P_{TX}}\mathbf{h}_2^T\mathbf{Q}\mathbf{H}_1\right)\mathbf{w}_{RS}. \end{aligned} \tag{18.23}$$

On the other hand, the noise power is calculated as

$$\begin{aligned} N &= E\left[\left(\mathbf{h}_2^T\mathbf{Q}\mathbf{N}_{RS}\mathbf{w}_{RS} + n_{RX}\right)^H \left(\mathbf{h}_2^T\mathbf{Q}\mathbf{N}_{RS}\mathbf{w}_{RX} + n_{RX}\right)\right] \\ &= \mathbf{w}_{RS}^H\mathbf{Q}'\mathbf{w}_{RS} \end{aligned} \tag{18.24}$$

where

$$\mathbf{Q}' = diag\left\{\frac{|h_{21}|^2 P_{RS}\sigma_{RS1}^2}{|h_{11}|^2 P_{TX} + \sigma_{RS1}^2} + \sigma_{RX}^2, \cdots, \frac{|h_{2K}|^2 P_{RS}\sigma_{RSK}^2}{|h_{1K}|^2 P_{TX} + \sigma_{RSK}^2} + \sigma_{RX}^2\right\}. \tag{18.25}$$

Defining $\mathbf{U}$ as

$$\mathbf{U}^H\mathbf{U} = \mathbf{Q}' \tag{18.26}$$

$$\mathbf{U} = diag\left\{\sqrt{\frac{|h_{21}|^2 P_{RS}\sigma_{RS1}^2}{|h_{11}|^2 P_{TX} + \sigma_{RS1}^2} + \sigma_{RX}^2}, \cdots, \sqrt{\frac{|h_{2K}|^2 P_{RS}\sigma_{RSK}^2}{|h_{1K}|^2 P_{TX} + \sigma_{RSK}^2} + \sigma_{RX}^2}\right\} \tag{18.27}$$

Equation (18.2) is rewritten as

$$N = \tilde{\mathbf{w}}_{RS}^{H} \tilde{\mathbf{w}}_{RS} \tag{18.28}$$

$$\tilde{\mathbf{w}}_{RS} = \mathbf{U}\mathbf{w}_{RS}. \tag{18.29}$$

Therefore, the *SNR* is given by

$$\frac{S}{N} = \frac{\tilde{\mathbf{w}}_{RS}^{H} \mathbf{R} \tilde{\mathbf{w}}_{RS}}{\tilde{\mathbf{w}}_{RS}^{H} \tilde{\mathbf{w}}_{RS}} \tag{18.30}$$

$$\mathbf{R} = \left(\sqrt{P_{TX}}\,\mathbf{h}_2^T \mathbf{Q} \mathbf{H}_1 \mathbf{U}^{-1}\right)^H \left(\sqrt{P_{TX}}\,\mathbf{h}_2^T \mathbf{Q} \mathbf{H}_1 \mathbf{U}^{-1}\right) \tag{18.31}$$

and maximizing (18.30) results in the following eigenproblem:

$$\mathbf{R}\tilde{\mathbf{w}}_{RS} = \lambda \tilde{\mathbf{w}}_{RS}. \tag{18.32}$$

The rank of $\mathbf{R}$ is one, so the optimal weight is given by

$$\begin{aligned} \mathbf{w}_{RS} &= \mathbf{U}^{-1} \tilde{\mathbf{w}}_{RS} \\ &= \mathbf{U}^{-1} \left(\sqrt{P_{TX}}\,\mathbf{h}_2^T \mathbf{Q} \mathbf{H}_1 \mathbf{U}^{-1}\right)^H. \end{aligned} \tag{18.33}$$

Consequently, from (18.33), the *k*th optimal weight is given by

$$w_{RSk} = \alpha_{RS} \frac{\sqrt{\Gamma_{RSk}\Gamma_{RXk}}\sqrt{\Gamma_{RSk}+1}}{\Gamma_{RSk}+\Gamma_{RXk}+1} e^{-j(\theta_{1k}+\theta_{2k})} \tag{18.34}$$

where θ_{1k} and θ_{2k} denote the phases of h_{1k} and h_{2k}, respectively, Γ_{RSk} and Γ_{RXk} denote the *SNR*s for the signal transmitted from the transmitter and received at the *k*th relay station and for the signal transmitted from the *k*th relay station and received at the receiver, respectively, which are defined as

$$\Gamma_{RSk} = |h_{1k}|^2 P_{TX} / \sigma_{RSk}^2 \tag{18.35}$$

$$\Gamma_{RXk} = |h_{2k}|^2 P_{RS} / \sigma_{RX}^2 \tag{18.36}$$

α_{RS} denotes the coefficient to let the weights satisfy (18.11), so it is calculated as

$$\alpha_{RS} = \left\{ \sum_{k=1}^{K} \frac{\Gamma_{RSk}\Gamma_{RXk}\left(\Gamma_{RSk}+1\right)}{\left(\Gamma_{RSk}+\Gamma_{RXk}+1\right)^2} \right\}^{-1/2}. \tag{18.37}$$

Finally, the maximal *SNR* at the receiver is given by

$$\frac{S}{N} = \sum_{k=1}^{k} \frac{\Gamma_{RSk}\Gamma_{RXk}}{\Gamma_{RSk} + \Gamma_{RXk} + 1}. \tag{18.38}$$

Equations (18.34) and (18.38) are the same as those derived by Larsson [2]. In addition, note that regenerative relaying with no decision errors at relay stations corresponds to the case with $\Gamma_{RSk} \to \infty$ and $\theta_{1k} \to 0$ in (18.34) and (18.38). Therefore, (18.38) leads to

$$\begin{aligned}\frac{S}{N} &= \sum_{k=1}^{K} \frac{\Gamma_{RXk}}{1 + (\Gamma_{RXk} + 1)/\subset_{RSk}} \\ &\to \sum_{k=1}^{K} \Gamma_{RXk} \quad (\Gamma_{RSk} \to \infty)\end{aligned} \tag{18.39}$$

Equation (18.39) is equivalent to the attainable *SNR* by normal maximal ratio combining [4].

18.4 Numerical Results

Figure 18.3 shows the layout of a cell for evaluating the performance of a non-regenerative incoherent and coherent transmission systems, where a transmitter and a receiver are located at $\left(-0.5/\sqrt{2}, -0.5/\sqrt{2}\right)$ and $\left(0.5/\sqrt{2}, 0.5/\sqrt{2}\right)$, respectively, and relay stations are located on a grid $\left(K = L^2\right)$. The channel gain for Link_1 and Link_2 is defined as

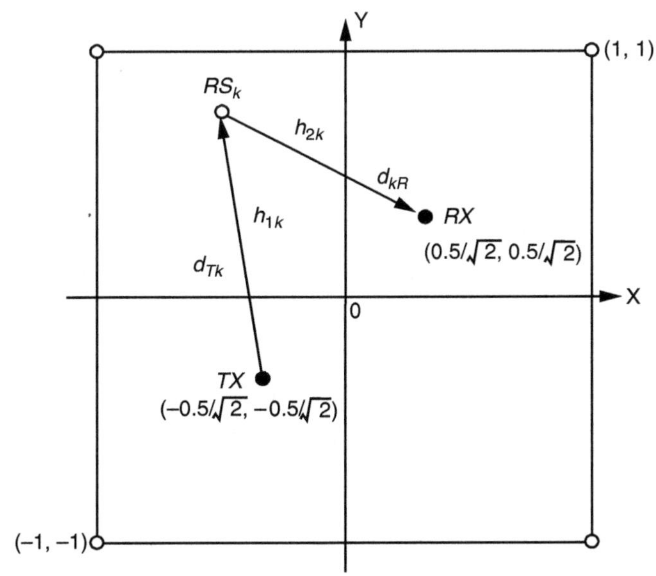

Fig. 18.3 Layout of a cell

$$h = (h_i + jh_q) \times (\alpha \cdot d^{-\beta}) \tag{18.40}$$

where h_i and h_q are mutually independent Gaussian random variables with mean of zero, respectively, and $(\alpha \cdot d^{-\beta})$ corresponds to the near/far effect with gain factor of α, distance of d and decay factor of $\beta = 1.5$.

Figures 18.4 and 18.5 show the spatial distribution of the average *SNR* for non-regenerative incoherent and coherent transmissions, respectively. Here, it is assumed that the transmission power from the transmitter is the same as the total transmission power from all the relay stations and the average SNR_{direct} (directly from the transmitter to the receiver) without relay is set to 0 dB. For the case of

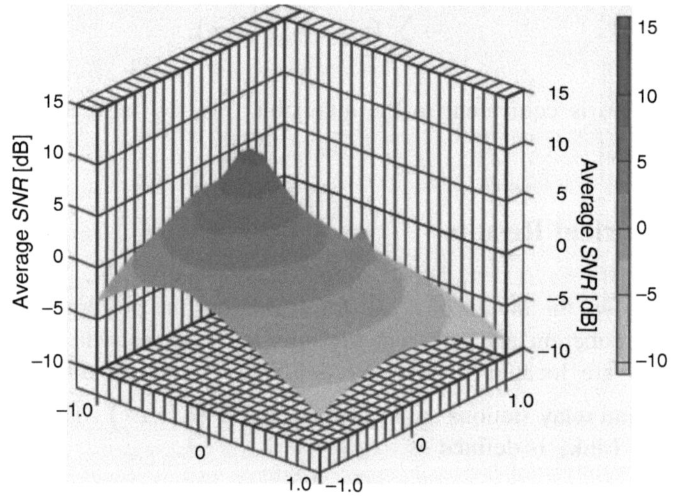

Fig. 18.4 Average *SNR* for the non-regenerative incoherent transmission ($K = 36$)

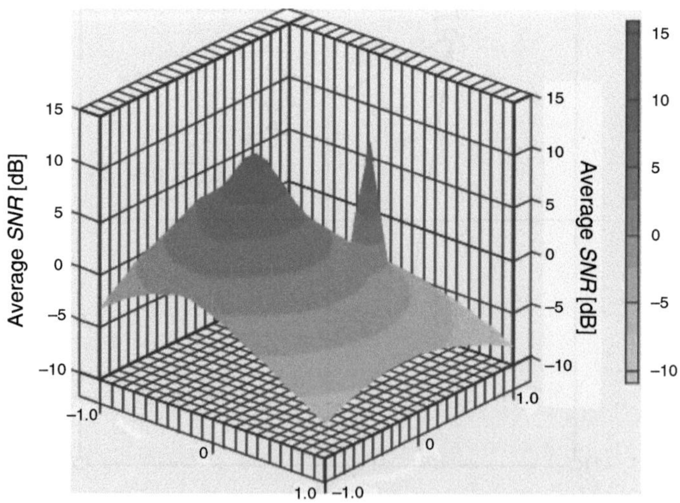

Fig. 18.5 Average *SNR* for the non-regenerative coherent transmission ($K = 36$)

incoherent transmission, the relay stations adjust the transmitted signals so as to compensate only for the near/far effect, so the average *SNR* is a little increased at the location of the receiver, but the value is still around the average $SNR_{direct} = 0$ dB. On the other hand, for the case of coherent transmission, the average *SNR* is much increased at the location of the receiver, and the value reaches around 13 dB. The *SNR*s at the relay stations located close to the transmitter are increased in the links so the signals forwarded from those relay stations to the receiver have higher *SNR*s. As a result, the average *SNR* is higher around the location of the transmitter.

For comparison purpose, Figs. 18.6 and 18.7 show the spatial distribution of the average *SNR* for regenerative incoherent and coherent transmissions, respectively,

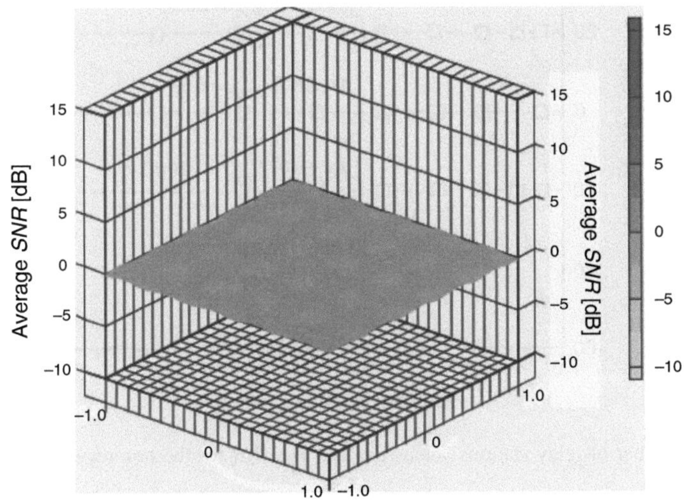

Fig. 18.6 Average *SNR* for the regenerative incoherent transmission ($K = 36$)

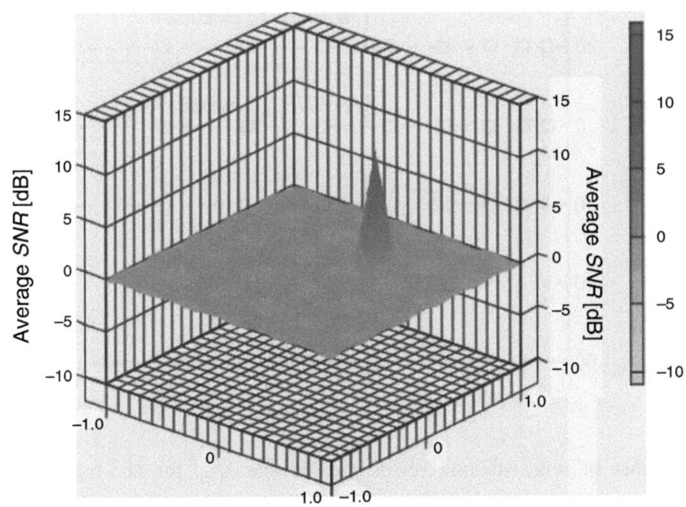

Fig. 18.7 Average *SNR* for the regenerative coherent transmission ($K = 36$)

where the decision errors at the relay stations are ignored. The coherent combining gives higher average *SNR* at the location of the receiver but the average *SNR* remains lower even around the location of the transmitter.

Figures 18.8 and 18.9 show the relationship between the number of relay stations and the average *SNR* for non-regenerative incoherent and coherent transmissions, respectively. The incoherent combining makes the average *SNRs* at most up to the average $SNRs_{direct}$, but the coherent combining makes the average *SNRs* much higher. The average *SNR* linearly increases as the number of relay stations increase.

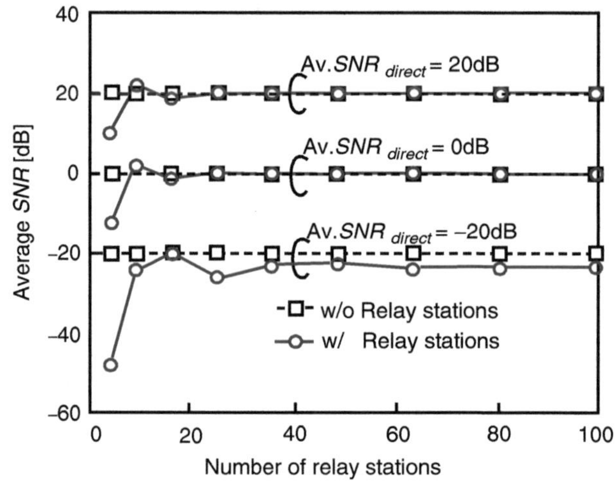

Fig. 18.8 Number of relay stations versus the average *SNR* for the non-regenerative incoherent transmission

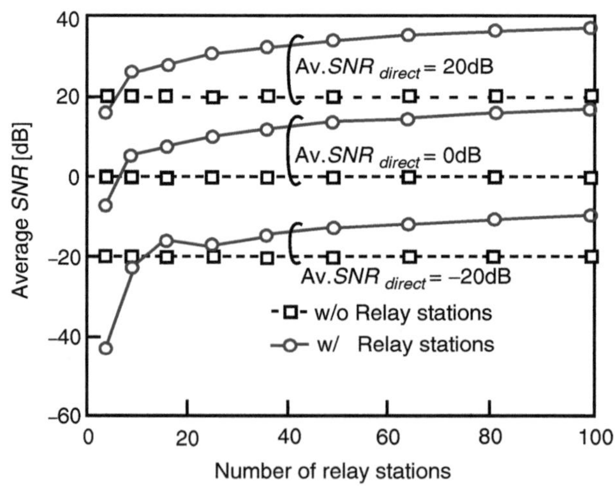

Fig. 18.9 Number of relay stations versus the average *SNR* for non-regenerative coherent transmission

18.5 Discussion and Technical Challenges

Figures 18.4–18.7 clearly show that the coherent transmission can give a high *SNR* only at the receiver, so this implies that, if the receiver is equipped with multiple antennas then spatial multiplexing can be easily achieved. Furthermore, there is no inter-cell-interference observed in those figures, so this also implies that the frequency reuse factor of one can be achieved, like code division multiple access (CDMA) in the current third generation (3G) mobile communication systems. Therefore, the coherent transmission can achieve not only lower energy consumption but also higher spectral utilization efficiency.

However, the performance evaluation has assumed "perfection", such as perfect synchronization in frequency and time among the relay stations, no frequency error and phase noise in the crystal of the local oscillator at each relay station, perfect estimation on the channel state information and so on. This perfection cannot be realized in reality, so the effect of releasing such perfection on the system performance needs to be realized first. Some technical challenges to realize the coherent transmission are as follows [4]:

- How to know and estimate the channel state information for the relay stations: some feedback information is required on the second link from the base station. The use of time division multiple access with time division duplex (TDMA/TDD) seems one of promising techniques for effectively estimating and transmitting the channel state information, but an efficient frame design is required
- How to synchronize in time and frequency among the relay stations: an efficient control signal design is required
- And so on

18.6 A Realization Scenario for Coherent Transmission

Figure 18.10 shows a conversation between a customer and a salesperson at a cell phone shop in the 2030s. Along this scenario, the number of relay stations, which can linearly reduce the energy consumption in cellular systems, can be increased. The relay stations must be powered by solar batteries with no emission of carbon dioxide.

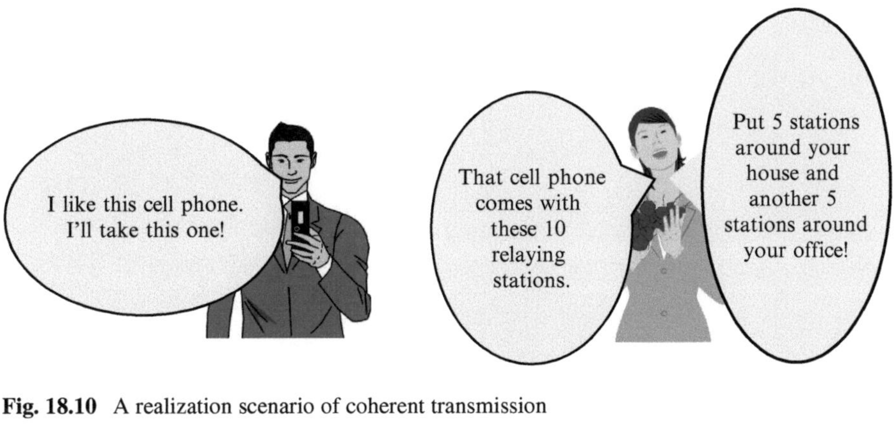

Fig. 18.10 A realization scenario of coherent transmission

18.7 Conclusions

This article has introduced coherent transmission as a "Green Wireless" for stopping global warming and climate change. It can achieve not only lower energy consumption but also higher spectral utilization efficiency as compared with current mobile communication systems. To realize the coherent transmission, there are many technical challenges, so some of them have been listed. Finally, a realization scenario of the coherent transmission has been suggested. The author personally believes that the coherent transmission-based double-hop transmission must be a core architecture in post-4G mobile communication systems.

References

1. Y.S. Tu, G. Pottie, Coherent cooperative transmission from multiple adjacent antennas to a distant stationary antenna through AWGN channels. Proceedings of VTC'02-Spring, in CD-ROM, Birmingham, USA, 4–9 May 2002
2. P. Larsson, Large-scale cooperative relaying network with optimal coherent combining under aggregate relay power constraints. Proceedings of FTC'03, Beijing, China, 2003
3. H.M. Tran, S. Hara, Y. Hara, A non-regenerative/coherent combining relaying system. Technical Report of IEICE, RCS2007-228, pp. 251–256, March 2008
4. M. Schwartz, W.R. Bennett, S. Stein, *Communication systems and techniques* (IEEE Press, Piscataway, NJ,. 1996)

Shinsuke Hara received the B. Eng., M. Eng. and Ph.D. Degrees in communications engineering from Osaka University, Osaka, Japan, in 1985, 1987 and 1990, respectively. Since October 2005, he has been with the Graduate School of Engineering, Osaka City University as a Professor. In addition, he is an International Scholar at Hyung Hee University, Korea and a Senior R&D Adviser of Synthesis Corporation, Japan. From April 1995 to March 1996, he was a visiting scientist at Telecommunication and Traffic Control Systems Group, Delft University of Technology, Delft, The Netherlands. He research interests include wireless communications systems and digital signal processing.

Chapter 19
Novel Services and Applications Demand Intelligent Software

Rasmus Løvenstein Olsen

19.1 Introduction

The idea of using context for enhancing application and service behavior toward humans is not new, and has its roots in the human mind and how it works, i.e. many decisions, perceptions and understanding of information we make as humans are based on the context we are in. A concrete example from the wireless communication world is how sometimes we find ourselves in need to exchange files between two mobile phones, and yet, having so much trouble finding the right settings to exchange this file when the devices are just there in front of us, if the file transfer is possible at all due to incompatible air interfaces!

It seems so simple to us humans, just think about exchanging an apple between two people. At first glance this requires only little thinking, but in fact a lot of things go on after all; a person first figures out he/she wants an apple, then this person notices the apple at the hand of the another person. Following this step, both person needs to communicate that they desire to exchange the apple (assuming the person does not just steal it or stretch it!) by some means, e.g. oral or by body language. Much of this part of the process does involve also understanding of the given context; e.g. does the person with the apple appears to be giving out the apple, selling it, or keeping it for personal use? Such observations easily simplify our decisions without even having to communicate then; say the person with the apple is eating it, or that the price of the apple is very high then maybe the person will not even ask for the apple, saving energy on the communication. Once the involved persons agree on the exchange, the actions follow by movements of hands. We do these types of processes everyday, and in fact our brains are so used to this that we do not even think about the involved steps for performing these types of actions.

But for the devices involved in a file exchange it is not so simple, in fact it may even be impossible if they do not agree on a common air interface. Imagine that the devices could figure out that they are close to each other, and they need to establish

R.L. Olsen (✉)
Aalborg University
e-mail: rlo@kom.aau.dk

R. Prasad et al. (eds.), *Globalization of Mobile and Wireless Communications: Today and in 2020*, Signals and Communication Technology,
DOI 10.1007/978-94-007-0107-6_19,

a connection for file exchange, relieving the user from the hassle of doing settings and experimenting which interface is the most appropriate to use. One may ask him/herself whether we would need systems like this in the future. Does it not increase complexity of systems? Endanger the user's privacy and security requirements? The answers to these questions are all *yes*! Why? We would need these systems to be context aware since with the growing technological capacity, communication means etc., bring an ever increasing complexity system which for users at some point (if it has not already happened) becomes impossible to manage by themselves. Think about searching for information on the Internet, for example. Without search engines, like Google, Yahoo, ..., or naming schemes like DNS or INS, there is absolutely no way we can find out ourselves where to find information, rendering the system basically useless to us. Take this thought and map it into a world where all our devices can communicate in different ways, with different air interfaces, with different protocols, with different settings, for different purposes at different times and places, it should be quite obvious that we need devices, services and applications to become aware of their situation to aid us in making decisions on what to do, where and at what time, i.e. to make them intelligent!

To combat increasing complexity, the wireless community has been working on cognitive radio to make communication between devices more intelligent for e.g., intelligent spectrum sharing. However, there is more to context awareness than simply observing a given spectrum, namely, the extraction of information from the environment, and the user and other relevant objects involved in the correct interpretation of a situation. Thus context aware systems must depend on many different types of information sources, where sensors are the key sources of context information. This means that activities with respect to (wireless) sensor networks are extremely important to context aware systems, but they build on top of system solutions because context aware systems are more focused on the interpretation and use of information than on the actual sensing part. But, as it will be later discussed in the paper, the gathering of information will have a significant impact on the reliability of information pertaining to context awareness, which must be taken into account to deliver enhanced experience to the end user.

In this chapter, we highlight the challenges and discuss the future of context information at a network level. We also present some key solutions to ensure a successful deployment of context sensitive services and applications.

19.1.1 Context Information and Context Awareness

The concept that applications and services are aware of the user's context and is capable of reacting upon it is known as being context aware. This is not a new concept, and was already studied in the 1990s, see e.g. [3]. However, due to the complexity of dealing with such a general topic, this is still being pursued. A very popular notion and definition of context is given [1] in his Ph.D. thesis, which states that:

> Context is any information that can be used to characterize the situation of an entity. An entity is a person, place, or object that is considered relevant to the interaction between a user and an application, including the user and application themselves.

The definition itself is quite wide, but hits the key point. It is in fact only the *relevant* information that we want our applications and services to apply to in order for them to behave as we would like them to. At one point, a user might want the phone to be in silent mode because he is in the cinema, but another time it may be okay for it to ring loudly because of an emergency in the family, or because he is in a private setting. It all depends on the context the user is in. Some may claim it also depends on content and other information types, but if we look at the definition of context, then we see that it is all covered by the overall term. This is, in its core, what makes context awareness such a strong concept, but it is also its weakness and why ***true*** context awareness is still somewhat difficult to exploit. The vast space of context information available to us is simply greater than the number of devices, sensors, and computers etc. existing today.

Being context aware [1, 2, 3] is defined as the capability of an entity to be aware of its situation, and be able to potentially react appropriately to the situation, or as minimum take a decent decision:

> Being context aware is the ability of an entity to be aware of the current circumstances and be aware of relevant information that may influence any decision and behavioral change based on context it may or may not take.

In practice, context awareness can be divided into two groups:

- ***Proactive kind***: Those entities providing proactive context awareness uses context information proactively to react on changes, e.g. by changing parameters and configuration to fit into the new situation or notifying a user of something, [24, 25].
- ***Reactive kind***: Those entities who are reactively using context information typically needs to know the current situation to make a decision here and now. An example is context aware service discovery; see [4, 5].

Whether proactive or reactive type of context awareness is to be used, context information must be made available to the application beforehand.

19.1.2 Scenarios and Examples for Beyond 2020

Technology beyond 2020 would need to be much more adaptive and responsive to the environment that a user is in. Some examples of how context awareness will benefit the user can be described by a few scenarios:

- Imagine that a user has gone to a cinema, and forgets to switch off his mobile phone and his mother in-law calls him. With today's technology the phone would ring, but imagine that the mobile phone can learn automatically that the user is at the cinema and does not want to be disturbed. Perhaps the user would like the phone to silently indicate an incoming phone call. However, if it was an incoming call about his father having a heart attack, then it would be a different matter, and he would have wanted to be disturbed even though he is at the cinema. Later in the evening, of course, the user does not want the phone to behave in this way, because he is now at a location where he does not mind being

disturbed by incoming calls, and finally when he goes to sleep, he may want unnecessary incoming calls to be redirected to his answering machine. It all depends on the context of the user how the phone should react.
- A traveling businesswoman arrives at an airport but is in need of a printout of some documents. She does not know the airport well, and does not know where the printers where she can print private and confidential documents. She uses her intelligent mobile phone to discover available printers, or the mobile itself is aware that she needs a printer and notifies her when she gets close to a printer that fits her current need of printing private and confidential information in a public place. The system detects nearby printers that are not in much use but fit her immediate needs, color, resolution etc. Later, when she arrives back in her office, her needs and requirements for printers are different.
- A user who has a mobile music player may want to play the audio to the stereo when he gets back home instead of using earplugs or the car stereo. The user may want this to happen automatically without him having to fiddle with settings and want to have the music follow him around the house. However, if he is not alone in the house, then it is not a good idea for the system to play loudly the heavy metal music he likes so much, while the other person in the house may be relaxing to classical music in the other room or simply like to be in a quiet environment.

The problem here is not related to if this can be done, but that it should happen automatically and at all places, because the involved devices would need to know of the users whereabouts, situation (personally as well as socially) and gain knowledge about the environment. And in some cases this information may simply not be accessible, be inaccurate or corrupted, e.g., due to large access delays, hacking, system failures, etc. Furthermore, as the above examples illustrate, it is really a large range of information types needed at different times for different applications that need to be provided to the applications before we can talk about a true context aware application and service environment, which lead to the main concerns and challenges to overcome for context aware applications and services to work properly, namely efficient context management. In the following, we look deeper into some of the most important issues that need to be solved to achieve this goal.

19.1.3 Other Typical Examples of Context Information Types

Even though context as on the definition is basically everything and nothing at the same time, we provide below a few concrete examples of what is typically considered context information today:

- *User context*: Any attributes describing a user's context, e.g., the activity, position, status and so on.
- *Device context*: Any attributes describing the situation of a device and its environment, e.g., battery status, OS environmental parameters such as available memory, CPU usage and so on.

- *Network context*: Any information regarding the network, e.g., link state, end to end delay and bandwidth, topology information, network type etc.
- *Environmental context*: Any information about the environment, e.g., the ambient temperature, light intensity, sound level, etc.

The above categories, although still quite high level, describe well what is relevant to networks of the future.

19.1.4 Increasing Competitiveness by Developing Intelligent Software

The trend that is being observed when migrating from wired communication to wireless communication is that user expects basically the same advantages of high data rates from mobile devices as from stationary devices. The same can easily be envisioned when discussing applications and services; as people get more used to interact with services and applications on a daily basis, the more they expect from the software. When working on stationary devices, adaptive services may not appear as attractive because the settings are mostly static, but when considering mobile devices and applications, the user would expect better experience as illustrated in the example above. In fact, software companies and service providers should take such requirements seriously, since they can easily become deciding factors for the consumer. For example, some of the most deciding factors for a successful deployment include:

- ***Level of security***: applications, services and systems involving sensitive and personal information related to the user's context are critical to be kept safe. The user should have trust that the software would not spread personal information, spam, or in any way jeopardize the user's identity.
- ***Level of correctness determining the adaptive behavior***: if the service or application does not act properly on the context, but appears to be random, this will surely annoy the user more than its help. If the reliability (or correctness) of the system is not robust, the advantage that the adaptive software renders would have just the opposite effect.
- ***Reaction times***: similar to the correctness, if the system does not respond to changes relatively quickly, the system may simply be rendered useless. Some users may even just override the functionality by setting the software manually instead of leveraging intelligence built into the software. Again, what was envisioned to be an advantage would turn into a disadvantage to the user.

Obviously, other metrics like power savings for mobile devices, processing capabilities, etc., are also important, but if the above requirements are not fulfilled to a satisfactory level, there would be no businesses in this area, see also [22]. However, there is hope! Context Management systems can help address the above issues, and provide platforms that serve providers and application developers to offer efficient, reliable and secured information access. Next, in the following section, Context Management will be the focal point, with more attention given to the issue of level of correctness and what it means to the system and to the user.

19.2 Context Management Systems

Context information is not always accessed from the device itself. In fact, in most cases access to other devices is necessary to get proper and valid context information, because the needed information is either not available, or not valid (e.g., a microphone on a mobile phone that has been put in a pocket does not provide the true ambient noise level), or simply not accessible due to an interface mismatch between the source and the client. To make context information available, a Context Management Framework is needed (Fig. 19.1).

The main purpose of a Context Management framework is to support context sensitive applications and services for easy access to information distributed in a network. Therefore, it is important that the framework can access any arbitrary information type and provide this information using uniform semantics. Furthermore, it is vital that the framework has an efficient distribution algorithm as quick access to the information, which typically is dynamic and changes value over time, is vital for context sensitive applications to react correctly in a given situation. As context information may not necessarily be measured directly by any single sensor or sensing application, context information may need to be inferred or derived by dedicated algorithms such as Bayesian Networks or Fuzzy Logic based algorithms. Thus a context management framework would need to support such type of inference logic, which poses the challenge to produce algorithms that are light enough to be able to run on low power, low processing capable devices.

However, such inferred information may also benefit from the distribution possibilities envisioned for Context Management system, as a single node may share

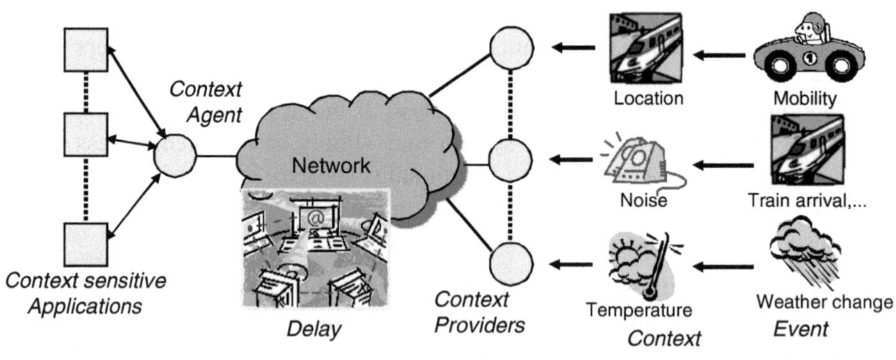

Fig. 19.1 The concept of context management, namely to ensure easy access to various information types distributed in a network

the inferred information with other devices, thus releasing the responsibility of doing hard computational tasks from low computational devices to high capability devices. Finally, since context relates very closely to a user, his privacy, security and privacy protection of context information are vital to ensuring that Context Management and context sensitive applications will eventually be successful. For year 2020 and beyond scenarios, a context management system is capable of gathering all required information and provides this to the application/services, and even manipulates the data so it matches perfectly to the client applications requirements. In case this information is not directly accessible, it will either use advanced search mechanisms or inference techniques to provide the information.

Context Management framework is a concept, hence there may be different ways to implement it. In Section 2.1 a concrete example is presented, but such a framework should be seen as a network concept. The major reason why it is a network concept is that a context management framework is supposed to provide access to all information available. Just relying on specific communication technologies, e.g., WLAN would limit a potential large group of other information sources, such as those from sensor networks, etc. Thus, a common platform is needed for these types of systems, which encompasses both IP and other network types, e.g., wireless sensor networks, which incidentally raises the issues of heterogeneous networks.

19.2.1 A Concrete Example of Context Management System for Personal Networks

Within the framework of MAGNET Beyond [16], a dedicated context management framework was developed in order to support context sensitive applications and context triggered actions such as establishment of certain cluster or PN configurations.

Figure 19.2 shows an example of a Context Management framework developed in MAGNET Beyond for Personal Networks.

This framework consists of a set of Context Agents, capable of collecting context information locally available on the node, and distributes this information into the network of Personal Nodes, [8]. The framework provides access to context information in this network through a dedicated query language, Context Access LAnguage (CALA) see [6], which effectively allows the client application to scope queries in either time or space. The internal working of a Context Agent is illustrated in Fig. 19.3.

At the local access level, the raw data is accessed through a set of so-called ***retrievers*** which is written for access to specific data types using any appropriate proprietary protocol. The retriever converts the native data format into an internal and uniform data format, which is designed for efficient communication between the internal blocks. When accessed through the Context Management Interface, the results are converted into a well defined XML formatted output (the CALA is also

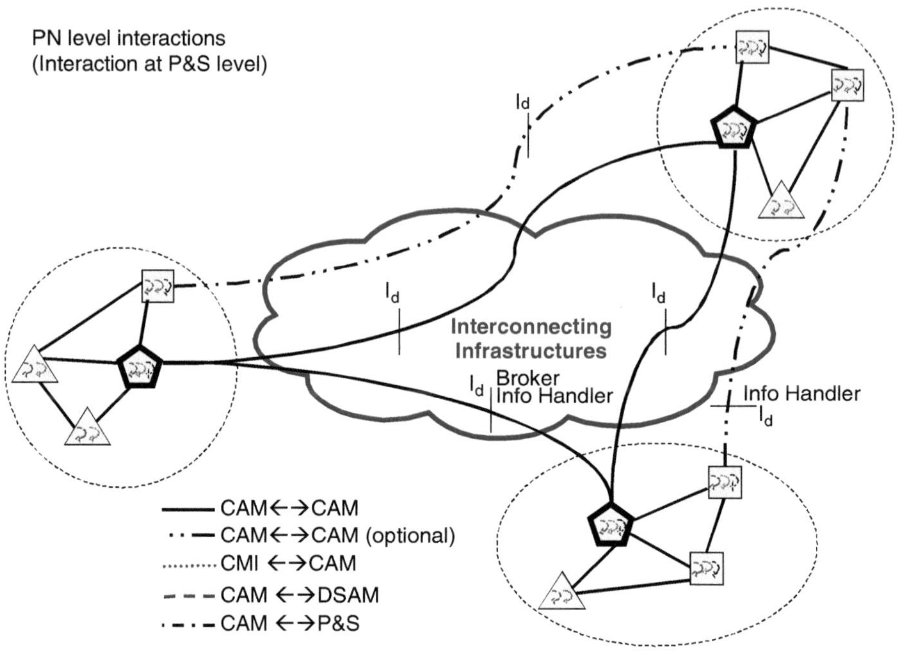

Fig. 19.2 Example of a Context Management Framework for Personal Networks, [7]

described in XML). Internally it is the Context Access Manager that ensures the distribution of context information, and controls the access to remotely located dynamic information. The Processing and Storage unit provides capabilities for added ***processing components*** (PU) allowing the Context Agent with possibility of inferring new types of context information, and in the same time it offers storage capacity for caching of information and storage of user profile information. Applications and services wishing to interact and use the service of a Context Agent would need to implement a CALA client, which is based on XML-RPC for this particular solution, and would then also need to implement reaction/reconfiguration mechanisms based on responses received from the Context Agent.

The presented framework is based on TCP/IP or UDP/IP based communication, and is implemented in Java. Thus all retriever and processing components would need to be implemented as java components. For retrievers, there is then no saying whether this would interface to other system components implemented in C, C++ or other languages. Since for each information element being accessed, a dedicated retriever is being written, there are no significant compatibility problems, except those limited to the retriever implementation themselves. An implementer of retrievers may choose to implement access to several versions of a system which provides information if required or choose not to. For example, a person implementing a retriever to get indoor positioning information from *System X v2.0* may also implement access from *System X v1.0*, or the person may choose only to implement support

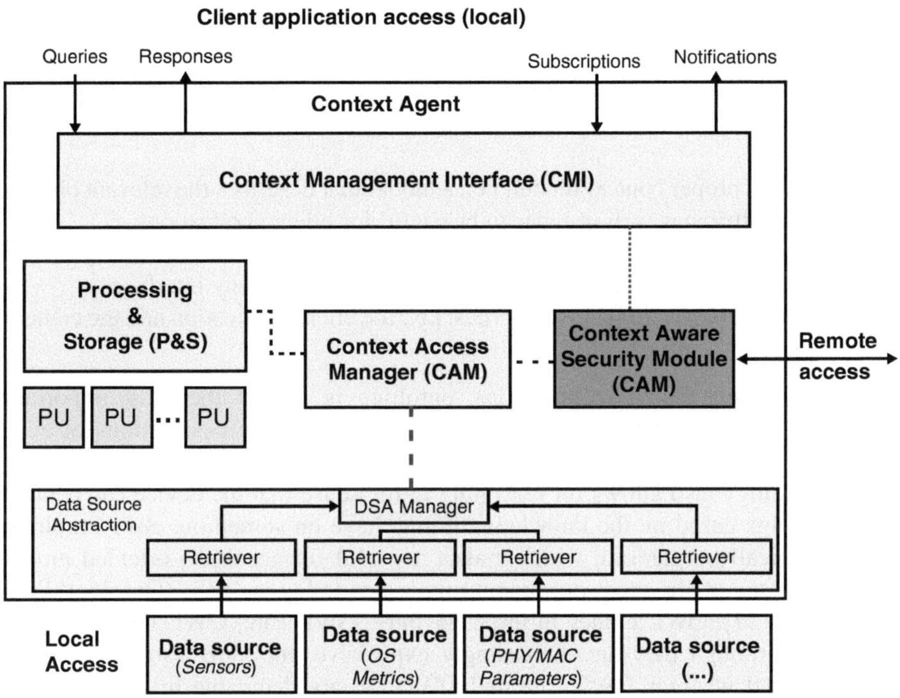

Fig. 19.3 A component view of a Context Agent capable of gathering, distributing, storing, processing and protecting context information [7]

for *System X v2.0.* The Context Agent doesn't care, as long as the retriever provides the information as specified by the given interface, but at the end of the day, if the user moves around in different environment where either *System X v1.0* or v2.0 is present, then obviously it is better to implement support for both, but at the cost of increased complexity. Hence compliance to legacy systems is entirely up to the developer of the retrievers to decide on which system to support.

Providing such a framework is a great challenge, and this work is advancing, not only in the European project as MAGNET Beyond, [16] but also in other European projects such as [9, 10, 11, 12, 13], etc. One of the real challenges to realizing such a framework is to ensure that it is lightweight and simple enough to be able to operate on any device at any time, while keeping security and processing requirements all in limits. Furthermore, the used data format and semantics of context information play a critical role in the final deployment of such systems. Therefore, standardization in this area is required.

19.2.2 Context Description and Semantics

A particular problem in dealing with context information is how to describe and model it. Obviously one can take any arbitrary approach if interaction with external systems is of no concern. But, especially for context, this is always not the case, since context as per definition, also constitutes information about the external world to an entity. This makes it necessary for the entity to understand how other systems describe and understand the world. It is important that context description is

- Based on a proper context model, i.e., a model that describes the relevant objects and their attributes well in order to be useful for client applications.
- Describable by a description language that is extensible, i.e., it is easy to add new objects to the description without having to change any interfaces.
- Understandable by all involved parties, i.e., the client application and the context provider.

Among all the existing technologies, ontology is one of the most important approaches to describing context. Ontology allows formal descriptions of concepts and attributes, which allow efficient modelling between instances of concepts. Ontology also allows for reasoning in the sense that the device can deduce certain things based on the knowledge it may have on something else. Ontology is hierarchically organised, which makes it useful for an object oriented model approach. One of the most prominent ontologies today is OWL (Ontology Web Language) [27]. OWL comes in three flavours, OWL Lite, OWL DL and OWL Full. Even though they are increasingly expressive, they are also increasingly complex, and whereas OWL Lite and OWL DL are decidable in time, the full version may have endless loops. OWL is based on XML and RDF (Resource Description Framework) which also makes this ontology better understandable for client applications. Notification 3 [28] is another ontology, which is basically a non-XML serialisation of RDF. Further work is also being done by 3GPP such as on standardization of user profiles [14], and the Dublin Core Meta data Initiative [15] has been working for some time now on standardization of interoperable meta data formats. Some examples of why and what challenges lies ahead of managing different types of context elements are described in the following section.

The challenge with semantics is not only to get to a common understanding of how to describe context, and relationships with other context elements needed for inference, reasoning and deduction techniques (*complexity*), but it is also a matter of striking a balance between a descriptive language that includes relationships with all types of context information and various amounts of Meta data (*completeness*) versus including all context information (which for context can be extremely large set of data) and what is absolutely needed (data *weight*).

What is needed may not be known in advance, hence it is not trivial to provide a useful language set in advance, and hence some intelligent pre fetching strategies may be needed. Furthermore, a useful language is required to be *extensible*, since new types of context information will appear over time without doubt. Thus a static structure is also not good, but if new information types appear, devices must be able to learn about this efficiently as well. The balance that needs to be found is illustrated in Fig. 19.4.

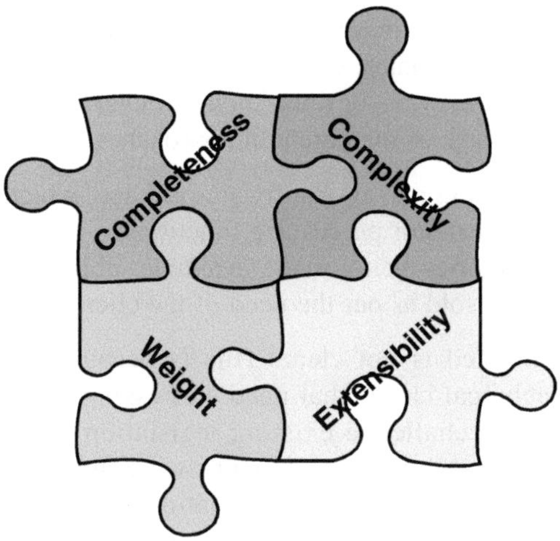

Fig. 19.4 The complexity of striking a right balance between useful semantic context descriptions including what is needed for a model

19.2.3 Appliance of Context Management Systems for the Global Business

As stated in Section 19.1, service and application developers must address some important requirements in order to successfully deliver adaptive and intelligent software to the end user. Context Management system aids in achieving this, by providing easy access to distributed information, and a range of functionality for ensuring, e.g., security and privacy control, fast and reliable response times and many other useful features. In the section, we discuss that it is much more complex than it may first appear.

From the developer's point of view, a context management framework leads to an easier development of adaptive services. However, in this respect there are several issues to be addressed:

- **Deploy and maintain the core framework**: Surely a core framework requires standardization effort for context access languages, semantics and models used to describe information prior any success for companies, but the question remains who will drive such work?
  - Large companies like Microsoft and Apple, who already have a large market share of the existing operating systems? An example could be UPnP [26], which in fact is an organization developing service discovery mechanism, such as service semantic standards.
  - Open societies such as the ones driving Linux?
  - Standardization bodies.
- **Deployment of active components for retrieving and processing context information**: As new information becomes available, either through new sensor

types, new algorithms for inference of context information, new components become available, need updates, or simply become outdated. How are components, including their life cycle, or reuse their retriever and processing power, are managed to be addressed. Some potential approaches include:

- Components becoming part of a software package, which would include the necessary retrievers and/or processing functions.
- Component repositories being made available, either for free access while being developed or sold as per the need of the client user.

Which solution is adopted is not clear. This is because context management also involves non-technical issues that need addressing such as the handling of personal data. This may challenge existing legislation in many countries, and the fact that such systems would operate worldwide, making the actual deployment of this technology very difficult. Legislation in one country on what type of information is allowed to be gathered and stored where, under given security requirements, may differ from one country to another, and the user may be traveling around the world. Should systems adapt to the legislation of each country, or should legislation of each country adapt to the technology? A solution for both will be very complex as it either requires countries in the world to agree to a common set of legislation, or a system which will need to adapt its internal mechanisms to each country's requirements. The sheer complexity and cost may significantly slow down the deployment or may make the solution infeasible.

Another fact also is that the core context management requires investment and participation from all partners (information/context providers, software developers, end users, etc.); hence a common collaborative effort is required to develop a workable solution. From a company's point of view, this would require capital expenditure to invest in underlying functionality, whereas for retrieval and processing components, it would require operating expenditure since these components would need to be included, removed or replaced as needed over time. Furthermore, the competition among companies will primarily be on the development of processing or retriever components, i.e., to provide the best quality information at the lowest cost.

19.3 Managing Distributed and Dynamic Context Information

What makes management of context information so difficult? What are the difficulties in technical deployment of context management? The fact that the context information space is so vast, and may be described in many (proprietary) ways is just one thing. But each single piece of information element has its own specific set of issues that need to be addressed. We discuss the above issues in detail in the next several sections.

19.3.1 Dealing with Heterogeneous Context Information Elements

In the following some examples of context elements and their individual characteristics that pose challenges are discussed.

19.3.1.1 Location Information

Geographical location has many potential uses, e.g., searching for nearby objects, showing direction to objects, indicating how other context should be perceived. However, these issues accompany many since this information comes from various sources:

- Location can be described either as textual, e.g., Room 203, or given as a set of coordinates. This implies that location given in different formats cannot easily be compared for example, distance calculation. A potential solution could be to put an attribute to the context information on where and how a location can be transformed. This would also allow transformation between different types of coordinate systems used.
- Bad tracking capabilities, e.g., relying on a GPS signal in an indoor scenario. If the system used does not provide the functionality to inform about this, how can the context management then figure out to change the source of information.
- Security and privacy issues related to providing the location to a client application. It may not be that simple due restrictions on the user's privacy or even legislation of a country (which may even change as the user travels around the world).

If there is any uncertainty about the location, this would have a severe impact on the user, e.g., a system directing a user in a wrong direction, finding objects that are hundreds of kilometers from their true location. Hence, it is important for a context management system to address these issues, and as a minimum to have a clear policy on what to do in such uncertain situations.

19.3.1.2 Timing Synchronization

Timing synchronization is yet another important aspect in context information that needs some attention. In many cases, reacting on context will almost always depend on the right time (and place). In most cases, the context agent would rely on the local clock on the device (if available), but there are some issues associated with that

- The clock may not be synchronized. If the device does not include automatic synchronization with some known source such as GPS, it will rely on the user having to set the clock manually. How does the system detect if the user did not do this properly? Or if it has switched to summer time? One way to address this issue is to check what other devices within a cluster consider as being the local time.

- The user may have switched time zone when traveling.
- Relative time relations may not easily be described as absolute values, such as "information must be less than 1 h old". What will happen if the information is 1 h and 2 s old, because of update delay? The consequence may be that the information is considered invalid, while in fact it may be not. One potential solution is to describe time relations using fuzzy membership functions. However, this requires an agreed definition of the membership functions that is available to all nodes within the network for such a solution to be used.

19.3.1.3 Network Context

The network state is fundamentally challenging context information to address. Since a network typically contains a broad range of network technologies (e.g. Personal Networks, [16]), using a plethora of communication protocol stacks with their own states for different purposes, with components working in different levels of the OSI stack model. Consequently, one would need to focus on a subset of information to be able to manage. It is, however, important to make the context agent extensible so that mapping new interesting information into the system would not be a problem. Two parameters that are commonly used are:

- Link and end-to-end bandwidth: What is available for the application may have an impact on what services to choose, e.g., video stream service. However, determining end-to-end bandwidth across a heterogeneous network such as a Personal Network (PN) remains a challenging problem to be solved.
- Latency: Similar to end-to-end bandwidth, end to end delay may not be easy to determine due to the heterogeneity in the network composition.

A context management framework may help to exchange information between nodes and clusters within a network, allowing a bigger picture of the network state in that network domain. This would, however, put additional requirements to the model used to describe complicated network structures. Furthermore, the variability within a cluster is rather fast compared to the time it may take to distribute the information; therefore, the context management designer must strike a right balance between the level of information distributed in the network and the delay of doing so with the changing nature of the information. If the changes are happening too fast to be efficiently distributed, then perhaps it makes no sense to distribute it at all, with the adverse consequence of the nodes within the network not having had a full picture of the network state.

19.3.1.4 Inferred Context Information

Inferred context information is defined as synthesized, inferred or derived information, which is based on one or multiple context information. Inference rules or other logical operations are used to figure out types of context information that sensors or other sources cannot provide. Some examples relate especially to user activity, such as sleeping, walking, in a meeting that is generally available or similar. Since this collection of information is based on other context information, some unique problems develop within this group:

- If one or more context information is missing, potentially out of date or not completely trustworthy, the output, i.e., the inferred information, will also suffer from not being reliable and accurate. If at least the probability of uncertainty can be estimated through the derivation, the application may be able to determine whether to trust the inferred information or not, and adapt accordingly.
- Depending on the composition and weight of different context information used, may it be individual and personal, i.e., depending on who the user is, where he is, what role the user currently has, a set of context information may be perceived in one way or the other. For example, how and what context information to determine if a user is busy or not, may depend on what function he or she performs.
- Since multiple information elements may be required, the time to access and get all the information may lead to performance issues. Also, the amount of computational resources may need to be carefully considered. On one hand, the computational performance is better achieved if done on the same node as the requesting application; while on the other hand, to achieve better resource utilization, computation may be a better option.

It is interesting to point out that profiles become an important aspect of inferring context information.

The key points in these examples are that the context agent would need to deal with all these different issues, and even more. Thus, additional information and rules are needed for the context agent to be able to handle all likely situations for different context information types. It is vital that the context agent is capable of providing unambiguous information to the client application to ensure that it behaves as expected. Since any type of information element is to be potentially handled by a context management framework, it is not possible to hard code any of such handling mechanism but it should be able to deal with as it becomes available. This pushes the requirements to not only increase the capability of the devices, but also to enable access to information providers, who besides the information should/could provide these additional information.

19.3.2 Dealing with Unreliable Information

Many types of context information have a certain level of dynamics, the information is with the user moving around and doing different things at different times. At any time, context may have changed when used by an application or service to invoke certain actions. This is normally an undesired side effect of using information that has been accessed remotely, or has undergone long processing delays before being used by an application. In the following, we present concepts to address these issues.

In [17, 18, 19] we investigated analytically three major metrics with respect to access strategies of remotely located dynamic information elements, namely, *network overhead*, *access delay* and *mismatch probability* (the probability of accessing outdated information), for different scenarios. The mismatch probability can be defined as the probability of accessing outdated information at a remote node (see Fig. 19.1), as given in Eq. (19.1).

$$mm\,\Pr = \Pr(E_{CA}(t) \neq E_{Source} \mid R(k) = t) \quad (19.1)$$

The problem, as mentioned, has already been studied for some time, and reveals that this probability depends to a high degree on which access strategy is used such as:

- ***Proactive, event driven***: The client has a subscription to the information element, and is informed at every significant change that happens. Updates may occur either *incrementally* (only the difference since last update is transmitted), or *fully* (the full data structure is transmitted every time).
- ***Proactive, periodic***: The client has subscribed to the information element to be updated periodically, with a specified time interval between updates.
- ***Reactive (with or without a cache)***: The client sends a request to the context provider when needed, and in return gets a response containing the current value. Potentially this information is cached for a certain time, from where subsequent requests obtain their responses.

Figure 19.5 shows the resulting mismatch probability using different access strategies, assuming Poisson event process and delay with (a) an increased delay rate ν decreased delay time and (b) increased event rate λ.

The impact of using different access schemes is clearly seen in Fig. 19.5a and b, e.g., for this scenario, the proactive, event driven (sending full information) and reactive without a cache (as these two approaches lead to equal mmPr [18, 19]) perform better over the investigated parameter range. The interesting part here is

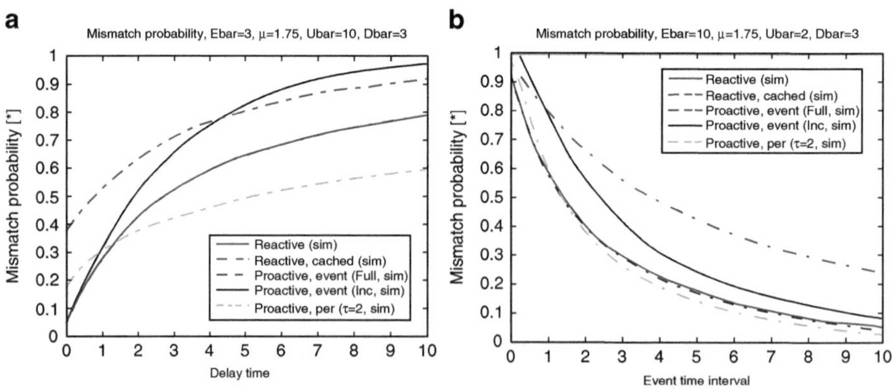

Fig. 19.5 Mismatch probability for a two node scenario with (**a**) varying delay and (**b**) event inter arrival times using different access strategies

that a context agent is in principle free to pick any of the above methodologies to access the remote information element, and furthermore, for the proactive periodic and reactive with a cache being applied, the context agent can control the level of mismatch probability achieved by adjusting (a) the update rate, or (b) the cache period C. In [18] we already showed that for the reactive caching strategy it is possible to meet mmPr requirements in most scenarios by adjusting the caching period, thereby enabling the context agent to provide a reliability indicator of the information element, and in [21] the same was shown just for the proactive periodic version with the limitation that the update rate was considered not to influence the delay via added network traffic.

The issue related to the example results shown in Fig. 19.5 is that even for a given scenario, a given access strategy may not be the best option in all circumstances. Thus, for context management framework, an adaptive scheme is required to keep the mismatch probability low, e.g., a proactive event driven approach leads to the lowest mmPr. However, achieving this is not an easy task for several reasons:

- The context agent has to become aware of the characteristics of the involved processes:
  - For delay, this means that end-to-end delay must be estimated under all circumstances, network configurations, topologies, etc.
  - For event processes, the characteristics must be known, e.g., event inter-arrival time distributions, if and how an event is recurrent (i.e. the process may return to a previous state and render a mismatch into a non-mismatch).
  - The request process, or more popularly, the information demand.
- It must know which of the involved performance metrics should be considered/prioritized.

The challenges indeed lay in the fact that these information types are not easily accessible, may not be available at all, be incorrect, or specified in a wrong data format. Even if the Context Agent tries to estimate values, in order to overcome reliability issues as done in [21, 23], the estimate may be biased or be very poor. This is one of the major reasons why it will take much research effort before context awareness can really be useful. Only when we have overcome these issues and are capable of ensuring that an application can request certain level of information quality, we can state that context awareness has arrived to be useful. Finally, not to underestimate, standards on how to describe this meta information and how to describe the general context information are needed.

19.3.3 The Business Aspects on Quality of Context

There are many clever engineers and software developers out there in the world. Many new technologies for sensing information, inference algorithms or other ways of dealing efficiently with the topics mentioned in Sections 3.1 and 3.2 are already being developed and implemented. To deal effectively with the side effects mentioned in Section 3.1, the reliability aspects in Section 3.2 and providing information that can actually be trusted (in terms of the information has not been hacked and tampered with), companies may offer different types and levels of *Quality of Context* depending on support systems and/or algorithms implemented. In this respect, the benefit from a context management framework as presented in this chapter is that the components implemented are completely modular entities in the core framework which only focus on distributing the information. This potentially protects the intellectual property that may be the foundation for a high quality context information element.

An example could be that one company develops positioning algorithms and support systems that guarantee accuracy of less than 1 m and a mmPr on less than 0.01 under certain circumstances (vicinity of the support system), while another company develops a less accurate positioning system with slightly higher mmPr that works in all situations.

19.4 Personalisation at Application Level

As already stated, any application or service that is personalized will need some profile on how this is done. In this section we distinguish between those profiles that are static and those that are adaptive. The difference between these two types will have an impact on the complexity of the data structures necessary to describe the profile, as the static type only depends on the profile information while the adaptive type also depends on context.

19.4.1 Static Personalization

By static personalization we mean the kind of personalization that does not depend on the context. This would typically consist of preference values, e.g.,

- Background color or image of an application screen.
- Default notification sound.

and other behaviors or description that are mostly static. These should of course be changeable to suit the user's will. What is primarily relevant for this kind of information is:

1. **To what client the data is relevant for**: It is necessary to know what client a specific set of profile data is associated with.
2. **A set of attribute names and values**: For each behavior a set of attributes and value pair is needed to instruct the client what behavior is being described and to what value it needs to be set.

Whenever a user decides to use an application in one way, on a specific device, he or she may want also the application to use similar settings on other devices, e.g.

an email client from where a user wants to have access to the same functionality, as the one shows the same buttons and colors as on the laptop at home. This requires access to a set of preference values at all times from whatever device the user is currently using, thus leading to requirements to distribution and potentially also concurrency control of preference values. Furthermore, this information may be learned over time by devices in a cooperative manner, so that the *experience* by each device that a user uses over time is shared and used to adapt the preference values and settings according to the user's habits and daily usage. In short, complex systems could be envisioned that make use of the observations made by devices during runtime, but here great care must be taken to respect the user's privacy.

19.4.2 Adaptive Personalization

Dynamic adaptation of services and application behavior are in many cases a desired functionality for the user, and will in most cases require context information, since the world we are living in is constantly changing. If the user has to always to manually setup the way a service or application should behave every time a parameter changes, he or she would need to do so constantly. Some examples of what adaptation of behavior could mean are:

- Adjusting the contrast level of the display automatically to the ambient light intensity and screen color so that the user can actually see what's on the screen. This is not necessarily an easy operation for the user to do.
- Change of notification methodology due to environmental restrictions, e.g. silence ringing when going to the cinema, which people sometime forget even when reminded, or increasing volume in noisy environments.
- Adjustment of video resolution when streaming; in cases where the network conditions are changing often as in the wireless domain, it may be preferable to shift to a lower resolution to maintain a smooth view.
- Output stream re-direction to more capable devices, e.g., in case of the video stream, which may be viewed originally on the user's mobile phone, may automatically be switched to TV screen or laptop for better viewing, performance, or due to lack of resources on the current device.

Common to these examples, and for all other cases, is that the adaptation is user and application specific and depends on the context information. In fact to achieve this kind of operation, a set of information needs to be available for the application. Some examples of these are

1. **What client this information is relevant for**: The following information must somehow be linked together with a specific client type, i.e. the context adaptation profile data for a calendar application would most likely not be the same as for a tourist guide application, simply because they focus on different objectives and require different inputs.

2. **What context to react on**: This depends on the service/application and to what extent this information is available.
3. **How to react on it**: This depends on the service/application and who the user is and what role the user has. In some cases, context information may not be available, or only related context is available, in which case some instructions are needed for the application to know what to do then.
4. **How to deal with uncertain information**: If context has been accessed remotely, there would always be some uncertainty, whether this information has changed during the update. How the application should react if the probability of using outdated context information is very high should also be specified. Such action depends not only on the service/application and their requirements to the context used, but also on what kind of person the user is.

In the above examples, adjusting and changing the video stream depends strongly on what the user really desires, how much the user wishes to be in control (which may also depend on how tech-savvy the user is), the current role of the user etc. Such information needs to be accessible by the client anywhere within the personal network (PN), and, furthermore, if two similar clients, e.g., two calendar applications are running on different nodes within the PN, they should adapt similarly.

19.4.3 Optimistic Versus Pessimistic Systems

Assuming that, based on different observed or given statistical information, the Context Agents are able to estimate the mismatch probability as discussed in Section 3.2, this meta data is valuable to any client application since it can be used to determine whether the information should be used or not. In this way an ***optimistic*** or ***pessimistic*** client behavior can be obtained. *Optimistic* meaning that the client may take chances on context information, while others would prefer or even require a more *pessimistic* system that requires the system to be very confident on the context information before being used.

To exemplify the difference, client A may require a pessimistic view on information such as location data for directing a blind person down the stairs in his or her home, the client may take a pessimistic view to avoid the person to be directed wrongly down the stairs. Client B, who needs network context to provide a link to a resource for the person's PN, may be more optimistic towards the information given, since the consequences of invalid information are less dangerous to the user. The loss in taking a pessimistic view on context is that client A will become more reluctant to adapt to a given context than an optimistic client. However, for client B, which takes an optimistic view of location data, risks that the adaptation is error prone, but will more likely adapt to the changes detected by the SCMF.

The level of *optimism* may depend on the individual application, but may also be a general system attribute, depending on what type of person and role of the user, that clients will take by default. The individual part will need to be a subset of a service profile, while the general system attribute can be a part of the user profile data.

Whether it will be per client or a global attribute depends on the level of *optimism* that a client or user wishes to have (which could also be based on the role of the user, and in the extreme based on other context information). Context information will come with an indicator of how much it is off by when accessed (the mismatch probability, see [18]). A simple methodology, as far as the user is concerned, is to define an *optimism indicator* in the interval [0:1], which simply describes the level of probability that the context information has to have before the system will consider it useful. A 0 indicates a completely pessimistic system (which can also be viewed as a non context aware system), and a 1 indicates a fully optimistic system. Using the Fuzzy Logic [20] operator, ***Not***, the level of *pessimism* is also defined as **not** (*optimistic*) which equals one minus *optimism*. A user's role would necessarily have to be somewhere in between zero and one to be really useful.

19.5 Conclusion and Outlook

Managing context information is not only about the distribution, but also a matter of ensuring that the output given to the client application is unambiguous and that the probability of any error is minimized. In this chapter we described some of the major challenges with respect to the concept of supporting context awareness, whereas some of the important ones discussed in this paper are summarized as follows

- Standardized description and model language for describing context information.
- Flexible and extensible context management framework which meets the set requirements.
- Efficient distribution and access to dynamic information elements in a potential dynamic network environment.
- Easy and quick access to error free, unambiguous, useful context information to client applications.
- Lightweight, but still powerful enough to infer new information types not directly measurable by sensors.
- Provisioning of distributed storage for user profiles for (a) personal application and service adaptation and (b) personalized interpretation of context.
- Enforcement of user privacy and security of context information.

To address all these challenges is a daunting task, but must be addressed if we ever wish to have user friendly context adaptive applications and services deployed. Once solved and if future devices are equipped with software components that offer all of above, integration of context sensitivity into applications and services becomes much simpler, especially interface to the information elements can be provided.

However, it should be noted that providing context aware applications is also not only a matter of providing the distributed, dynamic context information, but it involves the applications and services developers. Not only these need to learn of the coming possibilities of using context management systems, but also because application logic needs to be implemented so that it can automatically adjust for different situations with the probability that the information is already outdated when triggering e.g. state changes of an application. Certainly, this calls for awareness to aspects such that the developer cannot rely only on his/her opinion of how things should work, but need to take into account much more, namely what situation do we account for, how do we ensure user satisfaction in most relevant situation etc. Based on this, we might see a whole new types of businesses focused entirely on supporting intelligent software development and selling just context information which compete on their quality.

Finally, there is the most important issue of globalization of such frameworks and support for intelligent services and applications. Because of different countries having different legislations towards personal data gathering and distributions, there are a lot of potential issues of non-technical types that need to be taken seriously as well for a successful deployment. This problem is obviously closely related to the security requirements of systems not to leak out personal information to third parties, like governments, terrorists, hackers, or just the annoying and curious neighbor. Security systems for context management must fulfill their requirements from day one, otherwise users will quickly lose faith in such systems, and the deployment of intelligent software for future networks will face a very difficult future.

References

1. Dey, A.K., Providing architectural support for building context-aware applications, Ph.D. thesis, Georgia Institute of Technology, 2000
2. J. Pascoe, N.S. Ryan, D.R. Morse, Issues in developing context-aware computing, in *Proceedings of the International Symposium on Handheld and Ubiquitous Computing* (Springer-Verlag, Karlsruhe, Germany, Sept 1999), pp. 208–221
3. B.N. Schilit, N.I. Adams, R. Want, Context-aware computing applications, in *Proceedings of the Workshop on Mobile Computing Systems and Applications* (IEEE Computer Society, Santa Cruz, CA, 1994), pp. 85–90
4. R.L. Olsen, A. Nickelsen, J. Nielsen, H.P. Schwefel, M. Bauer, Experimental analysis of the influence of context awareness on service discovery in PNs,in *Proceedings of IST Summit 2006*, Mykonos, Greece, 2006
5. M. Khedr, A. Karmouch, ACAN – ad hoc context aware network, electrical and computer engineering, 2002. IEEE CCECE 2002. Canadian Conference on Volume 3, 12–15 May 2002, pp. 1342–1346 vol. 3 Digital Object Identifier 10.1109/CCECE.2002.1012948
6. M. Jacobsson, V. Prasad, M. Bauer, J. Stoter, R. L. Olsen, J. Lanza, L. Sanchez, J. Rico, Y. Gu, M. Ghader, M.G. Gene, D. Zeghlache, *PN Framework Solution and Performance*, D2.3.2, MAGNET Beyond, IST-027396
7. Martin Jacobsson, Venkatesha Prasad, Martin Bauer, Jan Stoter, Rasmus L. Olsen, Jorge Lanza, Luis Sanchez, Juan Rico, Yanying Gu, Majid Ghader, Marc Girod Gene, Djamal Zeghlache, *Interim PN Framework Solution and Performance*, D2.3.1, MAGNET Beyond, IST-027396

8. M. Petrova, M. Wellens, S. Oosthoek, D. Zeghlache, W. Louati, W. Louati, M.G. Genet, K. Masmoudi, L. Sanchez, J. Lanza, L. Muñoz, R. Olsen, H. Murakami, D. Kyriazanos, M. Argyropoulos, J.F., Yacine Rebahi, S. Mirzadeh, M. Jacobsson, V. Prasad, *Overall secure PN architecture*, MAGNET – IST-507102 Deliverable 2.1.2, October 2005
9. E-SENSE project website: http//www.ist-esense.org/
10. http://www.sensei-project.eu/
11. http://giove.isti.cnr.it:88/
12. http://www.ist-spice.org/
13. http://www.ist-daidalos.org/
14. S. Salsano, Presentation of the IST simplicity project, 3GPP SA1 meeting, Povoa de Varzim, Italy, July 2005, doc. num. S1–050725
15. http://dublincore.org/
16. http://www.ist-magnet.org/
17. R. L. Olsen, H. Schwefel, M.B. Hansen, Global telecommunications conference, CLOBECOM 2006 – IEEE. Electrical Engineering/Electronics, Computer, Communications and Information Technology Association, San Fransisco, California, USA, 6 p., 27 Nov–1 Dec 2006
18. H.-P. Schwefel, M.B. Hansen, R.L. Olsen, Adaptive caching strategies for context management systems, in *Proceedings of PIMRC'07*, Athens, Greece, 2007
19. M.B. Hansen, R.L. Olsen, H.-P. Schwefel, Probabilistic models for access strategies to dynamic information elements. Perform. Evaluation. Elsevier **67**(1) (2010)
20. L.H. Tsoukalas, R.E. Uhrig, *Fuzzy and Neural Approaches in Engineering* (Wiley-Interscience, John Wiley & Sons, New York, 1997), ISBN 0-471-16003-2
21. R.L. Olsen, Enhancement of wide area service discovery using dynamic context information, PhD. Dissertation, Aalborg University, 2008, ISBN 87-92078-37-0
22. M. Bauer, R.L. Olsen, L. Sanchez, J. Lanza, M. Imine, N. Prasad, Context management framework for MAGNET Beyond, invited paper, Joint MAGNET Beyond, e-SENSE, DAIDALOS and CRUISE IST workshop, Myconos, Greece, June 2006
23. R.L. Olsen, H.P. Schwefel, M. Bauer, Influence of unreliable information on context aware service discovery, in *Proceedings of Third Workshop on Context Awareness for Proactive Systems*, University of Surrey, Guildford, England, 2007, ISBN: 978-0-9556240-0-1
24. Sun, J.-Z.; Sauvola, J. Towards a conceptual model for context-aware adaptive services, Parallel Distr. Comput. Appl. Technol. 2003. PDCAT'2003. *Proceedings of the Fourth International Conference,* pp. 90–94, 27–29 Aug 2003
25. S. Stabb, H. Werther, F. Ricci, A. Zipf, U. Gretzel, D.R. Fesenmaier, C. Paris, C. Knoblock, Intelligent systems for tourism. Intelligent Syst. IEEE **17**(6), 53–66 (Nov/Dec 2002), Digital Object Identifier 10.1109/MIS.2002.1134362
26. Universal Plug'n'Play consortium: http://www.upnp.org
27. W3C OWL overview, http://www.w3.org/TR/owl-features/
28. Notation3 (N3) A readable RDF syntax, http://www.w3.org/DesignIssues/Notation3.html

Rasmus Løvenstein Olsen is currently holding a position as Assistant Professor at Aalborg University working in the Network and security group. Rasmus received his master degree from Aalborg University and received his Ph.D. degree on the topic of Context Sensitive Service Discovery and Context Management with focus on access to dynamic information in 2010, for which he was nominated to a Spar Nord research prize. He continuous his research within the field of distributed, intelligent and context aware systems, with activities in European projects; MAGNET, MAGNET Beyond and OPEN. He is currently active as work package leader within the working topic of network support platform. He has published several journal and conference papers on the topic of context management and context aware applications, services and support, with particular focus on reliability of access strategies to distributed and dynamic context information. Previously, Rasmus has been guest visitor at National Institute of Communication Technology in Yokosuka Research Park in Japan, co-operating with a team of researchers on next generation network technology, and has an active collaboration with industry, research institutes and universities within Europe. Finally, Rasmus is also engaged in teaching and supervision of students at the university, and has been giving several lectures in Context Management and Service Discovery, system engineering, algorithms and software architectures.

Chapter 20
Let's Assume the System Is Synchronized

Fred Harris

20.1 Introduction

It is amazing how many papers on radio systems, networks, error correcting codes, and related topics contain a version of the sentence "Let's assume the system is synchronized." Alright, let's assume the system is synchronized. But I have a few questions: Who did it? How did they do it? Who will do it in the next decades as many of us retire from the field? An important one is; where are they acquiring the skills required to negotiate and navigate the future physical layers?

This brings us to the question of "What do we mean by synchronize"? Its etymology starts with Chronos (also Khronos and Chronus) the ancient Greek Immortal Being (Χρόνος) personified in the modern age as Father Time. We thus form synchronize from the Greek prefix *syn*, meaning *"together with"* and *chronos* which we interpret as *"time."* With the industrial revolution and the ascendancy of the high speed railroad came the need to synchronize clocks in adjacent towns in order to maintain arrival and departure timetables. This helps protect trains from cornfield collisions. Today, the higher speed transport and commerce of communication signals places an even greater premium on the measurement of time and the alignment of remote clocks and oscillators.

When discussing the importance of synchronization in my modem design class I present the needle point shown in Fig. 20.1 and remind the students that Momma's middle name is synchronizer! If the radio is not synchronized no other system can operate! Not the matched filters, not the equalizer, not the detectors, not the error correcting codes, not the decryption, not the source decoding! At the waveform level, synchronization entails the frequency and phase alignment of remote oscillators for carrier acquisition and tracking, for modulation symbol timing, for chip alignment and hopping boundaries of spread spectrum modulation overlays.

What have we missed by assuming the system is synchronized? We skipped a challenging and most interesting part of the process. We have skipped the task of

F. Harris (✉)
San Diego State University, USA
e-mail: fharris@mail.sdsu.edu

R. Prasad et al. (eds.), *Globalization of Mobile and Wireless Communications: Today and in 2020*, Signals and Communication Technology,
DOI 10.1007/978-94-007-0107-6_20,

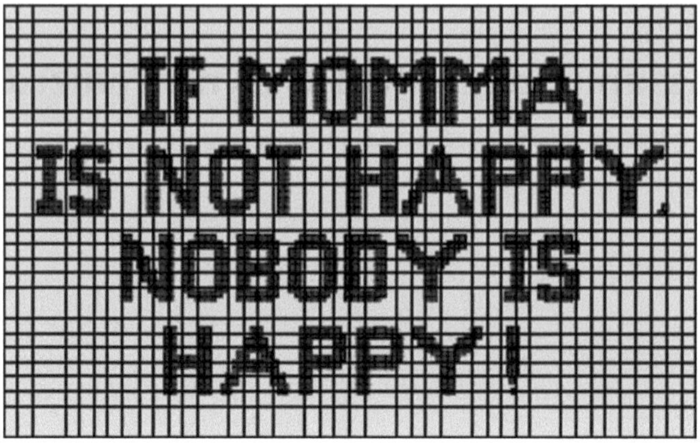

Fig. 20.1 The synchronizers' needle point

estimating, in a short time interval, the unknown parameters of a known signal in the presence of noise. We have replaced the task of processing a noisy wave shape for the easier task of processing a binary data stream with unknown errors. We have also skipped making Momma happy!

20.2 Source of the Problem

Figure 20.3 presents a very simple model of a communication system [1, 4]. Here input bits are subjected to a number of transformations at the modulator with matching inverse transformations at the demodulator. In total, the input bit stream to the modulator is converted to the radio frequency wave form at the modulator output. A communication system is seen through the eyes of the beholder and different eyes see different systems. Some see the system from the Shannon perspective, as a discrete modulator and demodulator connected to a discrete channel as shown in Fig. 20.2. In Shannon's model, the discrete modulator and demodulator perform a set of transformations on discrete sequences as shown in Fig. 20.4.

What Shannon did was brilliant: his model abstracted the system. His model separated the communication system from the physical system and the discipline blossomed under this separation. Truly remarkable advances in communication systems can be traced to developments in this model space. I fear that since there are no wave shapes in the Shannon model we have lost sight of their importance to the process. The model's success inadvertently relegates the wave shapes in the physical system to a secondary status in the academic community.

At the other extreme of the model space is the hardware model that emphasizes the signal conditioning and signal processing of communication waveforms exiting the modulator and entering the demodulator. Such a model is shown in Fig. 20.5, [3, 7]. In this model the discrete transformations emphasized in the Shannon model

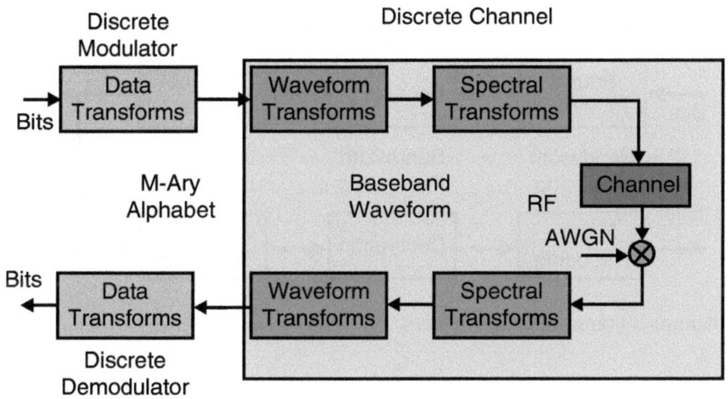

Fig. 20.2 Discrete communication system

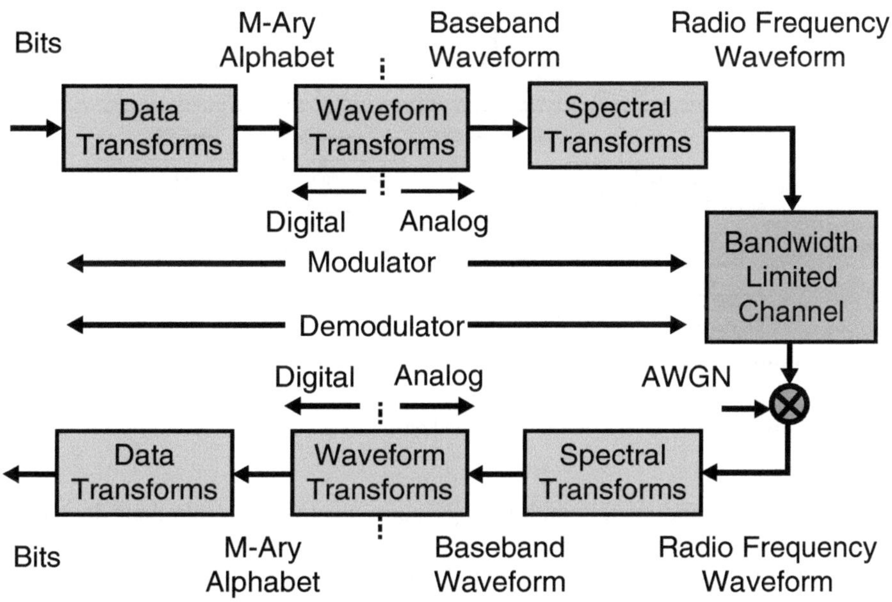

Fig. 20.3 Simple communication system

are located in the processing blocks titled Bit-Map at the input to the transmitter and at the output of the receiver. The transmitter of the physical layer model explicitly shows the shaping filter, the up-converter, and the output power amplifier which perform the base band spectral shaping, the linear RF spectral transformations, and non-linear spectral transformations respectively. We note the non-symmetry of the transmitter and the receiver. The receiver contains many more subsystems than does the transmitter. These subsystems are seen to be servo control loops that participate in the signal conditioning required to demodulate the input waveform. These loops estimate the unknown parameters of the known input signal and invoke corrective

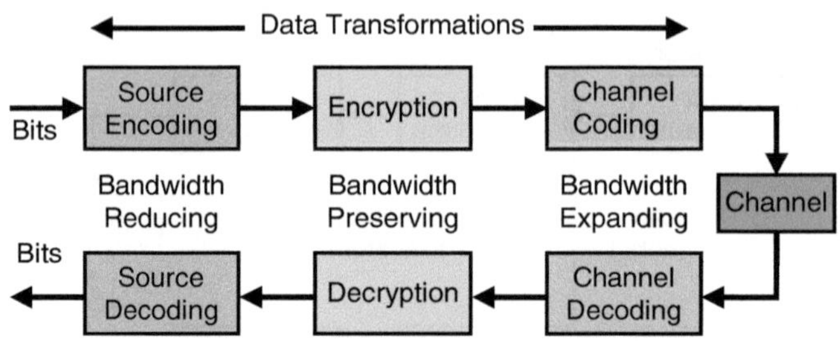

Fig. 20.4 Shannon's communication system

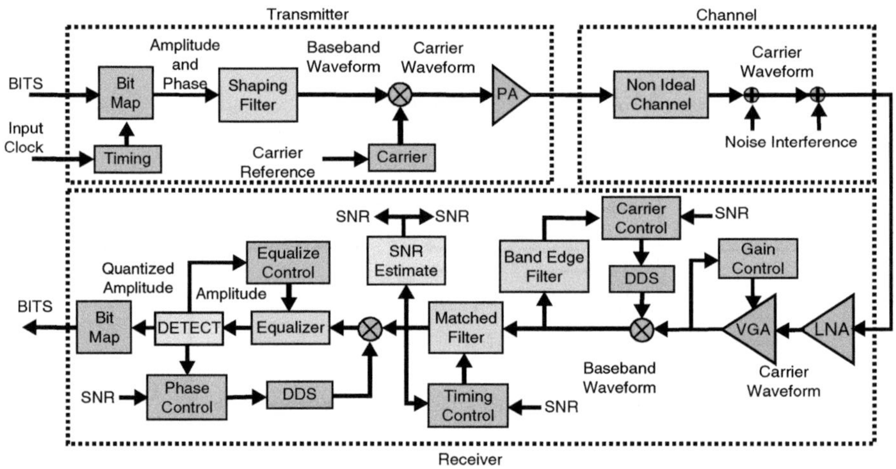

Fig. 20.5 Physical layer model of transmitter and receiver

signal processing and signal conditioning operations to ameliorate their degrading effects on the demodulation process [2].

A partial list of these processes includes the following subsystems: (i) an AGC loop to estimate and remove the unknown channel attenuation, (ii) a carrier recovery loop to estimate and remove the unknown frequency offset between the input signal's nominal and actual carrier frequency, (iii) a timing recovery loop to estimate and remove unknown time offsets between the receiver sampling clock and the optimal sample positions of the matched filter output series, (iv) an equalizer loop to estimate and remove unknown channel distortion responsible for intersymbol interference, (v) a phase recovery loop to estimate and remove unknown carrier phase offset between input signal and local oscillator, and (vi) an SNR estimator to supply important side information to the just enumerated subsystems.

The closing question in this section in this section is "Where do your students learn the science and engineering embedded in the sub-systems of Fig. 20.5? Where do they learn the new DSP based techniques that have supplanted legacy analog designs? We now examine an overview of some of the signal processing tasks performed during the synchronization process.

20.3 Phase Lock Loop, Timing Recovery

The function of the various loops in the receiver is to estimate the various unknown parameters of the received noisy signal. The estimate is based on observations of a sampled data sequence y(n) derived from the output of a matched filter that acts to reduce the effects of the additive noise. The estimate can be obtained from a bank of matched filters parameterized over the unknown variable such a time delay τ. This option is shown in Fig. 20.6. The outputs of the filter bank at specific symbol time increments nT are subjected to a detector and are averaged to obtain stable statistics The smoothed outputs are compared and the filter with the largest output magnitude is the one matched to the signal time delay τ_k.

In modern receivers, the filter bank is available to the receiver as the paths of an M-path polyphase filter. Rather than operate all the paths simultaneously, they are operated sequentially in response to side information which guides a state machine to the peak of the correlation function. This side information is the slope at the output of each hypothesized filter selection. The system selects any filter in the bank and tests the hypothesis this it is the correct filter. It does that by forming the derivative at the test point. In legacy receivers the derivative was estimated from early and late matched filters bracketing the test point in question. In modern receivers the derivative is formed by a derivative matched filter bank. The derivative at selected points of a correlation function is shown in Fig. 20.7 for positive valued and for negative valued correlations. Note that for a positive valued correlation a positive slope indicates the

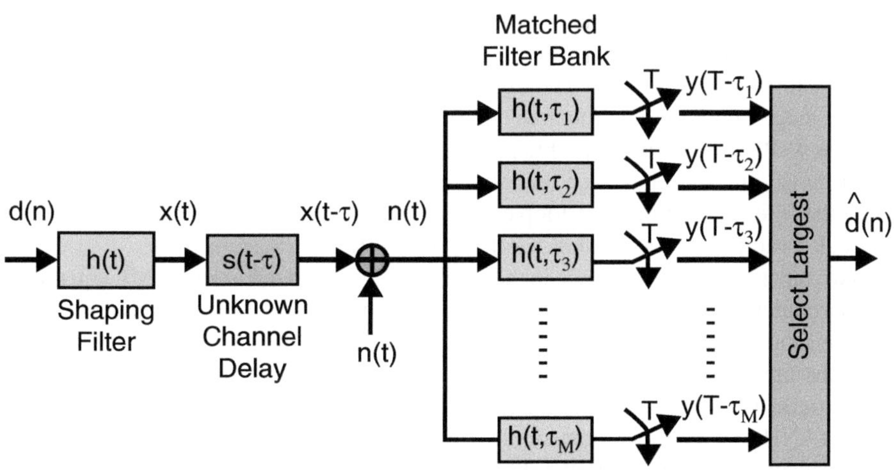

Fig. 20.6 Bank of matched filters parameterized over time delay variable τ

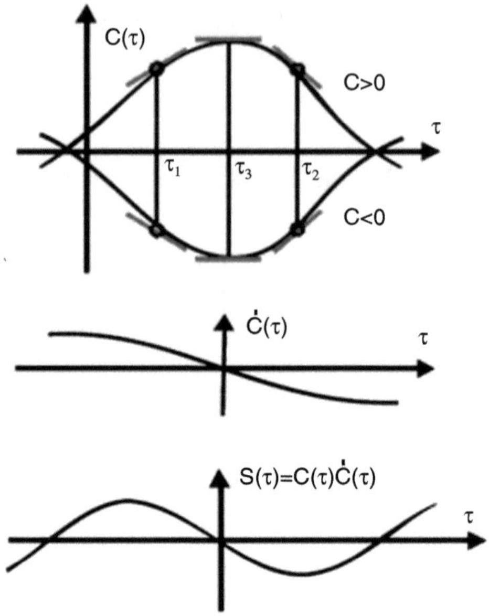

Fig. 20.7 Correlation function showing slope at various test points and detector S-curve formed as product $c(\tau)c(\tau)$

peak is ahead of the test point and a negative slope indicates the peak is behind the test point. Since the slope has the reverse polarity when the correlation value has a reverse polarity the information residing in the slope must be conditioned on the polarity of the amplitude. The amplitude conditioning of the slope is embedded in a detector S-curve formed as the product of the amplitude and the slope.

The state machine is designed to move the hypothesis test point in the direction that sets to $c(\tau)c(\tau)$ zero. Note this happens at two locations! If $c(\tau)$ is zero, we are at the peak of the correlation function and the system is a maximum likelihood estimator and if $c(\tau)$ is zero, we are at the zero of the correlation function and the system is denoted the Gardner or minimum likelihood estimator. The state machine operates as a servo system called a phase locked loop whose block diagram is shown explicitly in Fig. 20.8. The figure shows the standard components of the PLL timing loop. These include the matched filter bank and the derivative matched filter bank as well as the loop filter which averages through the received noise and modulation noise and the phase accumulator that selects the candidate hypothesis filter from the filter bank. Since the loop contains two-integrators it is a type-2 loop, able to track a ramp input, a frequency offset, with zero steady state error.

Two other components shown in the block diagram, derived from the ML equations, are particularly interesting. These are the SNR scale factor (2Eb/N0) that serves to tell the loop the quality of the signal it is processing. If the signal sample has a low SNR the input to the loop filter should be scaled in proportion to its quality. This of course reflects the philosophy of all matched filter processes. The second component is the TANH that conditions the amplitude from the matched filter as it interacts with the derivative. At high SNR the TANH faults to the sign of the input signal as a conditional correction to the derivative information.

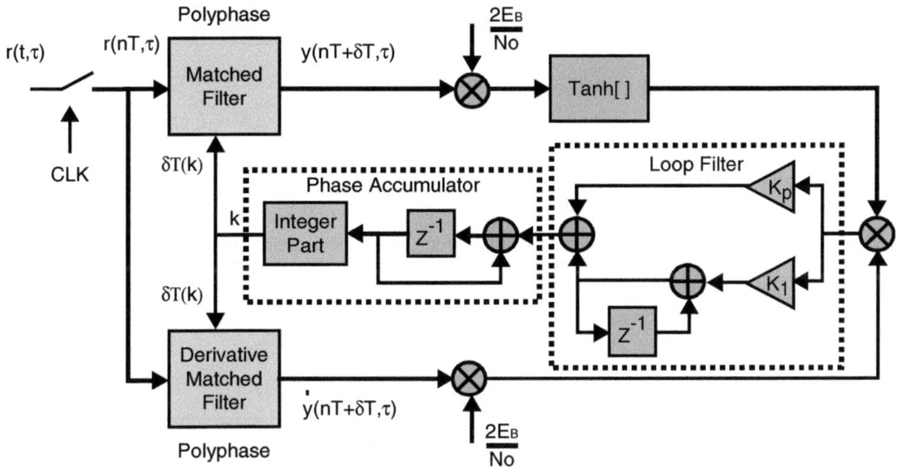

Fig. 20.8 DSP based polyphase filter bank maximum likelihood timing recovery loop

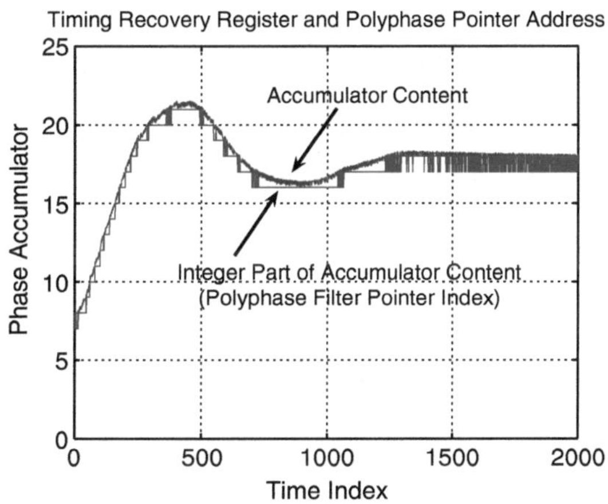

Fig. 20.9 Transient response of phase accumulator during timing acquisition

At low SNR the TANH defaults to a unity gain applied to the matched filter output to avoid possible errors in the sign decision of the amplitude as the conditional correction of the derivative. The SNR gain term in the loop filter throttles the loop bandwidth in response to the SNR. At low SNR the loop filter reduces its bandwidth so that it has to average over a longer interval to obtain stable control signals. Conversely, at high SNR the loop filter increases its bandwidth since it can obtain stable control signals by averaging over shorter time intervals.

The problem with these two components is that most receivers do not have real time background SNR estimators operating to feed the quality assessment of data samples to the loop filter. Thus most receivers replace the TANH with its small SNR gain and operate as suboptimal systems. We can do better than that! Figure 20.9

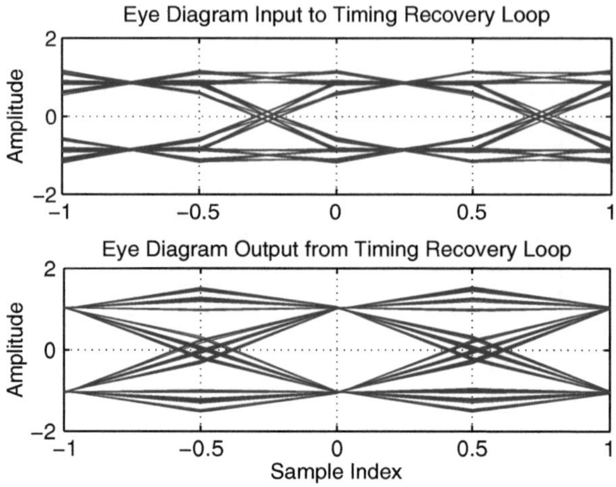

Fig. 20.10 Eye diagrams at input and output of timing recovery loop

shows the transient response observed at the output of the phase accumulator as it moves from an initial filter path to the correct filter path in the polyphase filter bank. The red curve is the accumulator content while the blue overlay curve is the integer part of the accumulator that defines the index pointer to the selected path of the M-path filter. Figure 20.10 shows the eye diagram of the signal at the input and output of the timing recovery loop.

20.4 Phase Lock Loop, Phase Recovery

In many modulators the signal formed by the shaping filter is amplitude and phase modulated in accord with the input bit mapping process. The amplitude and phase terms are represented in Cartesian coordinates and described as a complex base band signal. The quadrature components of the signal are upconverted or amplitude modulated on the quadrature components, the cosine and sine, of a radio frequency carrier. At the demodulator the process is reversed and the radio frequency carrier is down converted by a pair of cosine and sine quadrature sinusoids. The frequency and phase of the upconverter and the downconverter oscillators do not match by virtue of manufacturing tolerances, age and temperature related drift, and Doppler offsets due to velocity vectors between platforms.

The signal obtained at the output of the quadrature down converter is monitored and applied to a phase detector to obtain a measure of the phase misalignment. Here we quickly describe the process by which the phase lock loop aligns the local oscillator phase with that of the phase of the received signals' underlying carrier. First we examine the phase detectors for a binary phase shift key BPSK signal. What do we do with the samples of the I-Q pair, which are time aligned with the

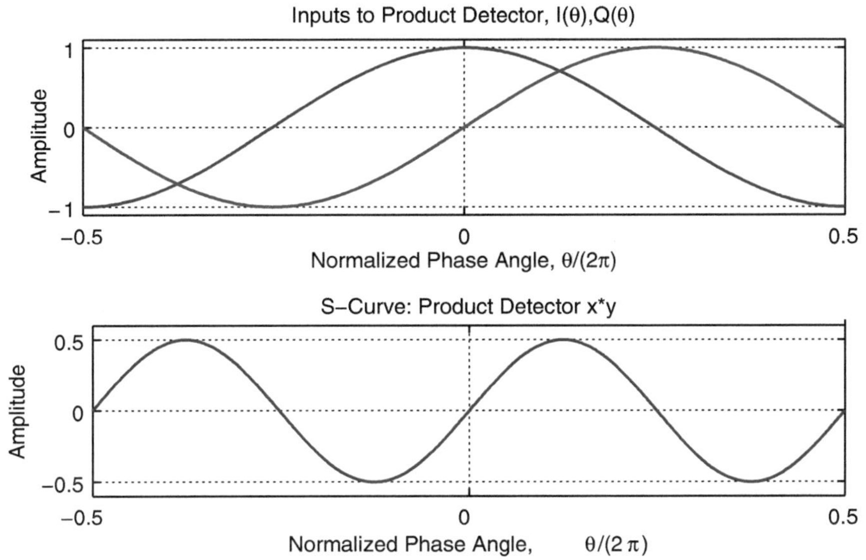

Fig. 20.11 I·Q BPSK phase detector and S-curve

correlation peaks of their matched filters, to obtain information about their phase θ? What we do is examine a legacy solution formed by their product I·Q. As seen in Fig. 20.11, the product is proportional to the sine(θ/2) which is the ex-pression for an S-curve phase detector. The detector has two zero value references at 0° and at 180° which of course is responsible for the two-fold ambiguity in the synchronized phase lock.

An alternate phase detector is the SGN(I).Q which is shown in Fig. 20.12 where we see a more linear S-curve spanning a wider range of input phase offsets. These correspond to the small signal to noise ratio and the large signal to noise ratio approximations to the TANH described in the previous section. Thus it will come as no surprise to see the structure shown in Figs. 20.13 and 20.14 of a BPSK and a QPSK receiver implementing Maximum Likelihood phase recovery.

Incidentally, we make an interesting observation with respect to Fig. 20.13 which is the ML phase estimator for the BPSK receiver. Suppose we leave off the SNR scaling factors from the two paths, and further replace the TANH of the upper path with a wire (or a unity gain path) and similarly replace the 2-to-1 down sampler on each path with a wire. What would we have? Give up? We would have a Costas loop phase recovery system. The Costas loop is embedded in many legacy receivers. Having discarded the items listed earlier we quickly conclude that the Costas loop is far from an optimum phase recovery process. In fact it is easy to verify that the sub-optimal loop performs quire poorly at low SNR. As we bravely move forward to the next decade, let us not carry legacy suboptimal solutions with us. Figure 20.15 presents the S curves for the ML phase detector of the QPSK receiver for a range of scaled SNR. Note the change in slopes at the zero crossings of the S-curve. These

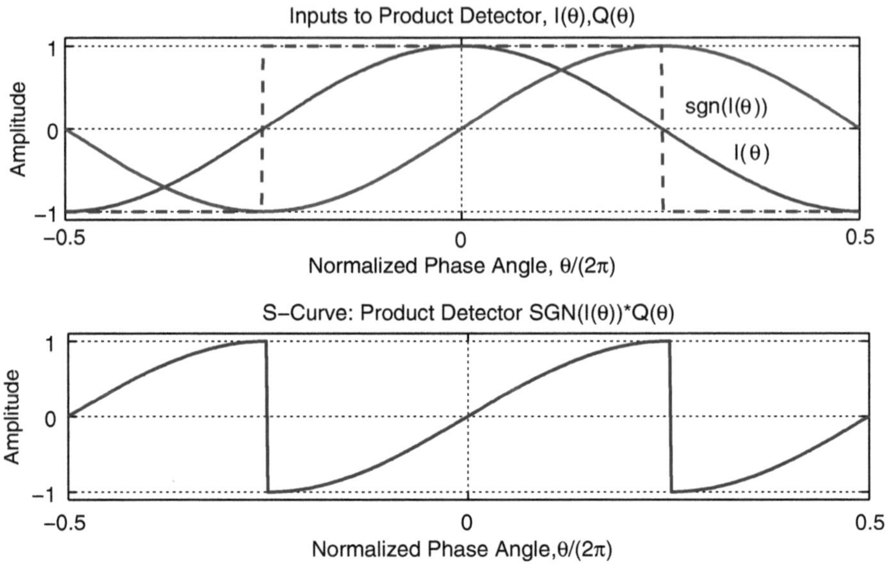

Fig. 20.12 SGN(I)·Q BPSK phase detector and S-curve

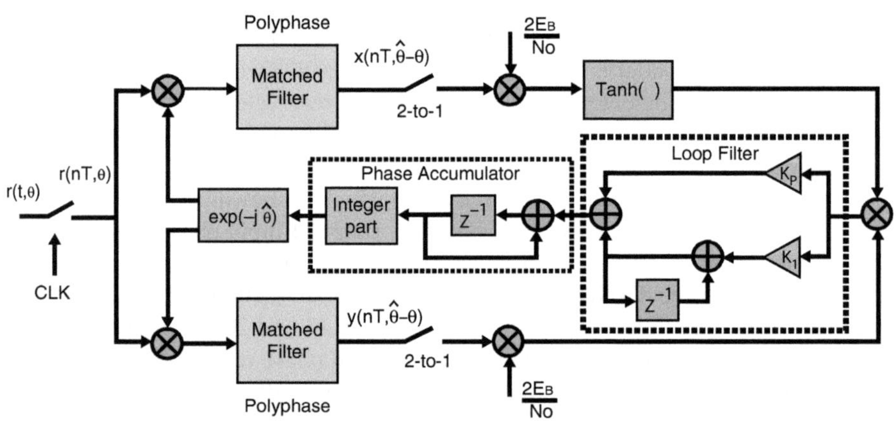

Fig. 20.13 Maximum likelihood phase recovery for BPSK receiver

slope changes alter the loop gain of the PLL and are responsible for SNR dependent loop bandwidth of the acquisition system.

Figure 20.16 shows the trajectory of the QPSK constellation during Phase acquisition. The trajectory has the appearance of a comet with its trailing tail because the figure plotted the constellation points near the end of the trajectory in red to emphasize the final state of the trajectory. As observed in the section on timing

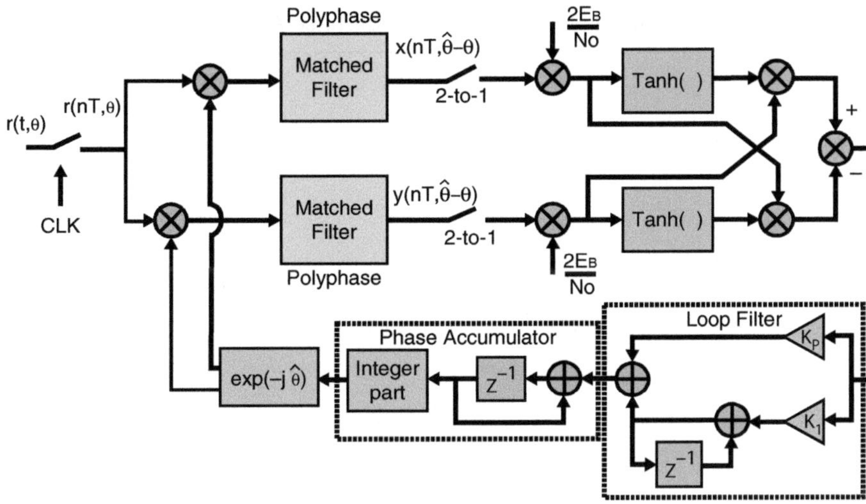

Fig. 20.14 Maximum likelihood phase recovery for QPSK receiver

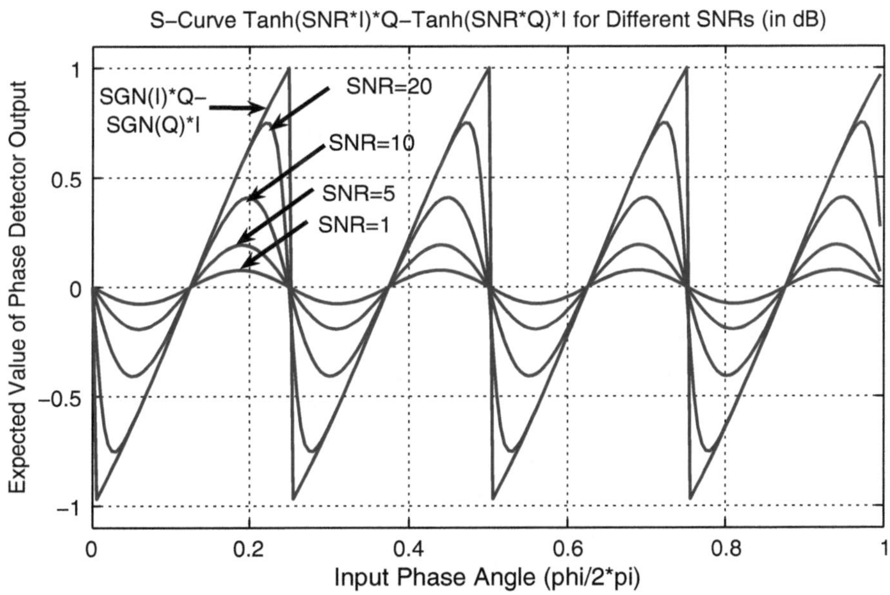

Fig. 20.15 S-curve for QPSK phase detector for range of input SNR

recovery, most receivers do not have real time background SNR estimators operating to feed the quality assessment of data samples to the loop filter. Thus most receivers replace the TANH with its large SNR gain and operate as sub-optimal systems. We again note that we can do better than that!

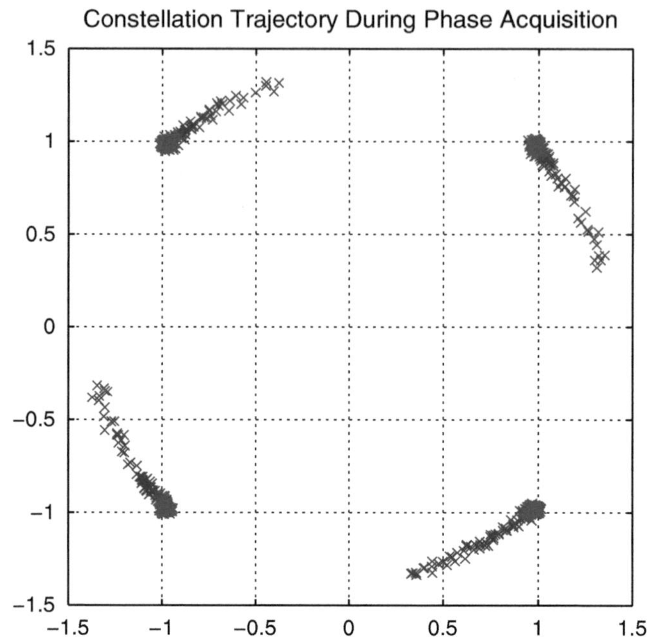

Fig. 20.16 Constellation trajectory while PLL acquires phase

20.5 Phase Lock Loop, Frequency Recovery

If the frequency offset between the local oscillator used in the final down converter and the center frequency of the input signal is sufficiently small the phase detector and the phase lock loop of the previous section can acquire and de-spin the input signal. On the other hand, if the frequency offset is significantly larger than the bandwidth of the PLL loop filter the loop will not successfully acquire and de-spin the signal. In this event an acquisition aid must be invoked to assist the phase lock loop.

In legacy receivers a lock detector is interrogated after a preset time-out to see if the acquisition aid should be invoked. If invoked, the local oscillator is slowly swept through the likely range of frequency offsets till the loop acquires the signal. The acquisition occurs when the frequency offset is small enough for the frequency error signal to pass through the narrow bandwidth loop filter. The lock detector terminates the sweep upon detection of the acquisition.

In modern receivers a maximum likelihood frequency estimators perform the task of frequency acquisition. To reduce phase jitter due to the frequency lock loop noise injection, this loop is disabled when the system acquires final phase lock. The frequency lock loop is based on a ML frequency estimator. When we take the derivative of the output of matched filter with respect to the unknown frequency offset we obtain the frequency derivative matched filter. The frequency derivative filter, often called the band edge filter, is a sensitive detector of frequency offsets of the input signal's spectral mass. How this is accomplished is visualized with the aid of Fig. 20.17.

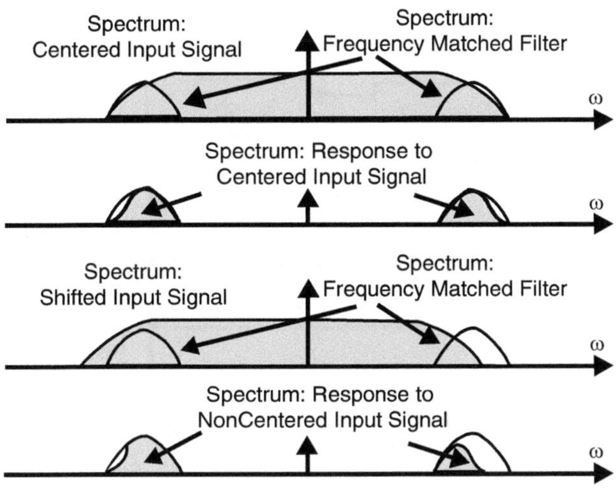

Fig. 20.17 Frequency centered and offset input spectra interacting with band edge filters

Here we see two scenarios; One in which the input spectrum is centered between the two band edge filters and one in which the input spectrum has shifted towards one and away from the other band edge filter. When centered, the two band edge filters collect the same energy from the input spectrum and their average energy difference is zero. When offset, the two band edge filters collect different amounts of energy from the input spectrum and their energy difference contains a DC term proportional to the frequency offset. The energy difference of the band edge filters is formed as the difference of the conjugate products of the time series from each band edge filter. This difference, proportional to the frequency difference, is the input signal to the frequency lock loop loop-filter. A block diagram of the frequency lock loop with its associated lock detector is shown in Fig. 20.18.

Figure 20.19 shows the phase profile of the frequency lock loop operating on an input signal with a frequency offset equal to 2.0% of the signal symbol rate. We see here the phase of the input carrier offset and the phase response of the loop. In steady state the two phase profiles have the same slope, hence same frequency but with a phase offset. We also see the band edge energy difference which returns to zero as the signal acquires frequency lock. The lower subplot of Fig. 20.19 shows the frequency offset of the loop DDS relative to the known input frequency offset.

20.6 Conclusions

We have commented on a gaping void in the communication sequences we each to the next generation of communications and signal processing engineers. A red alarm flashes every time I see a paper starting with the comment "Let's assume the system is synchronized." I think we are doing the next generation of engineers a

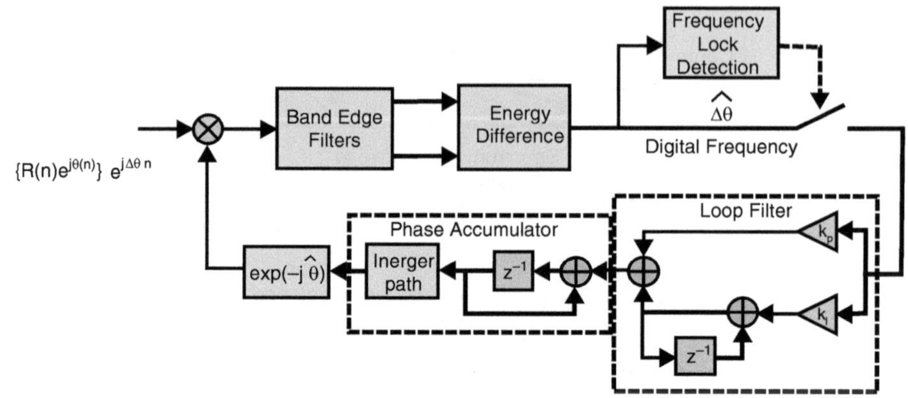

Fig. 20.18 Maximum likelihood frequency recovery for QAM receiver

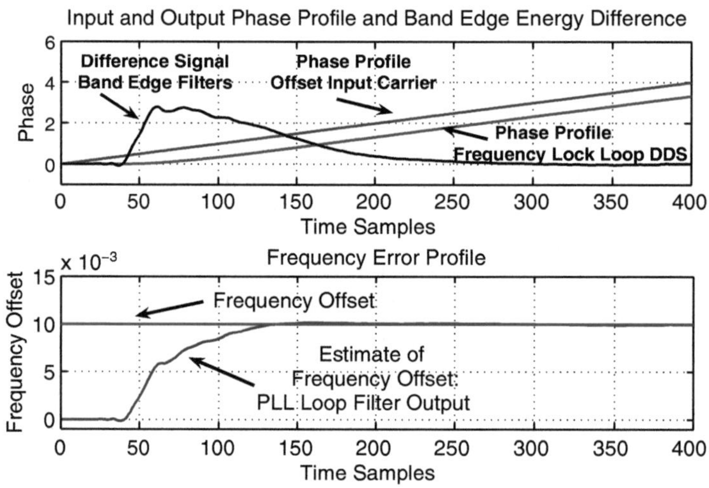

Fig. 20.19 Phase profiles of input and output of frequency lock loop and energy difference control signal and frequency error profile

disservice by not presenting an important aspect of communications residing in the physical layer.

We have presented a very light overview of the synchronization process performed in many radio communication systems. The particular processes we examined were timing recovery, phase recovery, and frequency recovery. We avoided derivations and emphasized concepts easily illustrated with figures and discussions based on senior level undergraduate electrical engineering curriculum. What we have not discussed would fill a few small textbooks. We did not ask nor answer the question where does the energy reside in the modulation signal that the synchronizer accesses to perform its tasks. Do you know? It resides in the excess bandwidth of the modulation spectrum. We might wonder what systems do not have excess bandwidth. We might and the answer is standard OFDM! How do we synchronize in OFDM? Answer: we acquire on preambles and track on pilots! What about other modulation formats such as Offset or Staggered Quadrature Amplitude Modulation (O-QPSK) [5, 6], or Vestigial Sideband (VSB), or Gaussian Minimum Phase Shift (GMSK), O-QPSK OFDM, or Shaped OFDM (S-OFDM), and on and on and on?

References

1. Rice Michael, Digital Communications: A Discrete Time Approach (Prentice-Hall, New Jesey, 2009)
2. Myer, Heinrich, Marc Monocle, Stefan A. Bechtel, Digital Communication Receivers: Synchronization, Channel Estimation, and Signal processing (Wiley, 1998)
3. Harris, Fred, Multirate Signal processing for Communication Systems (Prentice-Hall, New Jersey, 2004)
4. Proakis, John, Digital Communications, 3rd edn. (McGraw-Hill, New York, 1995)
5. Ramanmurthy, Arjan, fred harris, An All Digital Implementation of Constant Envelope, Bandwidth Efficient GMSK Modem using Advanced Digital Signal Processing Techniques (WPMC-06, San Diego, CA, 2006)
6. Ramanmurthy, Arjan, fred harris, An all Digital Implementation of M-VSB Modem Using Advanced Digital Signal Processing Techniques (SPAWC-06, Cannes, France, 2006)
7. harris, fred, Chris Dick, Michael Rice, Digital Receivers and Transmitters Using Polyphase Filter Banks for Wireless Communications, IEEE Trans. Microw. Theory, MTT **51**(4), 1395–1412 (2003)

Fred Harris holds the Signal Processing Chair of the Communication Systems and Signal Processing Institute at San Diego State University. He has taught various courses in Digital Signal Processing and Communication Systems in the same university. He holds 17 patents on digital receiver and DSP technology and lectures throughout the world on DSP applications. He has written over 170 journal and conference papers and is the author of the book Multirate Signal Processing for Communication Systems. He has also contributed to a number of other books on DSP applications including the "Source Coding" chapter in Bernard Sklar's 1988 book, Digital Communications and the "Multirate FIR Filters for Interpolation and Resampling" and the "Time Domain Signal Processing with the DFT" chapters in Doug Elliot's 1987 book Handbook of Digital Signal Processing, and "A most Efficient Digital Filter: The Two-Path Recursive All-Pass Filter" chapter in Rick Lyons 2007 book Streamlining Digital Signal Processing. He is a Fellow of the IEEE since 2003. In 2006 he received the Software Defined Radio Forum's "Industry Achievement Award." He is the Co-Editor-in-Chief of the Elsevier DSP Journal.

Author Index

Subject Index

R. Prasad et al. (eds.), *Globalization of Mobile and Wireless Communications: Today and in 2020*, Signals and Communication Technology,
DOI 10.1007/978-94-007-0107-6, © Springer Science+Business Media B.V. 2011

W

MIX
Papier aus verantwortungsvollen Quellen
Paper from responsible sources
FSC® C105338

If you have any concerns about our products,
you can contact us on
ProductSafety@springernature.com

In case Publisher is established outside the EU,
the EU authorized representative is:
Springer Nature Customer Service Center GmbH
Europaplatz 3, 69115 Heidelberg, Germany

Printed by Libri Plureos GmbH
in Hamburg, Germany